Vision, Illusion and Perception

Volume 3

The Vision, Illusion and Perception (VIP) book series publishes new developments and advances in the fields of Vision and Perception research, rapidly and informally and with a high quality. The series publishes fundamental principles as well as state-of-the-art theories, methods and applications in the highly interdisciplinary field of Vision Science, Perception and multisensory processes related to vision. It covers all the technical contents, applications, and multidisciplinary aspects of fields such as Cognitive Science, Computational and Artificial Intelligence, Machine Vision, Psychology, Physics, Eye Research, Ophthalmology, and Neuroscience. In addition, the series will embrace the growing interplay between the art and science of vision. Within the scope of the series are monographs, popular science books, and selected contributions from specialized conferences and workshops.

Nicholas Wade

Vision and Art with Two Eyes

Nicholas Wade
Department of Psychology
University of Dundee
Dundee, UK

ISSN 2365-7472 ISSN 2365-7480 (electronic)
Vision, Illusion and Perception
ISBN 978-3-030-77997-9 ISBN 978-3-030-77995-5 (eBook)
https://doi.org/10.1007/978-3-030-77995-5

This Springer imprint is published by the registered company Springer Nature Switzerland AG
The registered company address is: Gewerbestrasse 11, 6330 Cham, Switzerland

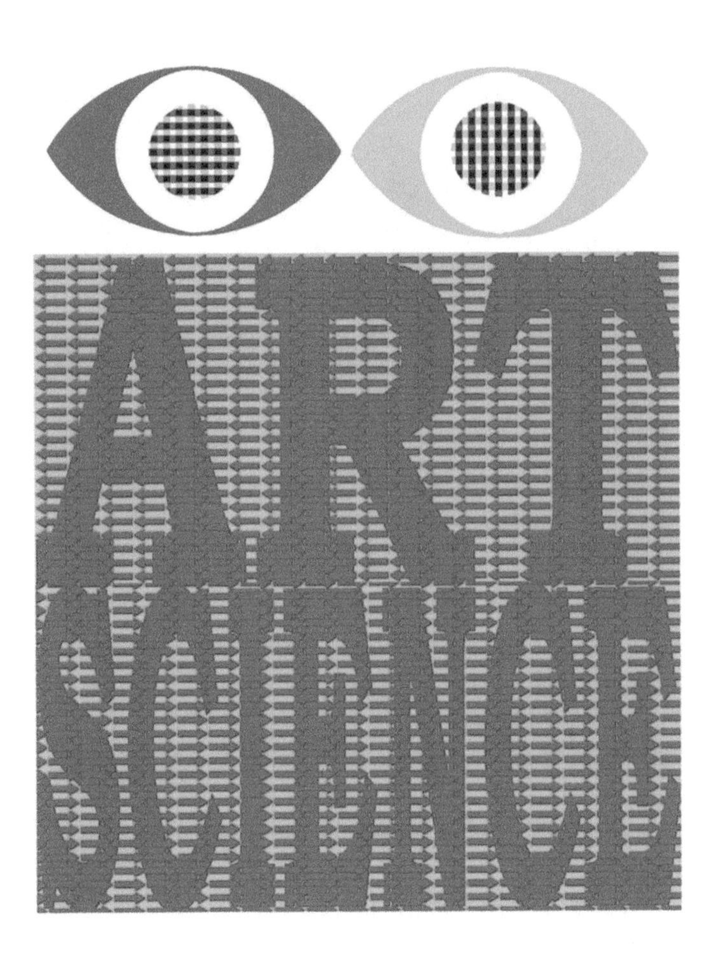
ART
SCIENCE

To my family with deep affection

Preface

Vision with two eyes has provided fascination for scientists but frustration for artists over the last two millennia. Our vision is naturally binocular: we use two eyes and see the world as single and in depth. The slight differences in what is seen with each eye were recorded in antiquity, but the manner in which those differences determined depth perception remained mysterious. This is not the case for the pictures produced: they appear much the same no matter which eye we use, and in fact they are better viewed with one eye alone. Since the early fifteenth century, when the principles of linear perspective were outlined, the ideal pictorial representation of space has been monocular. Despite this, Leonardo da Vinci stated the impossibility of a picture having the perceived depth of a scene viewed with two eyes. Pictorial art alludes to features of the world that are perceived but are not present, like the solidity of objects that are represented. The allusion is available to a single eye. This is not the case for binocular art: the allusion to depth is only available when two eyes combine two slightly different pictures.

The science of binocular vision was transformed with the invention of the stereoscope by Charles Wheatstone in the 1830s. Stereoscopes enable presentation of different pictures to each eye; if the differences are small, depth can be seen, but if they are large then the two pictures engage in rivalry. The tremors of this transformation stirred photographers but were barely felt by painters despite the fact that pictorial art is typically viewed with two eyes. However, the graphic arts are not binocular in the sense that the works require two eyes to appreciate them. Two-dimensional representational art works allude to depth that they do not contain; they are essentially monocular. The distinction between monocular and binocular art can be summarised succinctly in terms of the perception of depth and rivalry. Monocular pictures reveal to one eye what is concealed from two, whereas binocular pictures reveal to two eyes what is concealed from one. That is, monocular art alludes to aspects that are seen despite binocular information that the pictorial surface is flat; binocular art signals separations between objects in depth or rivalry that are unseen by each eye alone.

For most of us, most of the time, the world appears single and objects are seen at different depths or distances. This remarkable outcome occurs despite the fact that we have two eyes that are separated by about six centimetres. We tend to take our binocular vision for granted both in terms of the experience of a single world and the appearance of the solidity of objects within it. Through the binocular pictures in this book and the words that accompany them there will be an appreciation of just how remarkable the processes are that yield singleness and depth. Having two eyes also results in some objects that are not fixated upon having differences that are too great to be combined; rather than seeing them double we are unaware of this. When radically different images are presented to each eye our perception is unstable and dynamic; this is called binocular rivalry, and it is an expression of binocular competition in contrast to the cooperation that yields stereoscopic depth.

All the images have been made for this book to illustrate the nature of binocular vision, the scientific and artistic explorations of it, and the many ways in which binocular art can be expressed and extended. The images (with a few exceptions) are anaglyphs that have been made to be viewed with red/cyan glasses. Anaglyphs are simple stereoscopes that have an advantage over mirror and prism stereoscopes in that there are four possible outcomes from viewing them: with each eye alone to see the monocular images, with both eyes to see them in stereoscopic depth or rivalry, or without the red/cyan glasses where they can have an appeal independent of the binocularity they encompass. Reflecting and refracting (optical) stereoscopes do not offer this last possibility. Although anaglyphs do not separate the patterns to each eye as fully and equally as optical stereoscopes, they do have the advantage that the superimposition of red and cyan images creates a third design that can have an allure of its own. Many of the illustrations have titles, often based on wordplay. It seemed appropriate to combine the ambiguities of language with those of pictorial representation!

For some people the stereoscopic images can take some seconds for the depth to emerge, whereas others see the depth immediately. The time taken can also vary with the complexity of the carrier pattern as well as of the figure in depth. The same can apply to the rivalling images although some people find that they can be difficult to look at. Thus, art for two eyes is not restricted to cooperation between them yielding stereoscopic depth, but it can also draw upon their competition when markedly different patterns are presented to each eye resulting in binocular rivalry.

In visual science the perception of depth is often reduced to reliance on two sources of information, and they are often referred to as monocular and binocular. The former refers to the projective properties captured by a camera, and they are used by artists to provide the allusion to depth in a picture. Differences between the images in each eye (disparities) and the angle between the visual directions (convergence) are only possible with two eyes. With the invention of the stereoscope it was possible to produce two pictures with defined horizontal disparities between them to create a novel impression of stereoscopic depth. Stereoscopy and photography were made public at about the same time, and their marriage was soon cemented; most stereoscopic art then as now is photographic. Wheatstone sought to examine stereoscopic depth without monocular pictorial cues so that disparities alone determined depth. He was unable to do this, but it was achieved a century later by Béla Julesz with random-dot stereograms. The early history of non-photographic stereoscopic art is described in the book as well as reference to some contemporary works. Novel stereograms employing a wider variety of carrier patterns than random-dots are presented as anaglyphs; they show modulations of pictorial surface depths as well as inclusions within a binocular picture.

The chapters in the book cover the history of research on binocular vision as well as the artistic manipulations that can be made when different images are presented to each eye. Each chapter commences with an illustration containing its number in a binocular form. In some cases the number is presented to one eye only, whereas in others it emerges in stereoscopic depth or engages in rivalry. The chapters cover the phenomena of binocular vision, the history and science associated with them as well as the art of stereoscopic vision and binocular rivalry. Thus, the book is a celebration of the science of vision with two eyes and the ways in which interactions between them can be expressed in art.

My interest in the history of research on binocular vision is of long standing as is my involvement with graphic art. However, the fascination with binocular art is much more recent, and my enthusiasm for it will be evident in the illustrations contained within this book. The book started its life with a much narrower purview, reflecting my experimental research on binocular rivalry. The phenomenon is intriguing because all manner of percepts are reported when viewing unchanging but radically different patterns presented to each eye. I was surprised that rivalry had not been incorporated into contemporary styles (like op art) that amplify the vagaries of vision. I commenced by trying to fill this phenomenal void but in so doing became ensnared by stereoscopic imagery, both graphical and photographic. An important stimulus was participation in conferences concerned with the early history and contemporary expressions of stereoscopic photography. Examination of binocular competition should not proceed without engaging with its cooperation and indeed with the interactions between them. Thereafter, the graphical challenge to create patterns that were themselves of interest, independently of the stereoscopic depth they could display when viewed with red/cyan glasses, took hold. The hope is that the book will encourage others to explore the graphical delights afforded by vision with two eyes.

The production of the book has not been assisted by the pandemic, and I wish to thank the team at Springer for proceeding despite the constraints carried by COVID-19. Printing anaglyphic images poses challenges that are not so readily encountered when they are displayed on a computer monitor; I thank the printers for addressing this challenge. Finally, my family have humoured and encouraged me throughout this journey, and it is to them that the book is dedicated.

Newport-on-Tay, Scotland
January 2023

Nicholas Wade

Contents

Chapter 1
Setting the Seen

We view the world with two eyes and objects within it appear solid and three dimensional. In large part this is due to the slight differences in the images in each eye. This book celebrates binocular vision by presenting illustrations that require two eyes to see the effects of cooperation and competition between them. Pictures are flat but by printing them in different colours and viewing them through similarly coloured filters (included with the book) they are brought to life either in stereoscopic depth or in rivalry with one another. They are called anaglyphs and all those in the book display the ways in which the eyes interact. Thus, the reader is an integral element in the book and not all readers will see the same things. There are differences between vision with each eye and these can be marked for some people either in terms of cooperation or competition between them. The history, science and art of binocular vision can be experienced in ways that are not usually available to us and with images that have not been seen before. The history was transformed by the invention of stereoscopes comprised of mirrors, lenses or prisms in the 1830s. Anaglyphs are simple forms of stereoscopes that were developed a little later. They have an advantage over optical stereoscopes in that there are four possible outcomes from viewing them—with each eye alone to see the monocular images, with both eyes to see them in stereoscopic depth or rivalry, or without the red/cyan glasses where they can have an appeal independent of the binocularity they encompass. We tend to take our binocular vision for granted both in terms of the experience of a single world and the solidity of objects within it. Through the binocular pictures and the words that accompany them there will be an appreciation of just how remarkable the processes are that yield singleness and depth. Moreover, the opportunities for expressing these processes are explored with many examples of binocular art.

N. Wade, *Vision and Art with Two Eyes*, Vision, Illusion and Perception 3,
https://doi.org/10.1007/978-3-030-77995-5_1

Vision with two eyes. Viewing this pattern through the red/cyan filters results in the word STEREOSCOPIC appearing in depth, the alternate letters of DICHOPTIC being seen by different eyes, and the words BINOCULAR RIVALRY engaging in binocular rivalry. The apparently singular schematic eye exhibits some of these binocular effects

This book examines our vision with two eyes as it has been investigated in science and how it has been expressed in art. Binocular vision has been studied extensively by scientists but used sparingly by artists. The history of pictorial art has been concerned primarily with monocular vision: by trying to capture three-dimensional space on a two-dimensional surface it is as though the world was being viewed with one eye alone. The depth we see in a real scene is different from that depicted on the surface of a painting, drawing or screen. Even when we describe the apparent depth in a picture we do not see the surface as other than flat.

In normal binocular vision the two eyes cooperate to give us an impression of a single world in depth. The stereoscopic depth is based on the slight differences in the images projected onto each eye. This is illustrated in the figure below. Both eyes fixate on point F, so that its image falls on the centre of each eye (called the fovea). The images of points A and B, which are respectively further and nearer than F, fall on non-corresponding points. These differences in the projections from the two points are referred to as retinal disparities. If the disparities are not too great they can be processed in the visual system to provide information for relative distance, called stereoscopic depth perception or stereopsis. The sign of disparity depends on whether the non-fixated point is nearer or further than the fixation distance. Disparities can be described in terms of visual directions. Point A appears to the right of F when viewed by the right eye alone, and to the left with the left eye; this is called uncrossed disparity. Conversely, point B appears to the left of F when viewed by the right eye, and to the right with the left eye; this is referred to as crossed disparity. Suppose that A is a little higher than F and B is a little lower, then with the red/cyan glasses the stereoscopic depth would appear as in the small figure on the right. Despite the fact that the two eyes are in different locations in the head, producing these disparities, we still see a single world. This is not always so when the disparities are too large to be combined.

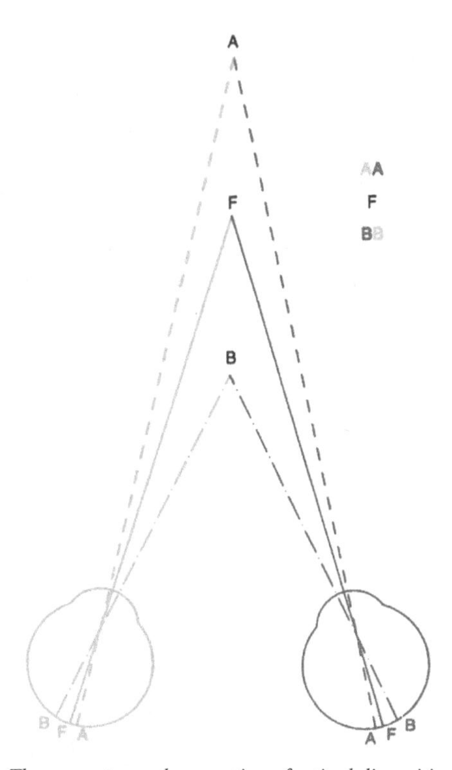

The geometry and perception of retinal disparities

The two eyes can also compete with one another to yield binocular rivalry which is experienced when dissimilar contours or colours are presented to corresponding regions of each eye. Rivalry occurs all the time when we use two eyes to view a three-dimensional world. It can be demonstrated very easily by looking at a finger held in front of the eyes and then closing one eye and the other. The same objects further away than the finger will appear in a different direction to each eye. It is a natural consequence of light travelling in straight lines to eyes that are separated from one another and it is illustrated below. When both eyes fixate on point F the object A is aligned with F in the right eye (RE) and object B is aligned with the left eye (LE). Normally F will occlude most of A and B but they are seen in different directions when using one eye at a time. Despite these differences we experience a single world when using two eyes. Not only does the fixated finger appear single but the background does also. In part this is because the more distant objects A and B are not in sharp focus, unlike the finger. However, there are other processes in operation, too: the rivalry between the competing images A and B is resolved in the brain. Two fingers can be seen when we combine A and B. Binocular rivalry is more compelling when different images are presented at the area of fixation. The situation is illustrated below both in terms of optical lines and perception. The red and cyan lines in the illustration will be visible to the left and right eyes when viewed through the red/cyan filters provided with the book. The red filter is usually in front of the left eye and the cyan filter in front of the right eye. Looking at images printed in equivalent colours results in the red image being seen as black with the right eye and the cyan image as black with the left eye. If the printed colours are matched with the filters then the red will be invisible through the red filter and the cyan invisible though the cyan filter. With both eyes, fixation on the dot F will result in three dots being visible. Maintain fixation on F then only two dots are visible when looking through each filter in turn—A and F with the left eye then F and B with the right eye. A and B are visible when both eyes are open because the dots in one eye are in rivalry with a white surface in the other; in general, contours are seen when in rivalry with evenly illuminated regions.

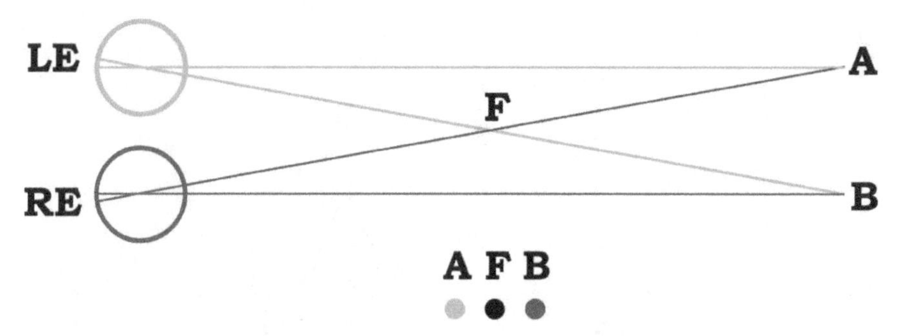

When the left and right eyes fixate on point F then more distant objects (A and B) will appear in different directions to each eye

Anaglyphs

Anaglyphs are displays in which the left and right eye images are printed in different colours, such as red and cyan, and they are viewed through filters of the same colours.

They have typically been used to present slightly different images to each eye so that they are seen in relief—stereoscopic depth. Anaglyphs can be used to investigate various forms of binocular combination, as is demonstrated with the following image. Viewing the upper pattern through the red/cyan glasses will lead to some perceptual instability because of the different images in each eye. If you close your right eye so that you are looking through the red filter with the left eye then the pattern you will see is shown in the lower left image (beneath the red rectangle). Closing the left eye to view through the cyan filter will result in seeing the pattern on the lower right. Anaglyphs are essentially simple stereoscopes and they can be used to present stimuli in five basic ways. A single image can be presented to a single eye even though both are open. This is the MONOCULAR condition shown in the top line. The simplest way of examining binocular vision is to compare vision using two eyes with that using one. The task is simple because one eye can be closed or covered, but with anaglyphs it is possible to control the eye to which the stimulus is presented without the observer's knowledge. If the same letters are presented to the same regions of each eye then it is BINOCULAR—like fixation on F in the figures above. Under natural viewing this is what happens when we look at distant objects—the projections to each eye are essentially the same. If the letters in the two eyes differ slightly in width, as in the second line then the word STEREOSCOPIC will appear to be in depth, with the letters slanting away to the right. These are due to small retinal disparities and stereoscopes are typically employed to view disparate images. It is also possible to present different but non-overlapping stimuli to each eye. This is referred to as DICHOPTIC stimulation and it is expressed in the fourth row of letters. Because there are no common features in the two stimuli the letters appear unstable and move with respect to one another. Finally, radically different displays can be presented to each eye, as is the case for the lowest line in which BINOCULAR is seen through the red filter and RIVALRY through the cyan filter. To add to the binocular rivalry the letters of the two words are formed from horizontal and vertical lines.

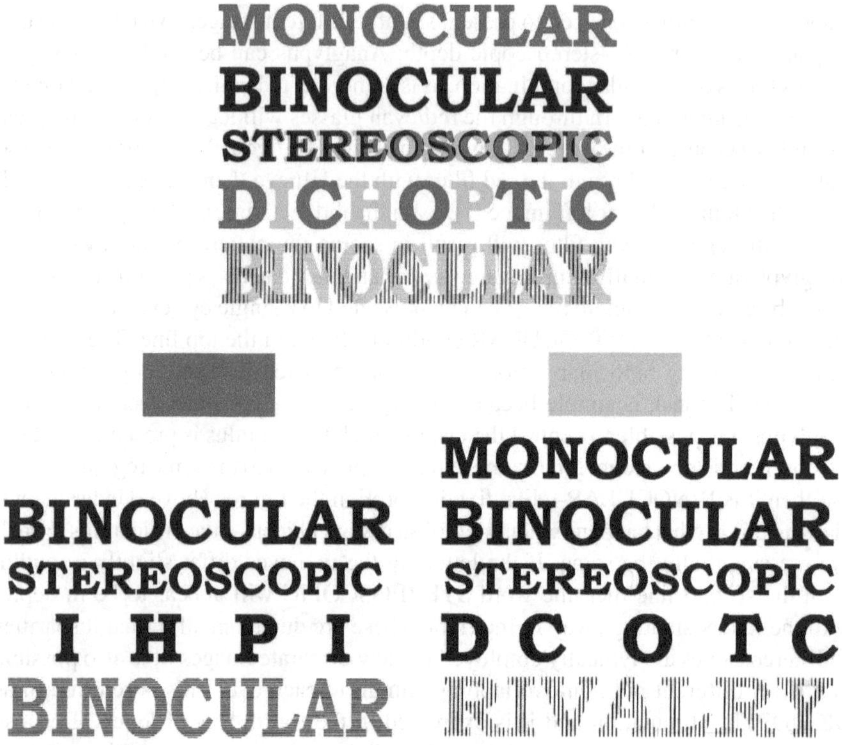

Viewing the upper words through the red filter with the left eye and the cyan filter with the right eye results in the black and white words shown below being visible to each eye

This book is about binocular vision—the history of the science behind it and the art of demonstrating it. In both science and art much more attention has been directed to stereoscopic vision than to the other aspects of binocular combination. This is somewhat surprising because the effects generated by binocular rivalry are more dramatic than the subtleties exposed by stereoscopic depth perception. This applies to both science and art. However, there is a great disparity in the ways binocular rivalry been examined in science and expressed in art. From the late nineteenth century scientists have embraced binocular rivalry but the same cannot be said about its expression in art. While stereoscopic photography has had an ardent following, rivalry has been slow to gather its exponents.

Strong binocular rivalry can be experienced with the patterns below in which concentric circles are presented to one eye and radiating sectors to the other. The patterns visible through the cyan filter are sectors on the left and circles on the right with the opposite pairing through the red filter. With both eyes open the patterns will change over time. Often a particular pairing is initially visible but with longer viewing the patterns change dynamically: sometimes only one component will be visible and at other times parts for the circles will be seen in one region and sectors in another with the regions changing. The fact that sectors can be seen in one region and circles in another indicates that the rivalry is not only between one eye and the other but also between one pattern and the other.

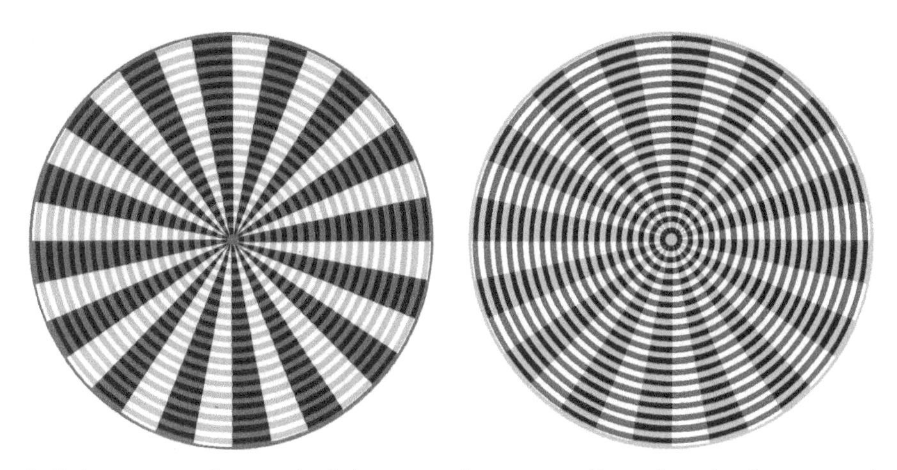

Radiating sectors and concentric circles presented to corresponding regions of each eye engage in binocular rivalry

The binocular rivalry experienced with the stimuli above is a consequence of the differences in the lines presented to each eye—either radiating from a central point or concentric circles. The ensuing changes in the appearance of the patterns is called binocular contour rivalry. Rivalry can also occur when different colours are presented to each eye and it can easily be experienced when looking at the pattern below through the red/cyan glasses. The single black circle is for keeping the eyes aligned; the same effect could be obtained by looking at a large white sheet of paper.

Looking at the centre of the white circle through the red/cyan glasses results in weak colour rivalry

Colour rivalry is by no means as dramatic as contour rivalry; the appearance changes from rather pale patches of cyan replacing pale reddish. This is one of the reasons why the majority of rivalling images in this book will be concerned with binocular contour rather than colour rivalry. Looking at coloured patches without filters for some time can also change their appearance. If you look at the two grey discs below while fixating the black dot between them, they will look much the same. This is not the case after you have looked at the black dot between the red and cyan discs for about 30 s, then look back at the dot between the grey discs. They will appear reddish on the right and greenish-blue on the left. Not only do we adapt to seeing the same colour over time but we also see an after-effect when viewing grey discs subsequently. Carrying out the same procedure while viewing the discs through the filters does not have the same outcome: both upper discs appear black and after prolonged fixation and then the lower discs both look a lighter grey rather than differing in colour.

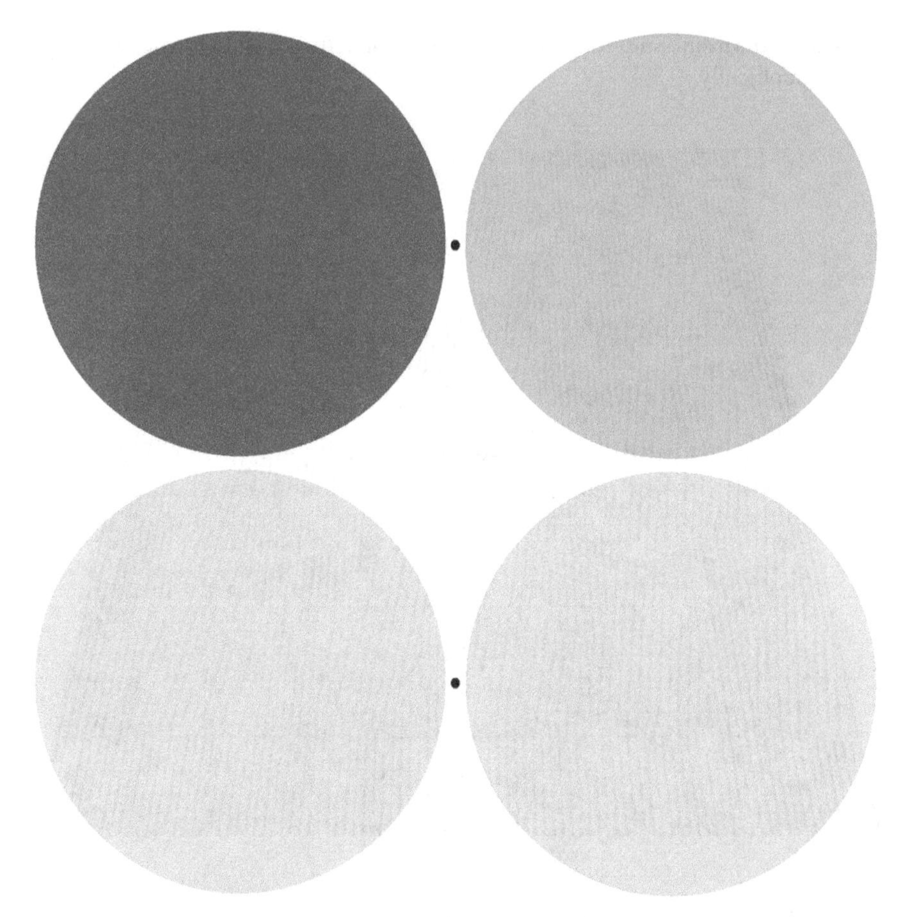

Colour aftereffects can be seen without the red/cyan glasses. The lower grey discs no longer look the same after fixation on the dot between the red and cyan ones for about 30 s—the lower left appears cyanish and the right reddish

Anaglyphs do not separate the patterns to each eye as fully and equally as optical stereoscopes but they do have some advantages. The superimposition of red and cyan images creates a third image that can have an allure of its own. The ease with which the components can be viewed alone (by closing or covering one eye) displays the monocular elements and (for rivalry patterns) acts as an efficient technique for presenting two patterns in a single image; this will be used extensively in the following chapters. Reversing the filters in front of the eyes is a simple means of reversing small disparities (for stereoscopic images) or large ones (for rivalry patterns).

The techniques for combining two images as anaglyphs have advanced considerably with the assistance of computers. Most of the figures in this book have been combined using a software package called StereoPhoto Maker.[1] It has been developed by Masuji Suto and it is freely available. Like many others, I wish to express

[1] http://stereo.jpn.org/eng/stphmkr/.

my thanks to Masuji Suto for creating such useful software and my appreciation is shown graphically.

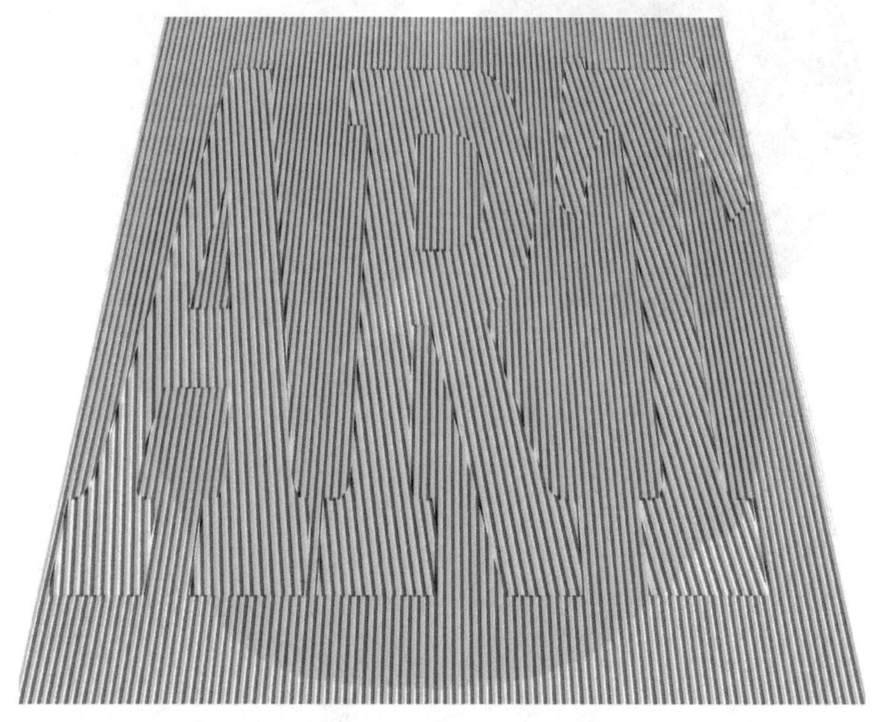

The art of Masuji Suto

Eye Dominance

It may be the case that not everyone sees the patterns in the way described. This could be for several reasons. One concerns the filters themselves which, although matched for colour with the printed patterns, are not equated for the amount of light they exclude. The red designs seen through the cyan filter tend to be of higher contrast than the cyan seen through the red filter. This can be seen in the radiating/circular rivalry pattern above: through the cyan filter sectors are seen on the left and circles on the right; through the red filter circles are on the left and sectors are on the right. When the patterns are in rivalry, the imbalance between the filters can result in one of them being visible for longer than the other. Another relevant factor is eye dominance. This can be checked by reversing the filters so that the pattern is viewed with the cyan filter/left eye, red filter/right eye combination. More will be said about eye dominance later in the book but an indication can be presented here. Normally one eye is able to maintain fixation more steadily than the other and it is usually designated as the sighting dominant eye. The difference between the stability of the eyes can easily be observed in the picture below which consists of two small discs, one above the other. When you look at them through the red/cyan glasses they will probably not appear constantly aligned one above the other: one will appear stable and the other will appear to move about. Reversing the glasses will probably reverse the disc that appears stable. Without any common contours that can be seen with both eyes the alignment between them breaks down. You are encouraged to experiment with the coloured filters in order to find which combination (red/LE, cyan/RE or cyan/LE, red/RE) produces the best depth and rivalry effects for you.

One of the dots might look more stable and in higher contrast than the other; this is likely to reverse when the filters are reversed

A small proportion of the population has had a squint in childhood that might have transformed their binocular vision; they are not able to use both eyes together and will be able to see only one of the component patterns. Squint (also called strabismus) is a deviation of the eyes so that the two eyes do not work together.[2] While people with a history of strabismus might not be able to see the component patterns in depth or rivalry they can see them individually by placing first one then the other filter in front of their dominant eye.

Stereoscopic Depth Perception

Anaglyphs are simple forms of stereoscopes: they provide a technique for presenting different patterns to each eye. A natural consequence of viewing the world involves such differences because the eyes are laterally separated by about 63 mm. This can easily be demonstrated by fixating a finger, held about 20 cm from the eyes, and then closing each eye alternately. You will probably be surprised at the large differences in alignment that are evident, and yet they pose few problems for normal perception. The small differences between the projections to each eye are combined to give us stereoscopic depth perception. Stereoscopes were devised to simulate what we naturally see with two eyes and they have been used to give us the impression of depth from two slightly different photographs as in the example below.

[2] Chopin et al. (2019) found that the incidence of stereo blindness in the population is about 7%.

Pierspective—Beneath the Tay Road Bridge

Unlike the effects with rivalry patterns, reversing the filters does not dramatically alter the depth seen in naturalistic scenes. This is largely because monocular cues to depth like occlusion are not changed when the sign of disparity is reversed. That is, nearer objects continue to occlude parts of more distant ones despite the reversal of disparities so that there is competition between different indicators of relative distance. The reversal of depth can easily be appreciated with simple stereoscopic shapes like discs—changing the filters in front of each eye changes the apparent depth of the dots above and below the central fixation dot.

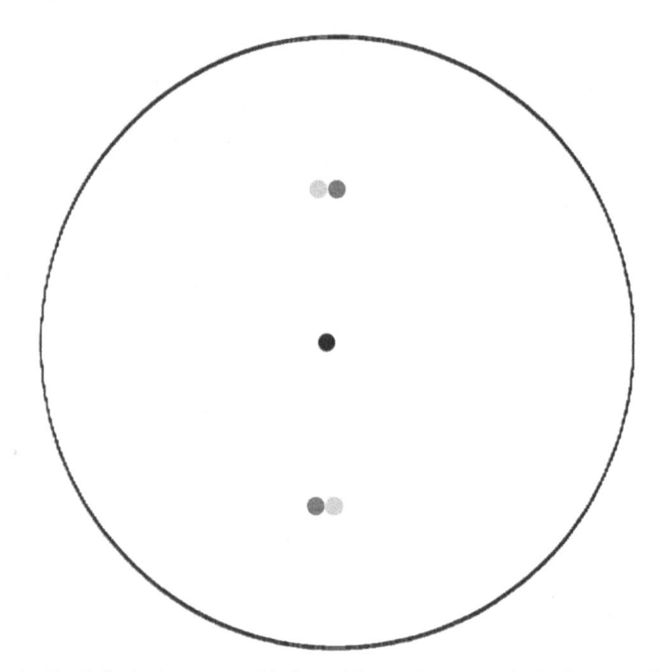

Fixating on the black dot in the centre with the red /cyan glasses results in the upper dot appearing farther away and the lower one nearer; these apparent depths are reversed when the filters are reversed

Drawings and photographs of objects give many signs of their solidity independently of stereoscopic depth. It is for this reason that scientists have tended to use simple patterns like dots, lines or more complex outlines. However, with the onset of computer graphics it became possible to generate designs in which depth only emerged when the component patterns are viewed with two eyes.[3] That is, stereoscopic depth could be investigated without any object recognition. An example is shown below in which each eye is presented with a pattern of hand-drawn dots and it is only with their binocular combination that circles can be seen in depth. As with all such effects it can take a few seconds before the depth is articulated. Reversing the filters in front of your eyes will reverse the apparent depth seen. Another feature of the design is that sideways head movement will result in the relative motion of the circles with respect to their surround: if the circles appear further away then they will appear to move in the opposite direction whereas if they look nearer then they will appear to move in the same direction as the head.

[3] See Julesz (1971).

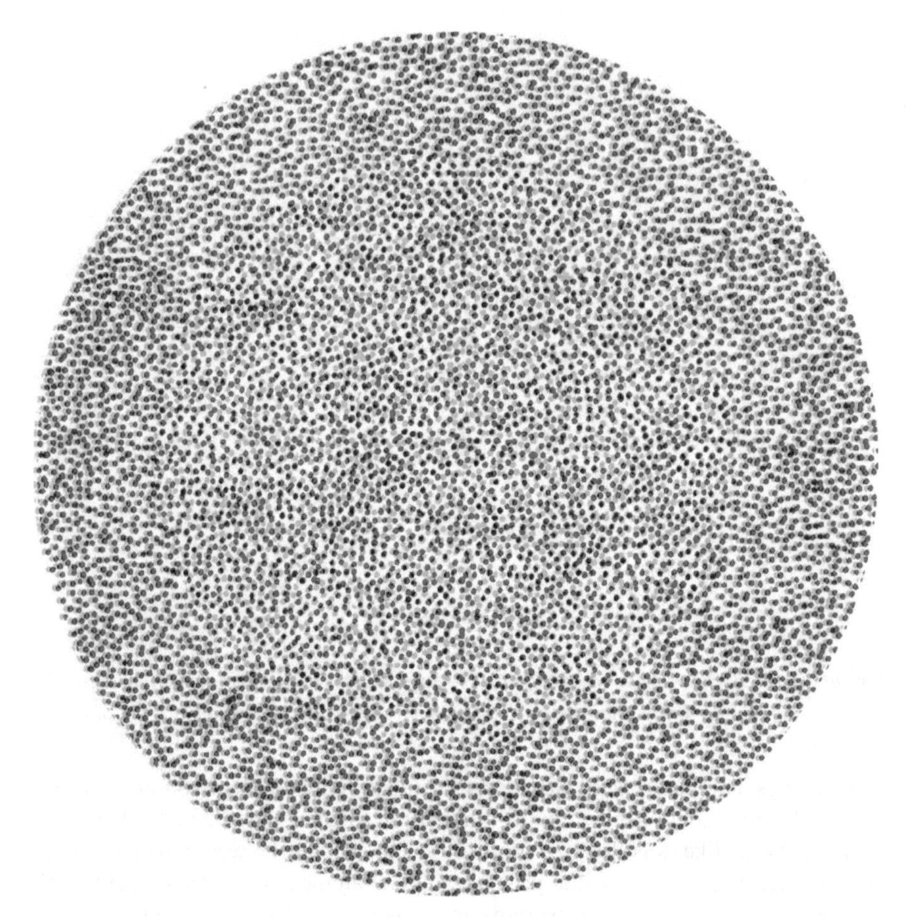

Random dot circles in depth

It is not necessary to use computer generated patterns in order to demonstrate stereoscopic depth without object recognition. The example below is derived from a photograph of autumn leaves lying on the ground; it has been folded over on itself to be symmetrical about a vertical axis. When viewed with two eyes and the red/cyan filters two circles will be seen in depth, one to the left and the other to the right of the centre, one apparently more distant and the other closer. Another fascinating feature of stereoscopic vision can be noted when viewing this stereo-pair: although the two circles are the same size, the one that is apparently more distant appears to be larger. This can readily be checked by reversing the filters in front of your eyes—then the apparent difference in sizes also reverses. Thus, circles of the same physical size appear larger or smaller as they are seen further away or nearer. The dependence of apparent motion and size on stereoscopic depth indicates its significance in the perception of space. Many more examples of stereoscopic depth in photographs and designs will be shown in the chapters to follow.

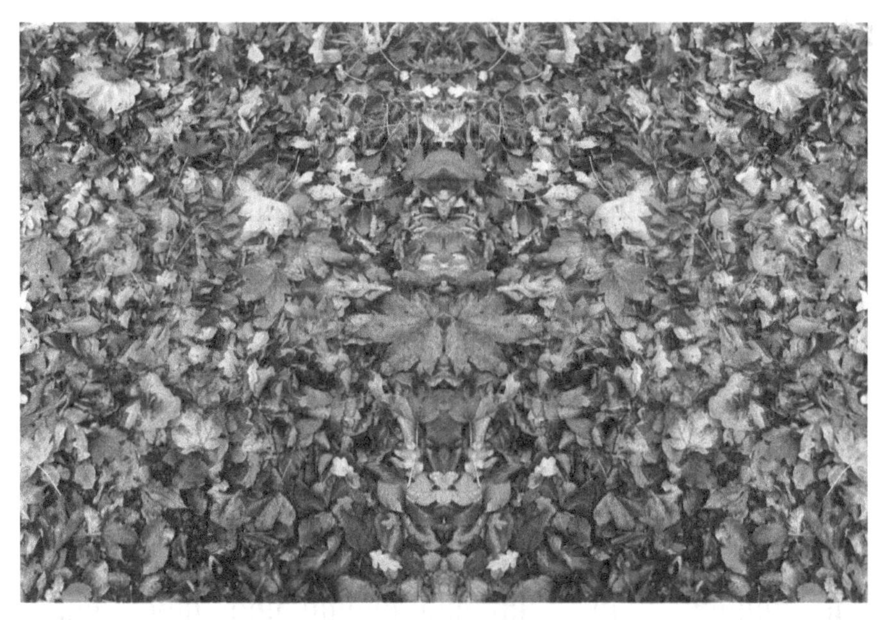

Leaf circles

Stereoscopic depth is not the only aspect of vision with two eyes that warrants attention of vision scientists and artists. Depth is visible because of small lateral displacements of the images in each eye. When the differences between the images are too large then the images in each eye no longer cooperate to yield depth but compete to produce rivalry.

Binocular Competition

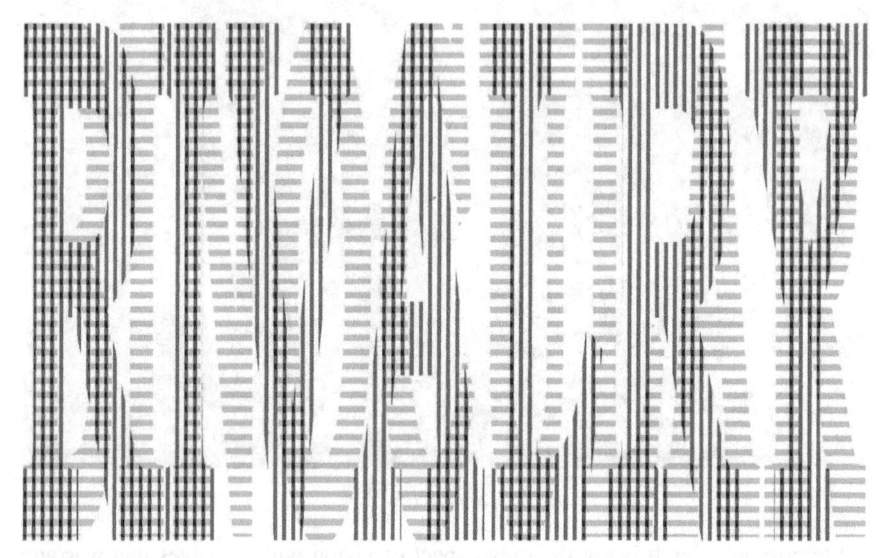

Binocular rivalry. The word BINOCULAR defined by horizontal lines is visible when viewing through the red/left filter alone and RIVALRY in vertical lines is seen through the right/cyan filter. With both eyes open the words or fragments of them compete for visibility

Stereoscopic depth perception is an instance of cooperation between the two eyes: slightly different signals are combined to give the impression of three-dimensionality. Binocular rivalry is an instance of binocular competition. It is a natural consequence of our binocular interactions with the world; rivalry is a resolution of conditions that apply to most of what we see when using two eyes. It occurs when the differences between the images in the two eyes are too large to be combined, and stereoscopic depth cannot be extracted from disparity. When we fixate with both eyes on part of an object most of what is projected to the peripheral retina is too disparate to yield depth; since the peripheral stimuli arise from different depths their retinal images also tend to be out of focus. We are not generally aware of this binocular rivalry as both visual resolution and attention are associated with the fixated object rather than peripheral ones. If attention shifts to a peripheral object then both eyes generally also move to fixate it. Binocular rivalry is rarely examined under these conditions of natural stimulation. It is typically studied with different patterns presented to corresponding central regions of the two eyes—as if we are looking at two different objects in the same place and at the same time. As noted earlier, this can be seen most readily when contours having great differences in their orientations are presented to corresponding regions of the eyes, as in the patterns below. Either simple line designs at right angles to one another can be combined or more complex curvilinear contours can compete.

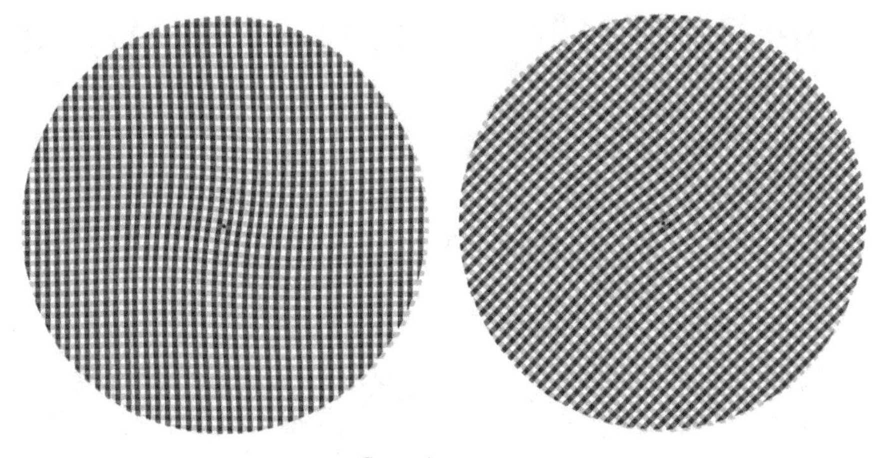

Competing contours

Competing curves

Binocular competition and co-operation can coexist. That is, the same pattern pairing can exhibit both stereoscopic depth and binocular rivalry. The illustration below is based on a symbolic representation of the two eyes and their nerve pathways to the brain.[4] The pattern visible when viewing through one filter is the negative of that through the other and so they engage in rivalry when viewed through both filters at the same time. However, the disparities in the upper and lower parts of the patterns yield stereoscopic depth so that the central eyes either bulge or recede—depending on which way round the filters are. This is an example of binocular art because the stereoscopic depth and the binocular rivalry can only be experienced when two eyes are used to view the illustration.

---

[4] See Wade (2007, 2009).

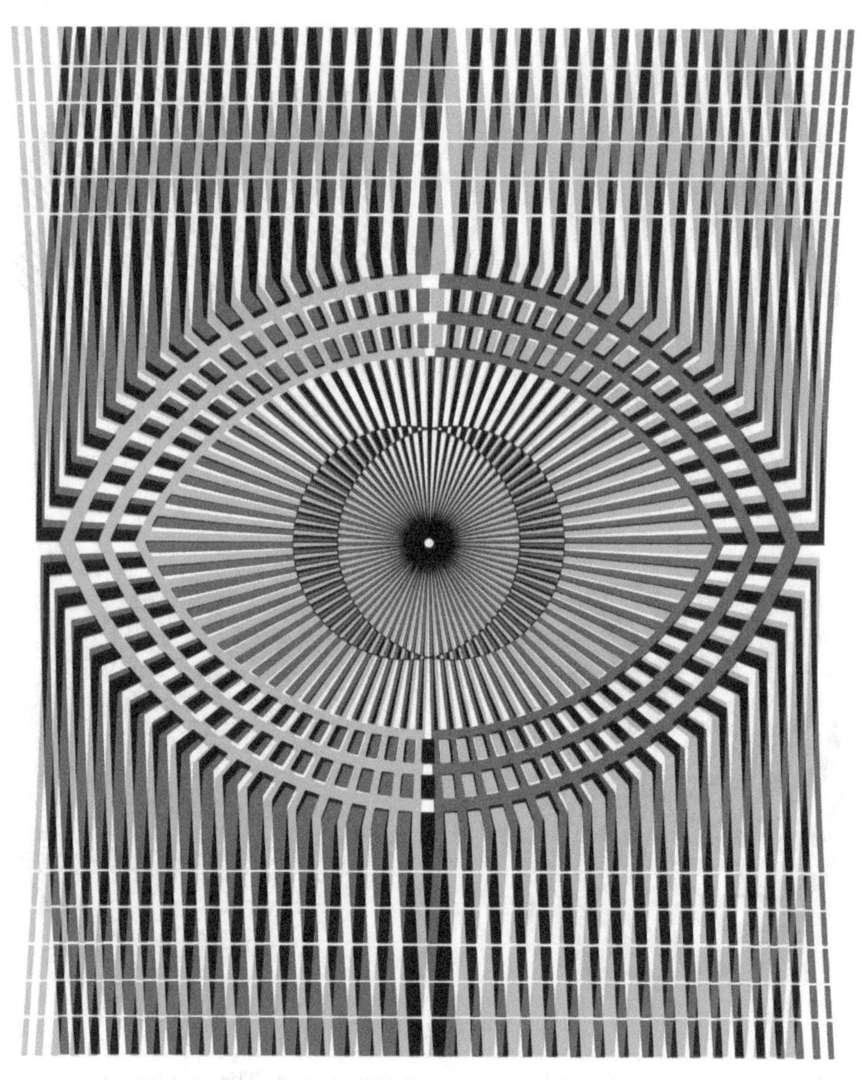

Symbolic representations of the eyes and the pathways to the brain

Binocular Art

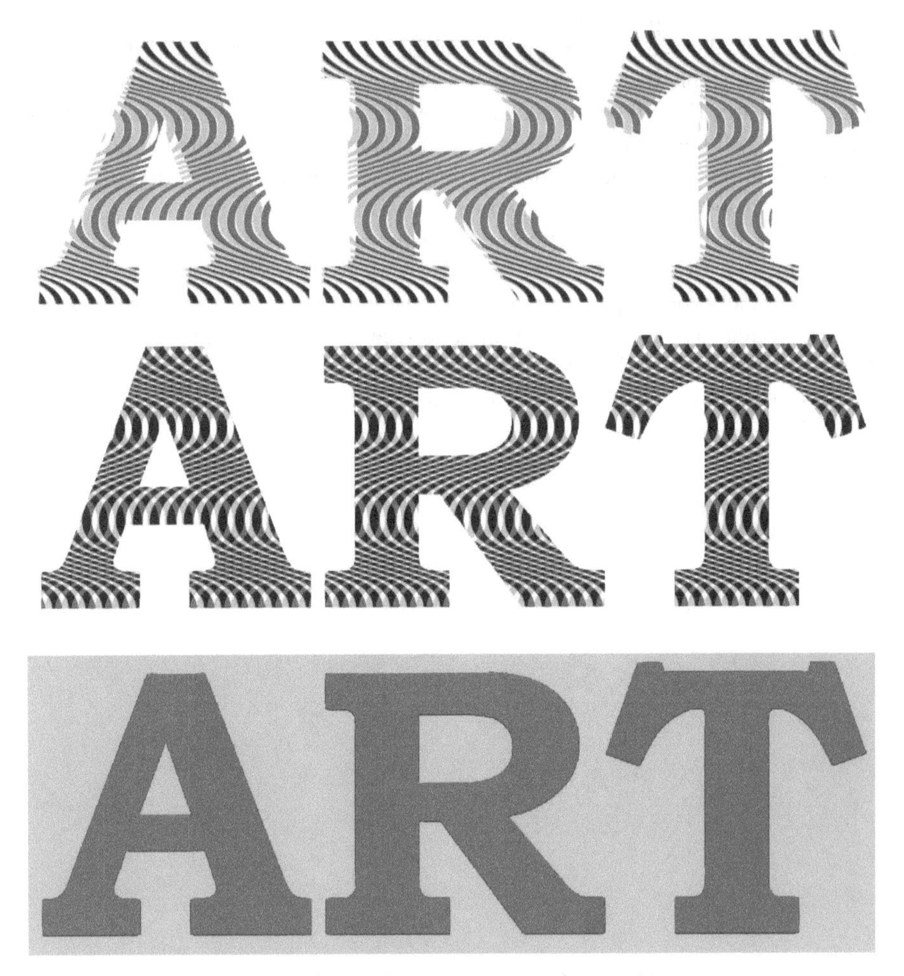

The art of binocular co-operation and competition

Art works or designs that are called 'binocular' are those that require two eyes to express the properties they contain. That is, they provide perceptions that are not available with conventional works. It often takes time for these aspects of our perception to be experienced and so binocular art is not cursory, it cannot be appreciated by the quick glance. Moreover, binocular art draws on several perceptual processes, not just stereoscopic depth perception. Designs can also compete with one another in terms of the contours they contain or the colours comprising them. These points can be made with the aid of the illustration above. Viewing the upper word with the red/LE, cyan/RE arrangement for some seconds transforms it from a flat pattern to one articulated in depth, with the central region appearing more distant than the upper and lower extremities of the letters. This is an example of stereoscopic depth perception. The patterns in the letters of the central word display binocular rivalry: with longer viewing the curved contours will seem to compete with one another for dominance and the two components will vie for perceptual visibility. The lower word ceases to look stable but takes on the appearance of a metallic patina that is called binocular lustre. The effects will change when the red/cyan glasses are reversed: the upper word will bulge rather than recede, the central one will probably display the opposite dominance to that experienced previously, as will the lower one which will be seen as positive and negative when viewed with each eye in turn.

We rarely look at pictures with one eye alone and so in this sense most pictorial art is binocular. However, the predominant technique for representing space since the Renaissance—linear perspective—is fundamentally one-eyed. The dimensions of solid objects are represented on a flat surface. This was delightfully illustrated by Robert Fludd (1574–1637) in his text on perspective.[5] The disembodied eye is mounted in front of a grid, the cells of which capture parts of the scene from one viewpoint.

[5] Fludd (1617).

The perspective of Robert Fludd

Linear perspective involves capturing visual angles from a single defined location. It provides a powerful illusion of depth but this is in conflict with the equally powerful perception of the flat surface on which the marks are made. A comparison between the two methods of creating the illusion of depth can be seen below. The perspective illusion of the third dimension is shown alone in the upper illustration and enhanced by stereoscopic depth in the lower one.

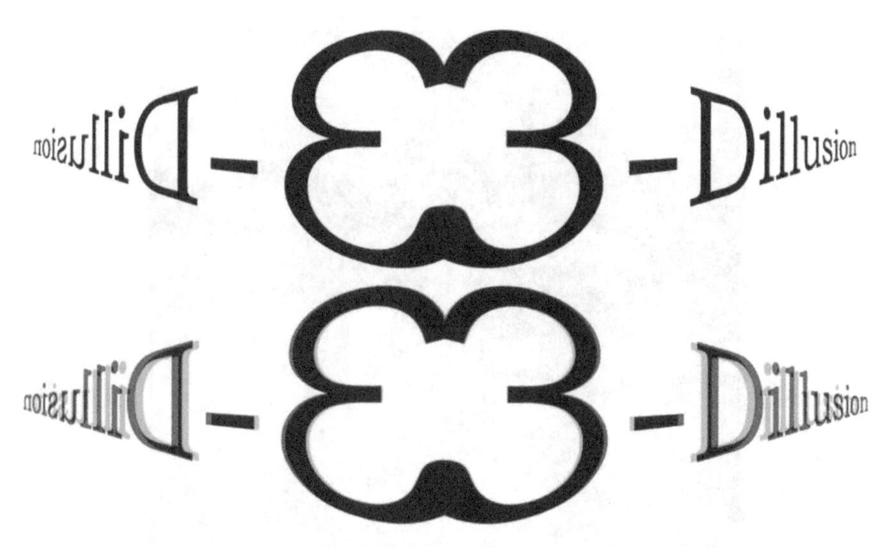

Delusions of depth

Following the invention of stereoscopes in the nineteenth century there has been a growing interest in producing binocular art, either from paired photographs or less commonly from paired drawings or paintings. Much less interest has been given to binocular art involving rivalry.

Art and Science

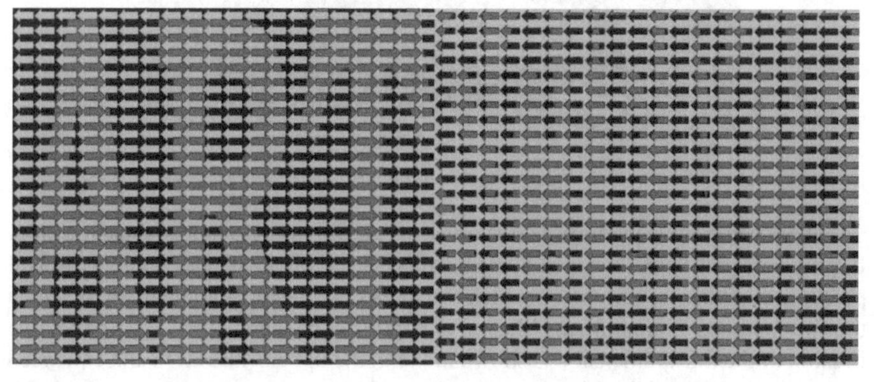

Art and Science

Unlike the neglect of binocular vision in art, visual scientists have embraced the phenomenon as it is seen as a tool for understanding more about how the two eyes interact. Many excellent books have been written about the science of binocular vision but I single out a three volume work on perceiving depth by Ian Howard together with Brian Rogers because of the breadth of its coverage.[6] In it can be found the history and experiments behind the aspects of binocular vision covered in the chapters of *Vision and art with two eyes*. A simple stereoscope is included in *Perceiving in depth* so that some of the paired illustrations can be viewed appropriately and Ian Howard can be seen in depth in the picture below. The volumes cover the science of binocular vision in detail but they do not deal with the art of vision with two eyes.

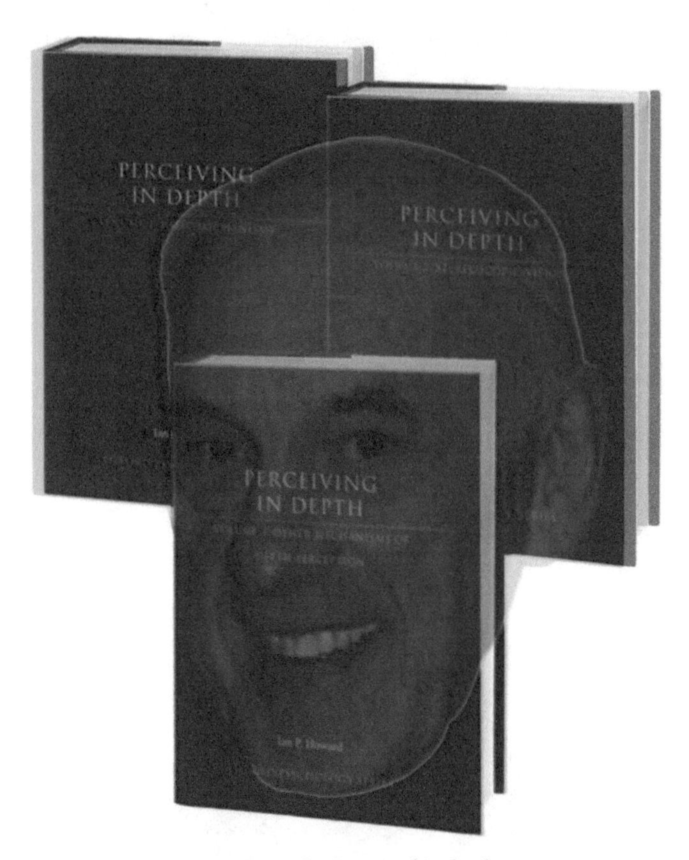

Ian Howard—Perceived in depth

Scientists tend to use simple patterns to investigate vision so that they have control over the stimulus variables at play, as in the illustration below. In each of the circles a grey patch is presented to one eye and a very low contrast grating (parallel vertical lines) in the other. The gratings are quite difficult to see on their own. With the cyan filter in front of the right eye the sequence from top to bottom is grating, grey, grating and the opposite sequence is visible with the red filter in front of the left eye. Viewing

[6] The three editions of the book have been published with different titles and with a growing number of volumes: Howard and Rogers (1995, 2002, 2012).

with both eyes results in varying clarity because of the rivalry between the gratings and the grey circles.

Grey/grating rivalry

Not all visual scientists have been content with simplifying visual experience and two examples will be mentioned, both of whom have examined binocular phenomena. Jacques Ninio has produced computer-generated patterns of stunning symmetry which are illustrated in his books and articles.[7] One of his books is entitled *Stéréo-magie* in which he has also adapted his skills to create stereoscopic pairs as well as autostereograms. His portrait in depth is combined with a symmetrical pattern derived from a photograph of tree branches.

Jacques Ninio—artist and scientist of stereo and symmetry

Depth can be seen in some pictures based on colour alone—red usually looks closer than blue but some people experience the opposite. The phenomenon is called chromostereopsis and the science behind it has been reviewed by Akiyoshi Kitaoka[8] and he has added considerably to its art. He is shown below displaying the phenomenon.

[7] See Ninio (1994, 1998, 2001, 2011).

[8] http://www.ritsumei.ac.jp/~akitaoka/index-e.html and Kitaoka (2016).

Chromostereoscopic Akiyoshi Kitaoka

In contrast to the simplicity of patterns used in visual science, artists revel in the complexities of pictures and manipulate them in myriad ways and this has applied to the limited examples of binocular art, too. In the example below, each pattern consists of radiating wavy lines that have an eye at the centre and an almost undiscernible face. Looking with either eye alone results in a puzzling picture that seems three-dimensional. However, the apparent depth is not consistent throughout the pattern as the same contours on one side can look like humps and on the other side hollows. The perceptual complexity is enhanced when both eyes are used—the normal cooperation between the two eyes is disrupted by competition between them.

Seeing eyes

The approaches to stereoscopic depth and rivalry in art and science will be reflected in the pages that follow. The illustrations have been designed to be viewed with the red/cyan glasses and they provide novel ways of presenting the history, science and art of vision with two eyes. For example Chap. 2 surveys the history of approaches to binocular vision and portrays those who have added to our understanding of it, together with some motif which represents their work. The motif can reflect a title page of a book, text or illustration from it, apparatus they have devised or an idea they have advanced. Rather than displaying the motif and portrait separately they are integrated so that each component is seen by one eye only.

References

Chopin A, Bavelier D, Levi DM (2019) The prevalence and diagnosis of 'stereoblindness' in adults less than 60 years of age: a best evidence synthesis. Ophthalmic Physiol Opt 39:66–85
Fludd R (1617) Utriusque cosmi maioris scilicet et minoris metaphysica, physica atqve technica historia: in duo volumina secundum cosmi differentiam diuisa. (Illustration Tomus 1, p 307)
Howard IP, Rogers BJ (1995) Binocular vision and stereopsis. Oxford University Press, Oxford
Howard IP (2002) Seeing in depth. Volume 1 Basic mechanisms. Porteus, Thornhill, Ontario
Howard IP, Rogers BJ (2002) Seeing in depth. Volume 2. Depth perception. Porteus, Thornhill, Ontario
Howard IP (2012a) Perceiving in depth, vol 1. Oxford University Press, USA, Basic mechanisms
Howard IP, Rogers BJ (2012) Perceiving in depth, vol 2. Oxford University Press, USA, Stereoscopic Vision
Howard IP (2012b) Perceiving in depth, vol 3. Oxford University Press, USA, Other mechanisms of depth perception
Julesz B (1971) Foundations of cyclopean perception. University of Chicago Press, Chicago
Kitaoka A (2016) Chromostereopsis. Encyclopedia of Color Science and Technology. https://doi.org/10.1007/978-1-4419-8071-7_210
Ninio J (1994) Stéréomagie. Seuil, Paris
Ninio J (1998) La science des illusions. Odile Jacob, Paris
Ninio J (2001) The science of illusions. Cornell University Press, Ithaca, NY, Trans Philip F
Ninio J (2011) L'empreinte des sens. Odile Jacob, Paris
Wade NJ (2007) Artful visions. Spat vis 21:27–53
Wade N (2009) Allusions to visual representation. In: Skov M, Vartanian O (eds) Neuroaesthetics. Baywood, Amityville, USA

Chapter 2
A Little History

Seeing with two eyes has posed puzzles for philosophers and physicians for many centuries. It was appreciated that slightly different things are seen with each eye and these differences were often illustrated. However, the problems they addressed concerned how we see a single world with two eyes rather than how we see it in depth. In this chapter the journey from descriptions of how we see to experiments examining the processes of perception will be outlined. In order to understand binocular vision it was necessary to have knowledge about the optics of the eye, its anatomy and microanatomy as well as how it responds to the action of light and how the brain processes the signals from the eyes. The major transformation in this history was the invention of the stereoscope by Charles Wheatstone in the early 1830s. He named the instrument and described experiments he conducted with it establishing that stereoscopic depth is based on retinal disparities. Thereafter, more precise experiments could be conducted so that the relationship between retinal disparities and depth perception could be established. When retinal disparities were small the eyes cooperated to yield the impression of depth; when they were large, the eyes competed with one another to yield rivalry. The competition was not restricted to the eyes but could involve local features of the patterns, too. Paradoxically, binocular rivalry was described before stereoscopic depth perception because it could be examined easily without the aid of a stereoscope. With the development of rudimentary experiments a more precise terminology for binocular processes was required and an account is provided of the origins of the terms now in use. These include stereoscope, stereoscopy, stereographs, stereopsis, dichoptic, cyclopean eye, binocular lustre, rivalry and colour stereoscopy. One of the many aspects of vision with two eyes that emerged was that there are differences between them. This is referred to as eye dominance and its history is surveyed. There are three principal aspects of differences between the eyes: eye movements, visual acuities and suppression as evidenced by binocular rivalry.

N. Wade, *Vision and Art with Two Eyes*, Vision, Illusion and Perception 3,
https://doi.org/10.1007/978-3-030-77995-5_2

In this chapter some of the history of investigations into binocular vision will be outlined. Not all were conducted by scientists as artists also mused about the relationship between perception and pictures. Before this some considerations are directed to the anatomy and physiology of the paired senses as well as the terminology applied to binocular vision.

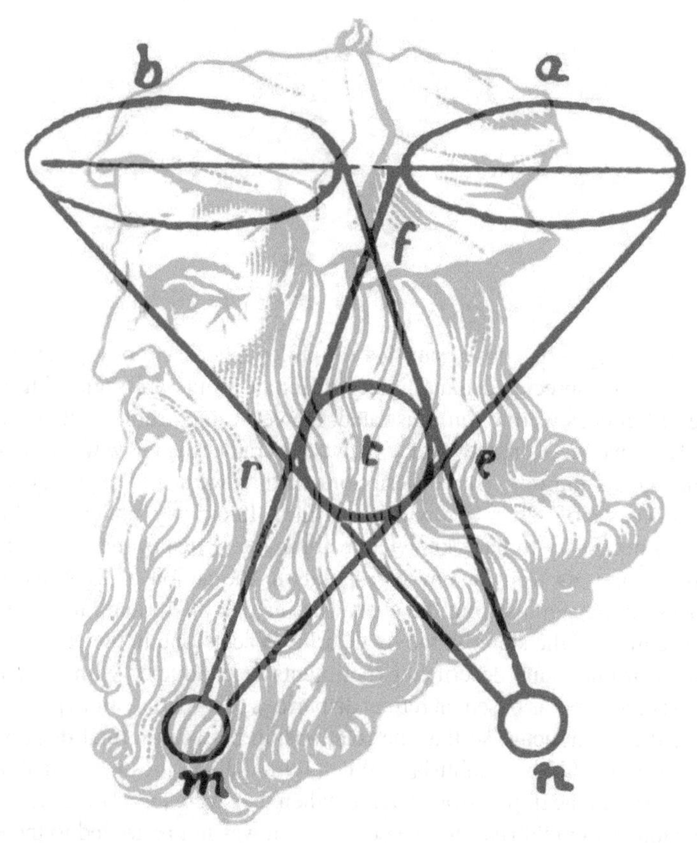

Leonardo da Vinci and his diagram of viewing a small sphere (t) with two eyes (m and n). The regions of the background obscured by the sphere (a and b) did not overlap so that the whole of the background could be seen with both eyes but not with one alone

For much of the long history of investigations on vision, speculations were that seeing with one eye was superior to that with two. In large part this was a consequence of the theory that visual spirit passed from the brain to the eyes.[1] According to this view, the visual spirit was more concentrated when it was confined to a single eye. Among those who disputed this theory was Leonardo da Vinci (1452–1519) who struggled long and hard with the contrast between monocular and binocular vision.[2]

[1] See Wade (1998a, b).

[2] See McMurrich (1930), Kemp (1989), Strong (1979), Wade et al (2001).

Linear perspective provided a monocular match between a picture and a view of a scene from a single point. But what happens when two viewpoints are adopted? Leonardo examined this many times in the context of a small object lying in front of a background. He returned to the issue repeatedly as indicated by the many diagrams he made of it. In each instance, vision with two eyes was optically and phenomenally different from that with one. The example he used, of viewing a sphere with a diameter less than the distance separating the eyes, reflected one condition Euclid analyzed, but Leonardo added the characteristic of seeing the whole background.

Every time Leonardo returned to the issue, he came to the same conclusion that he could not depict correctly on canvas everything he saw with two eyes. He was unable, in the terminology of virtual reality, to simulate what he saw with two eyes. In the early fifteenth century Leon Battista Alberti (1404–1472) described a procedure for conveying visual angles from a scene to a picture plane, now called Alberti's window: viewing a scene through a window with the eye in a fixed location and tracing the contours on the transparent surface.[3] This simulates the monocular visual world on a canvas, but not that of the binocular visual world. For example, Leonardo's drawings represent both binocular and monocular observation of a small sphere and the accompanying text emphasizes the differences between viewing a scene and a painting of it in terms of perceived depth and the amount of the background that is visible. Viewing an object that is differentially obscured in the left and right eyes provides an impression of depth that is now called Da Vinci stereopsis.[4]

[3] Alberti (1435/1966).

[4] Nakayama and Shimojo (1990).

Da Vinci stereopsis

In many ways the history of science is reflected in the history of sight. For example, the first approximately accurate diagram of the nerve pathways from the eyes to the brain were drawn (but not published) by Isaac Newton (1642–1727); he is shown below together with a diagram based on his notes.[5] Newton was able to draw upon more accurate knowledge about the anatomy of the eye than was available to Leonardo and he was able to incorporate putative processes in the brain, too. The individual nerves from each eye project to both hemispheres and separate (decussate) at the optic chiasm. A very brief summary of Newton's conclusions was given in the XVth Query of his *Opticks*,[6] but a fuller account, perhaps a complete one, was included in Harris's *Treatise of Optics*. Newton described the pathways eloquently, thus:

> Now I conceive that every point in the retina of one eye, hath its corresponding point in the other; from which two very slender pipes filled with a most limpid liquor, do without either interruption, or any unevenness or irregularities in their process, go along the optic nerves to the juncture EFGH, where they meet either betwixt G, F or F, H, and there unite into one pipe as big as both of them; and so continue in one, passing either betwixt I, L or M, K, into the brain, where they are terminated perhaps at the next meeting of the nerves betwixt the *cerebrum* and *cerebellum*, in the same order that their extremities were situated in the retina's. And so there are a vast multitude of these slender pipes which flow from the brain, the one half through the right side nerve I L, till they come at juncture G F, where they are each divided into two branches, the one passing by G and T to the right side of the right eye

[5] Newton's sketch was redrawn and published by Harris (1775) and Brewster (1855).

[6] Newton (1704).

A B, the other half shooting through the space E F, and so passing by X to the right side of the left eye $\alpha\beta$. And in like manner the other half shooting through the left side nerve M K, divide themselves at F H, and their branches passing by E V to the right eye, and by H Y to the left, compose that half of the retina in both eyes, which is towards the left side C D and $\gamma\delta$.[7]

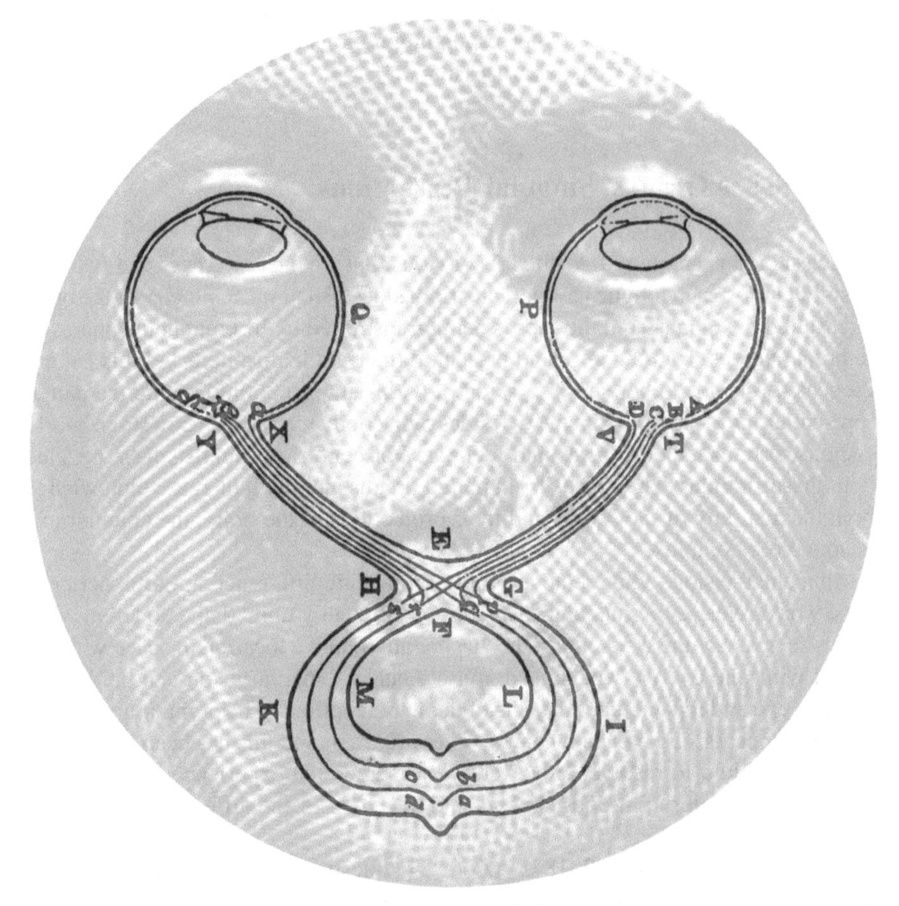

Isaac Newton and a diagram (derived from his notebook drawing) of the nerve pathways from the eyes to the brain

In the course of this description Newton gave a possible basis for binocular rivalry: "Why, though one thing may appear in two places by distorting the eyes, yet two things cannot appear in one place. If the picture of one thing fall upon A, and another upon a, they may both proceed to p, but no farther; they cannot both be carried on the same pipes *pa* into the brain; that which is strongest or most helped by phantasy

[7] Harris (1775), p. 109.

will there prevail, and blot out the other".[8] Unfortunately, he was wrong in his anatomy—the two 'pipes' (nerve fibres) in the optic nerve do not unite in the optic chiasm but remain separate. Nonetheless, Newton was seeking to examine binocular vision within the framework of his science and to give an interpretation of visual phenomena in mechanistic terms.

Modern research has built upon the foundations laid by Newton. Function is related to structure. In the case of binocular rivalry, the dynamic shifts in visibility of different patterns presented to corresponding regions of each eye is related to the known neurophysiology of visual processing.

Paired Sense Organs, Singular Perceptions

Our sense organs are paired. While it is obvious that we have two eyes and two ears the singular nose and tongue have left and right sides. Bilateral symmetry of the body has resulted in signals from the senses being transmitted to different hemispheres of the brain. Nerves from the left hand project to the right hemisphere and this pattern is repeated for most other senses. Vision provides an exception. Despite the paired representations of signals from the two sides of the body we experience a single world of sights, sounds, smells, tastes and feelings. For centuries scientists have puzzled about these singular experiences from paired organs and, with increased knowledge about the anatomy of the senses and their projections to the brain, the conundrums became more complex.

Monocular vision yields relatively little immediate difference in our experience of the world relative to binocular: perception is typically unified and coherent, with objects seen as having a specific size, shape, colour, and a location in space whether viewed with one eye or two. Indeed, a great deal of spatial vision can be derived from one eye and this includes aspects of depth perception.[9] The point of note is that the world appears single even when two eyes are used, and this has been seen as the principal problem to resolve. However, while perceptual experience changes little when one eye is closed, it does change, and the study of the differences has a long history.[10]

Charles Wheatstone (1802–1875) wrote "No question relating to vision has been so much debated as the cause of the single appearance of objects seen by both eyes".[11] Binocular single vision has been discussed at least since the time of Aristotle (384–322 BC), and, since five hundred years later, it has been examined experimentally when Claudius Ptolemy (ca. 100–170) defined lines of visual correspondence for the two eyes. Many of the statements about binocular single vision are reflections of its breakdown and the experience of binocular double vision. This was the case for

[8] Harris (1775), p. 110.

[9] See Vishwanath (2014).

[10] Wade (2018).

[11] Wheatstone (1838), p. 387.

Aristotle's description of one of the most common ways of inducing double vision—gently pushing one eye with the finger. The involvement of eye movements was stressed by Ptolemy, who suggested a functional advantage of double vision, namely, to bring the two eyes into register with regard to the objects under inspection.[12]

Perceptual experience is typically unified and coherent: objects are seen as having a specific size, shape, color, and a location in space. This obtains under conditions of binocular and monocular vision. Contrary to the evidence accumulated since the late seventeenth century, it was long believed that vision with one eye was superior to that with two. Aristotle took this to be self-evident, interpreting it in terms of eye movement control. The source of much subsequent comparison was driven by the theory of the visual spirit circulating between the ventricles and the eyes. It could be transmitted to one or two eyes, and thus was thought to be more concentrated in monocular viewing. This opinion was repeated over the following centuries, and it was held as late as the seventeenth century. Many statements were made about tasks that were more difficult to perform with one eye rather than two, or stimuli that were more difficult to see.

At the end of the sixteenth century Giovanni Battista della Porta (1535–1615) presented a delightfully simple solution to the problem of single vision with two eyes—we only use one at a time.[13] It is clear from the diagrams in Porta's book that his knowledge of ocular optics was wanting. A decade later Johannes Kepler (1571–1640) elucidated how images are brought to a focus on the retina and this transformed approaches to both monocular and binocular vision.[14] An alternative to Porta's theory of binocular single vision was contained in the book by François d'Aguilon or Franciscus Aguilonius (1567–1615) published in 1613.[15] He proposed that the signals from the two eyes combined or fused in order to see the world as single. Some of the finest illustrations of the study of binocular vision are to be found in the *Optics* of Aguilonius. The six frontispiece engravings were designed by his friend, Peter Paul Rubens.[16]

[12] See Wade (1998a, b) for a review.

[13] Porta (1593).

[14] Kepler (1604).

[15] Aguilonius (1613).

[16] See Ziggelaar (1983).

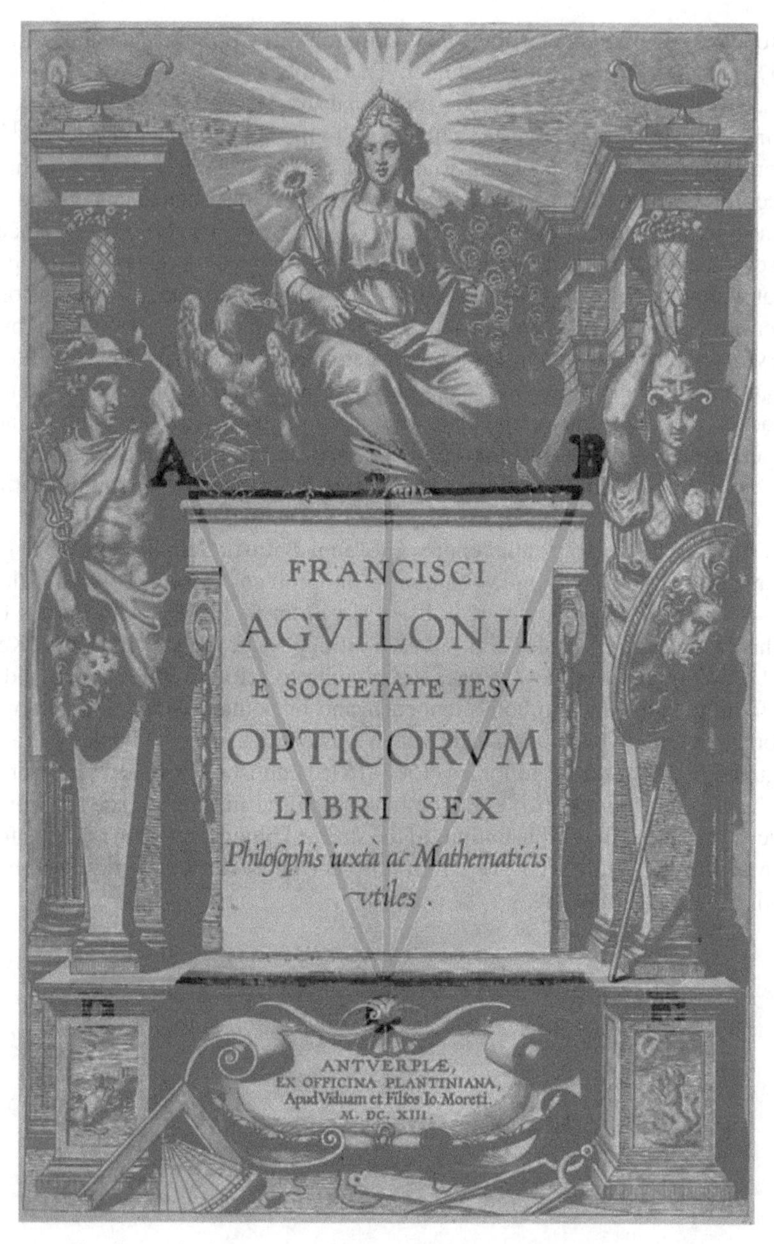

Frontispiece for the six books on optics by Franciscus Aguilonius together with his binocular diagram: "If the common radius intercepts the horopter at right angles, the horopter makes the angles equal with the axes. From the two eyes A and B the optic axes project AC and BC and the common radius FC intercepts the horopter DE at right angles. I say that those angles which the horopter makes with the axes, namely ACD and BCE are equal to one another."[17]

[17] Aguilonius (1613), p. 147.

Leonardo's often quoted comparison between viewing a painting of a scene and the scene itself was an implicit contrast between vision with one eye or two. It is most instructive because the concept of relief or depth is taken to be the distinguishing characteristic of binocular vision. Wheatstone, with his knowledge of binocular depth perception, enquired why more errors are not made in monocular viewing. His answer was that many cues for depth and distance exist, and that alternatives are used when one source is not available. Most particularly, he proposed that motion of the head (motion parallax) was a potent substitute for retinal disparity.

Anatomy of the Eye

Initially anatomists dissected the singular sense organs in order to describe their gross structures before they tried to trace the nerves to specific sites in the brain. Some knowledge of ocular anatomy must have been available to those specialists who carried out eye operations in ancient times. However, the records that have survived usually relate to the fees they charged and the penalties they suffered for faulty operations. Their skills and understanding would have been passed on to Greek scholars, who both developed and recorded them. The initial Greek speculations about the anatomy of the eye were fuelled by philosophy: the four elements of earth, air, fire, and water led to the proposition that there must be four coats to the eye. At about the same time the optic nerve was described and it was thought of as a hollow tube, enabling humors to pass from the ventricles of the brain to the eye. The dominance of philosophy over observation was reversed for one school of Greek medicine that emerged in the fifth century BC, of which Hippocrates (ca. 460–370 BC) was a member. Naturalistic observation superseded superstition, but the examination of anatomical organs was still eschewed. The moral strictures of the time did not countenance dissection of dead bodies, although this was soon to change with the Platonic dissociation of the body from the soul. It is known that Aristotle did dissect the eyes of animals, and he is believed to have written two books (now lost) on the eye. Drawing on the evidence from dissection marked the dawning of more exact knowledge of the structure of the eye. The anterior chamber of the eye was separated from the posterior by a membrane, and the lens was assigned the faculty of vision. This notion was to survive for many centuries. Vision was considered to reside in the lens and Claudius Galen (ca. 130–200) supported this assertion when he noted that clouding due to cataract resulted in blindness.[18] Thus we find a clear attempt to assign the function of vision to a structure in the eye. Galen based his anatomy on dissections of animals, particularly monkeys, but most of his ocular anatomy was derived from dissecting the eyes of freshly slaughtered oxen. The restrictions that were subsequently placed on dissections of humans and animals resulted in a reliance on Greek works on anatomy, and they were recounted until the

[18] See May (1968).

time of Andreas Vesalius (1514–1564) over one thousand years later. His portrait and diagram of the eye are shown below.[19]

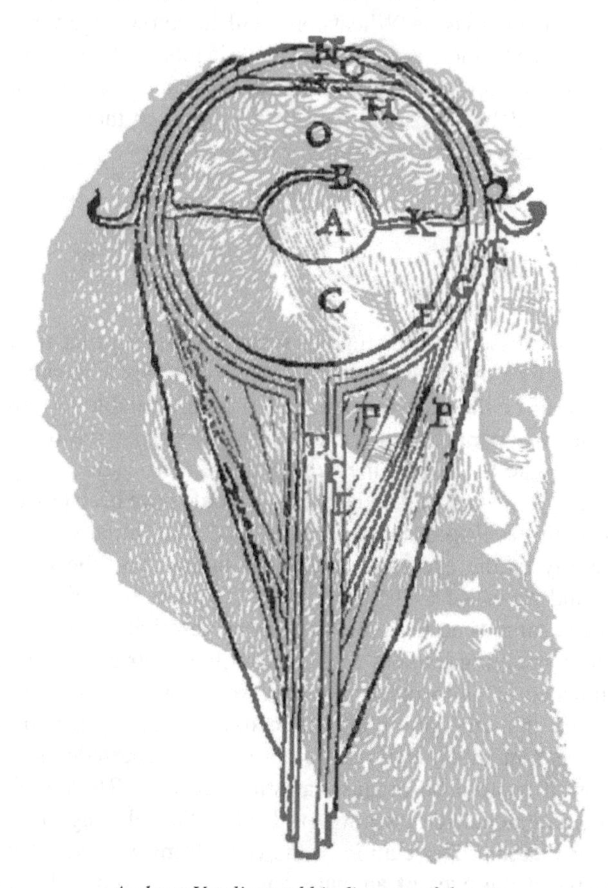

Andreas Vesalius and his diagram of the eye

The journey from Galen to Vesalius was not straight. The disinterest in science and medicine after the sacking of Rome in the fifth century left Europe in the so-called 'Dark Ages', but Greek wisdom was retained by Islamic scholars, who translated many books into Arabic and eventually transmitted them to late medieval students. Arabic accounts of the eye drew on Galen for inspiration, but their illustrations reflect a greater concern with geometry than anatomy. Scholars in the late Middle Ages derived much of their knowledge from manuscript translations of Arabic texts into Latin, and the diagrams of the eyes showed a similar preoccupation with geometry. Printed figures of the eye were published from the beginning of the sixteenth century. However, they are unlikely to have been based on observation of actual eyes, but

[19] Vesalius (1543).

derived from earlier manuscript drawings. Shortly thereafter, the genius of Vesalius was brought to bear on the topic, and the modern era of anatomy was founded.[20]

Following the relaxation of sanctions prohibiting dissection of human bodies in the fourteenth century, knowledge concerning anatomy in general slowly began to be based on more secure ground, although the descriptions were not always accurate and observation often remained a slave to Galenic dogma. The dissecting skills of the anatomist were critical, and the major advances came with practitioners like Leonardo and Vesalius. Leonardo's detailed drawing of dissections did not make any immediate impact because they remained both in manuscript form and in private hands.[21] Unlike his anatomical drawings of the musculature, those of the eye reflected a conflation of dissection and dogma: his rather crude drawings reflected a reliance on Galen, even though he did prepare the excised eye (by boiling it in the white of an egg) for dissection. The lens was represented as spherical and central in the eye, and the optic nerves passed to the cerebral ventricles. The renaissance of anatomy is associated with Vesalius, who published his book *On the Structure of the Human Body* in 1543. Vesalius presented an account of anatomy that was almost free from the legacy of Galen. While Vesalius could examine the structure of the eye with his own rather than Galen's eyes, he did not pay too much attention to it. His drawings did not match the detail or accuracy of those for the skeletal musculature and internal organs: a symmetrical lens was still located in the centre of the eye and the optic nerve was situated on the optic axis. Felix Platter (1536–1614) moved the lens towards the pupil and recorded the differences between the curvatures of its front and back surfaces.[22] Christoph Scheiner (1571–1650) gave the first accurate diagram of the mammalian eye; the lens and its curvatures are appropriately represented and the optic nerve leaves the eye nasally.[23] This figure has frequently been reprinted, but it is rarely acknowledged that it is not a human eye. Scheiner stated that he did not have the opportunity of dissecting a human eye, and so his evidence was based on the eyes of domestic animals; he applied reasoning in extrapolating the structure to the human eye. Scheiner is shown below together with his illustration of the mammalian eye.

[20] See Wade (1998a, b).

[21] See Clayton (2019).

[22] Plater (1583).

[23] Scheiner (1619), Daxecker (2004).

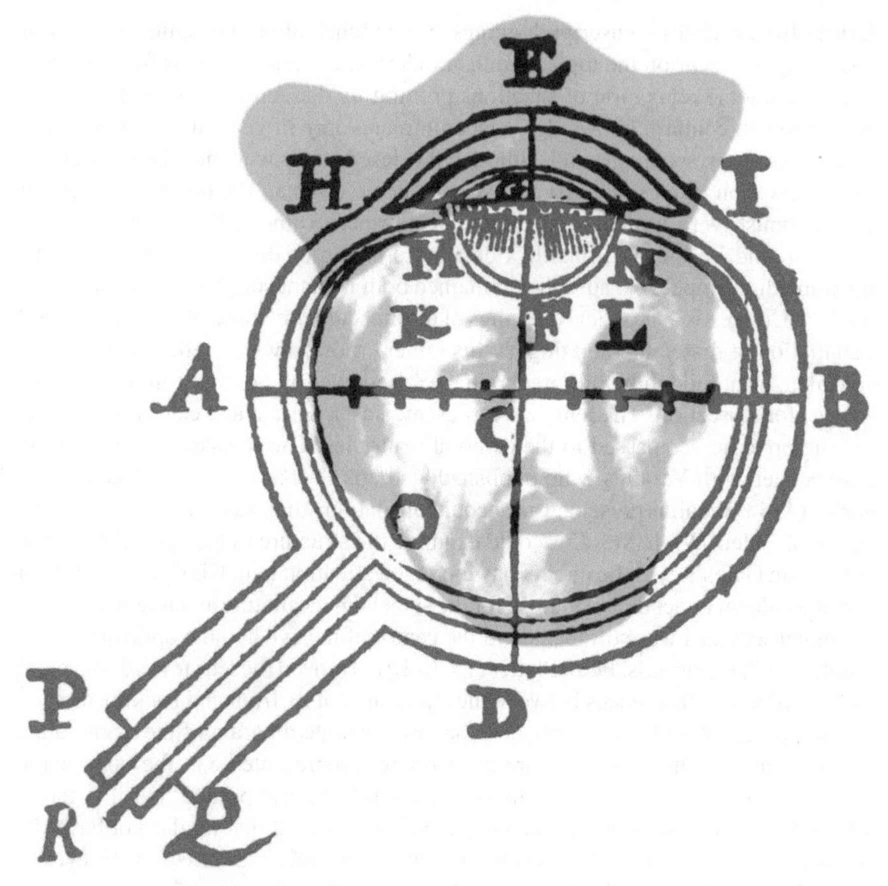

Christoph Scheiner and his diagram of the mammalian eye

Scheiner made an explicit analogy between eye and camera indicating how both operated according to similar optical principles. He presented a pictorial analysis of optical image formation in the camera and the eye—with both inverted and upright images due to the addition of convex and concave lenses. He noted that an upright retinal image resulted in inverted vision. Furthermore, Scheiner described how an image could be seen on the exposed surface of an excised animal's eye, an experiment "he had often performed". It is clear from Scheiner's analysis that the equation of eye and camera raised issues of focusing or accommodation within the optical system. The most familiar illustration of image formation in an excised eye is that from René Descartes's (1596–1650) *Dioptrics*, in which the cosmic observer ponders on the spatial arrangement of the inverted and reversed retinal image.[24] Descartes is shown below with his figure. He was able to provide a more precise analysis of image formation because he could apply Snell's law to the refractions taking place in the eye.

[24] Descartes (1637/1902).

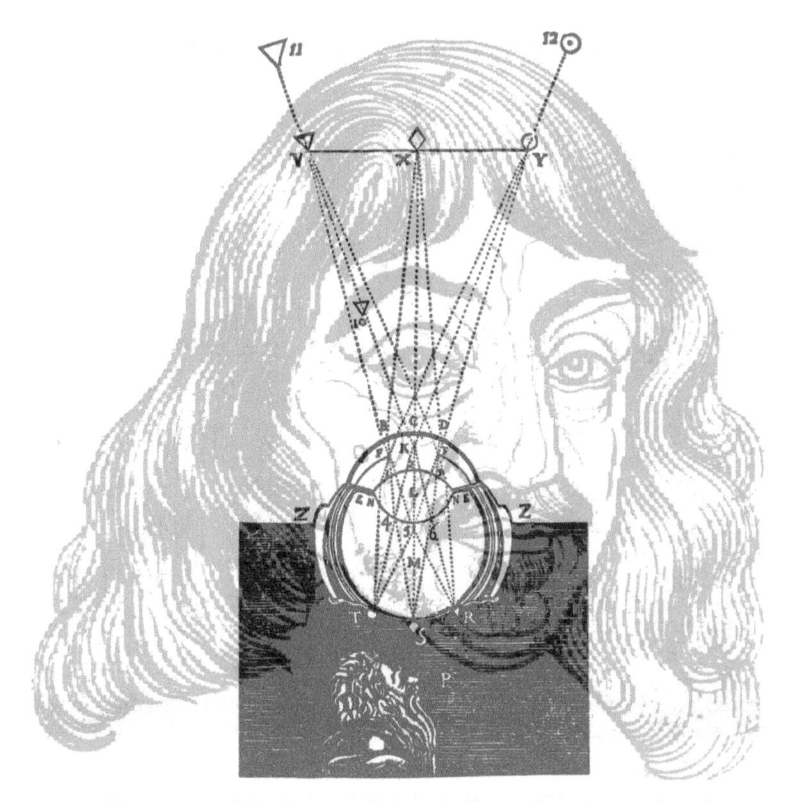

René Descartes and his diagram of observing image formation on the retina

With the introduction of achromatic microscopes in the 1830s the microanatomy of the retina was revealed.[25] When René Descartes was discussing the eye he considered that the retina consisted of nerve endings and that their dimensions defined the limits of visual resolution. Since these were of a particular size, he argued that no object smaller than a fibre ending could be resolved. When achromatic microscopes were directed towards retinal fibres in the 1830s a greater degree of structure was discerned, but the nature of the receptors was not immediately apparent. Gottfried Reinhold Treviranus (1776–1837) was a comparative zoologist who did much to establish biology as an independent discipline within Germany, and to provide support for the cell doctrine. From 1833 Treviranus measured the dimensions of many nerves in sensory systems and in the brains of a variety of animals. He considered that the brain was comprised of cylindrical cells arranged in parallel.[26] In his posthumously published volume on the inner structure of the retina, Treviranus presented drawings based on vertical and horizontal microscopic sections of cells in the visual systems of many species. His diagram of the crow's retina indicated a wider variation in retinal structure than had previously been represented, and the

[25] See Schickore (2007) and Wade (2019).

[26] Treviranus (1837).

layers within it are clearly shown. Treviranus can be seen in his diagram of the crow's retina.

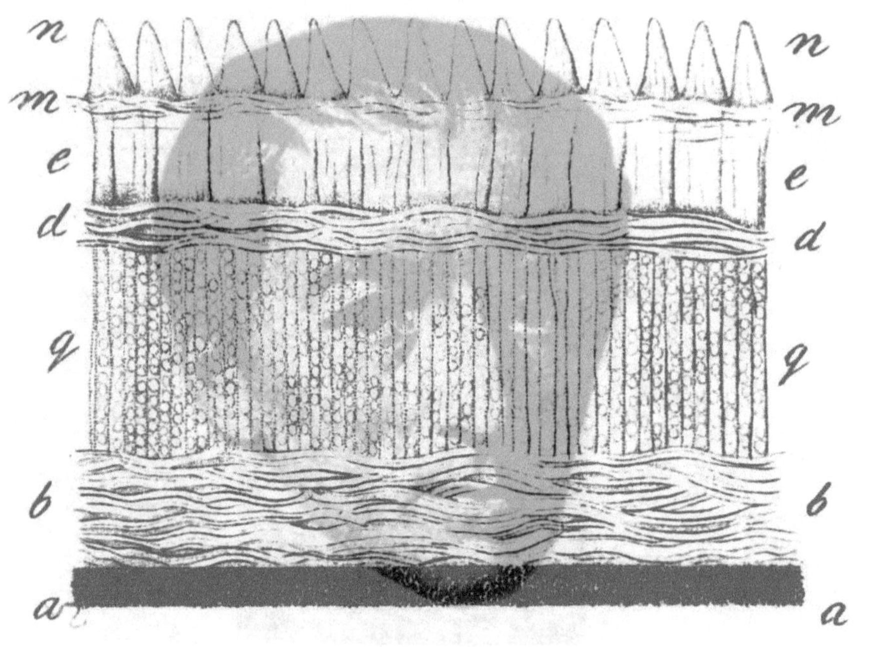

Gottfried Treviranus and his diagram of the cells in the retina of a crow

In the years following 1840 rapid advances in fixing, sectioning, and staining microscopic preparations were introduced and this applied particularly to the retina. Max Schultze (1825–1874) gave order to the many layers of cells in the retina and he also examined the complement of rods and cones in a variety of animals.[27] He suggested that rods and cones function under different levels of illumination—which became known as duplicity theory.

[27] Schultze (1866).

Max Schulze and the human retina flanked by individual rods and cones

The detailed structures of the retina became more visible with the introduction of cell staining methods, particularly the 'black reaction' of Camillo Golgi (1843–1926). In 1873 he devised a novel technique for staining nervous tissue; it consisted of hardening the preparation in potassium bichromate and then impregnating it with silver nitrate. The subsequent black reaction exposed the networks of nerves in grey matter in a manner that had not been possible previously.[28] One of Golgi's students, Ferruccio Tartuferi (1852–1925), applied the silver staining technique to the retina in 1887, and his beautiful illustration of it is shown below. The horizontal and amacrine cells are clearly represented, as are the rods, cones and bipolar cells.

[28] See Mazzarello (1996, 2010), Wade and Piccolino (2006).

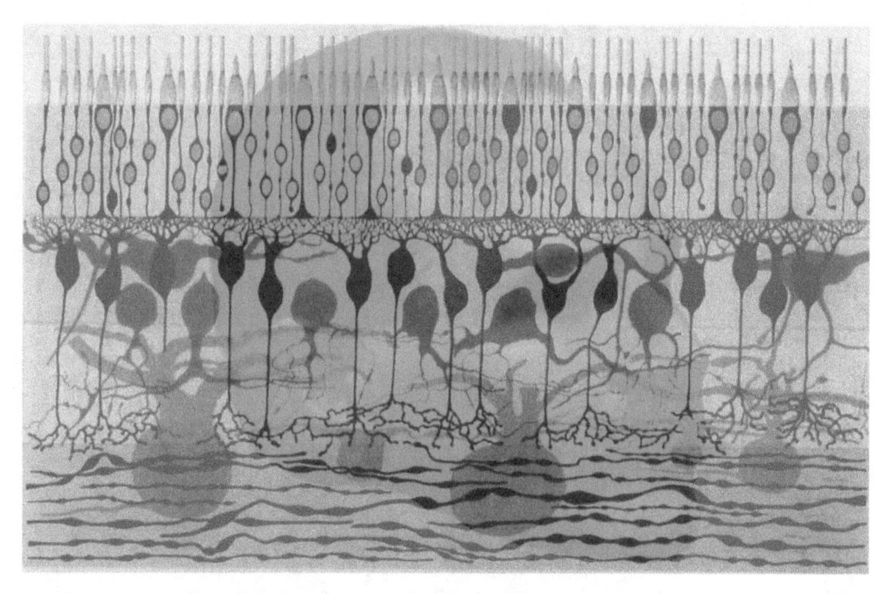

Ferruccio Tartuferri and his illustration of retinal structure

Anatomy of Visual Pathways

The ignorance of the anatomy of the eye in antiquity was multiplied with respect to the pathways from the eyes to the brain. Indeed, the brain itself was often considered of relatively minor importance. Hippocrates did locate the pleasures, sensations, and thoughts in the brain, but Aristotle did not follow him in this speculation: the heart was the centre of sentience. Galen believed that the origin of the visual pathways was located in the anterior ventricle of the brain, where the animal spirit could interact with the visual spirit, borne by the optic nerves. The optic nerves themselves came together at the optic chiasm, but each of the nerves remained on its own side. This error was to be repeated by Vesalius, and it was integrated into Descartes' analysis of vision, as is shown below. Descartes did stress the correspondence between points on the object, those on the retina, and their projection to the brain, but it is unlikely that he was addressing the issue of corresponding points in the two retinas. His analysis of binocular vision was by analogy to a blind man holding two sticks and it was not physiological. The union that was depicted in the pineal body reflected an attempt to match singleness of vision with a single anatomical structure.

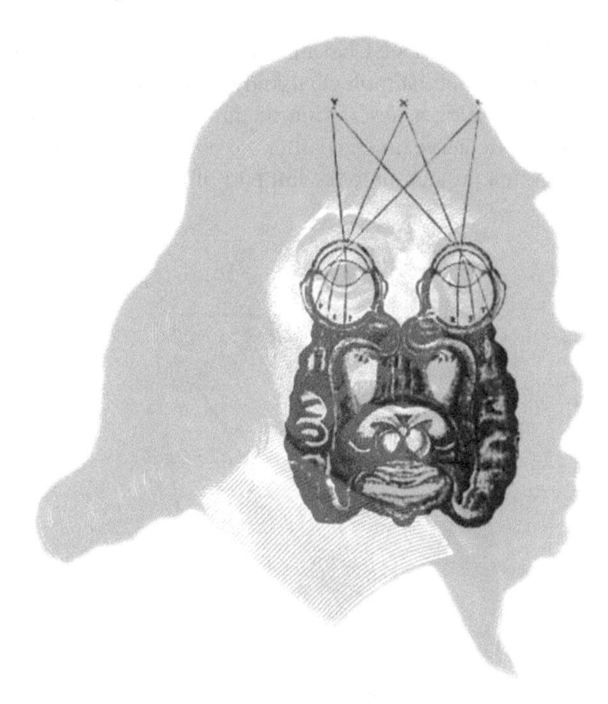

René Descartes and his diagram of pathways from the eyes to the brain

Newton made the first representation of partial decussation at the optic chiasm (shown at the head of this chapter) and proposed a theory of binocular single vision based upon it. The subtlety of Newton's analysis was not, however, widely disseminated. In 1704 he did make passing reference to it in Query XV of his *Opticks*, together with a telling reference to species differences, but the manuscript itself was not published. Newton was almost correct in his analysis: partial decussation was appropriate, but the nerves themselves united at the chiasm. That is, optic nerve fibres from corresponding points on each eye form single fibres in the optic tract. This detail was rectified by John Taylor (1708–1772) in an accurate representation of the partial crossing over and independence of the nerve fibres.[29] Newton's conclusion that the fibres united in the optic chiasm had important implications for his theory of binocular vision. Indeed, Newton used binocular rivalry to support his proposal that fibres from the two eyes were united in the optic chiasm because it was not possible to see two objects in the same location.

Taylor, who is shown below, was a most colourful character. His opinion of himself was high, and his manner was flamboyant: the tours he made around the cities of Europe were more like circuses than surgeries. Samuel Johnson said of him that he was an instance of how far impudence will carry ignorance! It is also possible that he was responsible for blinding more of the nobility of Europe than any other single person. He published books in several languages, usually ennobling himself with

[29] Taylor (1738).

such titles as Ritter von Taylor or Chevalier Taylor.[30] Nonetheless, he must have divined some optical and ophthalmological knowledge through his travels because he did provide the first correct description of the partial decussation at the optic chiasm: fibres in the optic nerve diverged after the optic chiasm, with those from the left halves of each retina projecting to the left part of the brain, and vice versa.

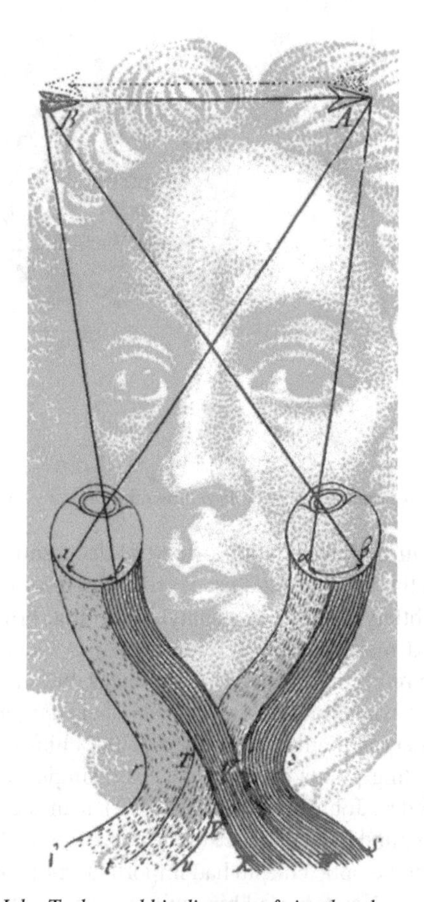

John Taylor and his diagram of visual pathways

There was much debate and controversy surrounding partial decussation at the optic chiasm.[31] It was resolved by Hermann Munk (1839–1912) who pursued the pathways from the eyes to the brain and provided confirmations of the speculations of the errant oculist John Taylor. Disputes about partial decussation at the optic chiasm had raged in the century and a half that separated the two but Munk's painstaking dissections and beautiful illustrations established the routes taken from

[30] See Wade (2008).

[31] See Duke-Elder (1961), Polyak (1957).

eyes to brain.[32] He is shown below in one of his figures. Munk went on to examine the visual cortical functions in greater detail. He found that cortical lesions in the visual areas produced two types of blindness which he called psychic and cortical. Psychic blindness resulted in the experimental animals (dogs and monkeys) behaving appropriately to objects (by avoiding collision with them, etc.) but showing no evidence of recognising what they were. Cortical blindness, on the other hand, reflected total absence of vision, and it typically followed complete removal of the primary visual cortex.

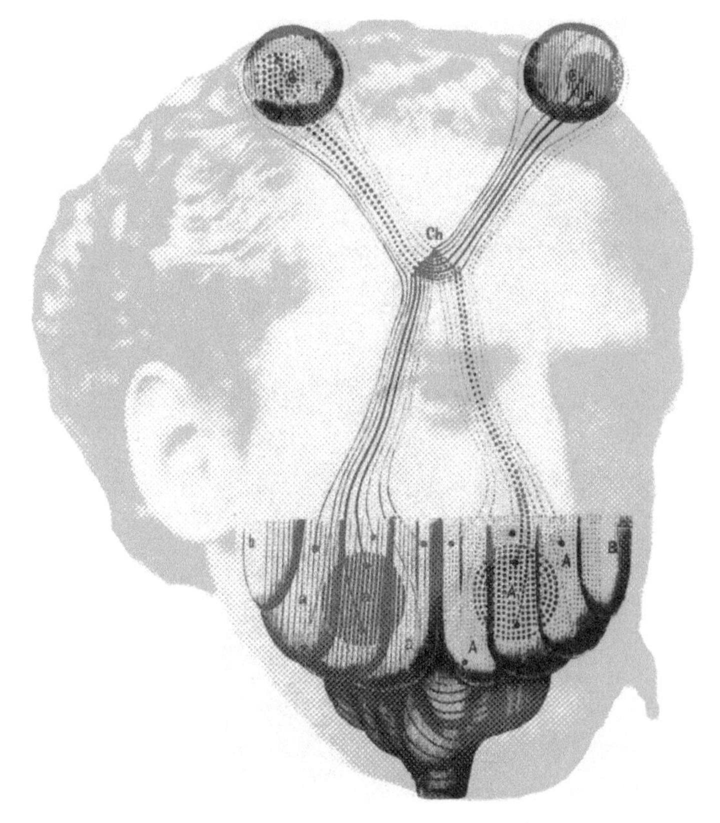

Hermann Munk and the projections of the nerves from the eyes to the brain

The great anatomist and artist Santiago Ramón y Cajal (1853–1934) represented the consequences of such partial decussation to the passage of patterns from the eyes to the brain.[33] The stimulus he used was the traditional one of an arrow viewed from the side so that half was projected to one hemisphere and the other to the opposite side of the brain, as is shown below. The principle of projection featured strongly in Cajal's analysis of vision and his detailed staining studies of pathways in the brain did

[32] Munk (1879).

[33] Cajal (1893).

much to hasten a clearer understanding of neuroanatomy. It is mainly due to Cajal's study of retinal architecture (culminating in 1893 with his monumental memoir on the vertebrate retina) that the photographic paradigm dominated physiological investigations of the retina for the first half of the twentieth century. In 1906, Cajal and Golgi were jointly awarded the Nobel Prize for medicine. After having been first exposed to Golgi's process in 1887, Cajal adopted and adapted it, and in the next year he published the results of his initial investigations in the cerebellum and in the retina.[34] In the cerebellum he was able to establish the individuality of the cellular elements and, moreover, to recognize the presence of an ordered pattern of connectivity.

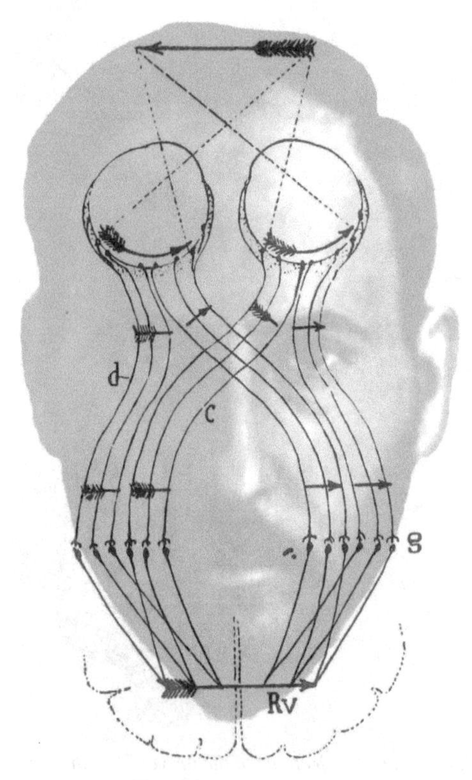

The arrow in Cajal's eyes

[34] Piccolino (1988).

The Visual Brain

Vision, like all other aspects of experience and behaviour, is mediated by activity in the brain. It also depends upon activity in highly specialised cells in the sense organs called receptors. A full understanding of vision will include an appreciation of the neurophysiological processes that are initiated by the action of light on receptors in the eye. These involve the modification of light energy into nerve impulses and their transmission to the areas at the back of the brain where they are analysed. However some measure of the receptor responses to light was required and this was provided by a Swedish physiologist in 1866. Frithiof Holmgren (1831–1897) wrote "It would be of great importance to find a method that would give, if possible, a direct and objective expression for the effect of light on the retina. The following is an attempt to solve this problem".[35] This was the first index of what was to be known as the electroretinogram and eventually it led to electrical recordings of nerve activity at subsequent stages in the visual system.

HOLMGREN, Fr., Studien ueber Athembewegungen, af *Rosenthal* (Ref.) . . . . 62.
» Undersökningar rörande iris' rörelsemechanism med tillhjelp af kalabar och atropin . . . . . 64, 160.
» Centripetal ledning af nervretning i motoriska nerver 76.
» Parasitdjur i ryggmärgen hos grodan . . . 84, 109.
» Anatomische Untersuchungen über die Blutgefässe des menschlichen Auges af *Leber* (Ref.) . . . . 157.
» Method att objectivera effecten af ljusintryck på retina . . . . 177.
» Nyare undersökningar om gangliecellernas byggnad. Efter *Beale, Arnold, Fromann* . . . . 302.
» Spottkörtlar och deras afsöndring . . . . . . 307.
» Lehrbuch der physiologischen Chemie, af *Kühne* (Anmälan) . . . . 314.
» Zur Anatomie und Physiologie der Zonula Zinnii, af *Heiberg* (Ref.) . . . . 433.
» Von den Folgen des beschleunigten Blutstroms für die Absonderung des Speichels, af *Giannuzzi* (Ref.) . . . . 436.
» Rättelser till *Hoppe-Seylers* Handbuch der chemischen Analyse . . . . 440.
» Rhodanreaction i spott . . . . (313) 441.

Holmgren's portrait is combined with the titles of the articles and book reviews that he contributed to the 1865–1866 volume of Upsala Läkareförenings Förhandlingar

[35] Holmgren (1866), p. 178. The article has been translated into English by Kantola et al (2019).

Tremendous advances have been made in our knowledge about the neural processes underlying vision since the nineteenth century. Experimental evidence of electrical activity in nerve cells was to await technological advances in recording and amplifying small electrical signals. This was provided in 1928 by Edgar Adrian (later Lord Adrian, 1889–1977) who was able to record action potentials from individual cells in the visual pathway and then determine the stimulus properties that would excite them. Adrian coined the term 'receptive field' to refer to this. He is shown in a recording of optic nerve responses to light.[36]

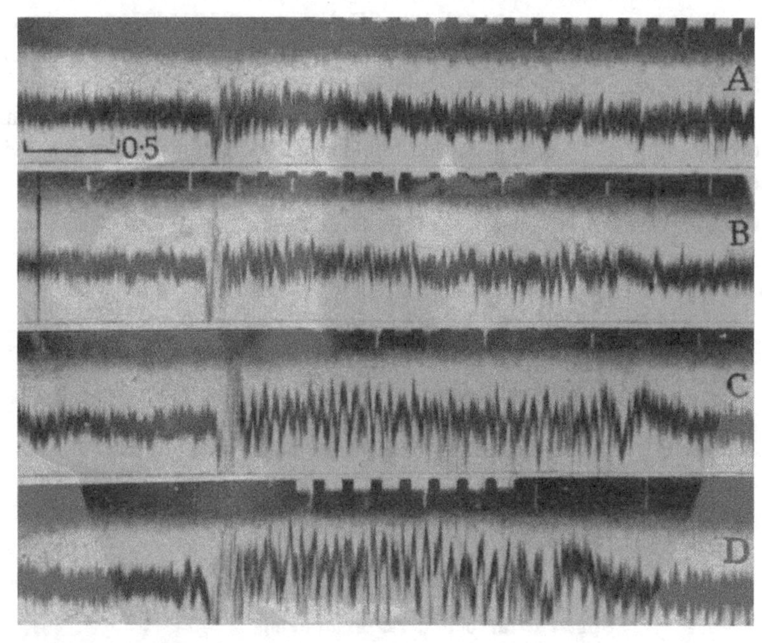

Adrian's record. Lord Adrian and recordings of electrical activity in the optic nerve of water beetle in response to different illuminations from feeble (A) to intense (D)

Haldan Keffer Hartline (1903–1983) later applied the concept of receptive fields to describe the region of the receptor surface over which the action of light modified the activity of a neuron. Hartline studied the responses to light in the horseshoe crab (*Limulus*) and found marked lateral inhibitory interactions between neighbouring receptors. The compound eyes of arthropods are comprised of independent facets (ommatidia) which have an optical and receptive function. The dioptric (conical lens and crystalline cone) concentrates light on to photosensitive cells in the rhabdom; the ensuing electrochemical changes are conducted by neurons with lateral as well as central axons. The eye of *Limulus* contains around 800 ommatidia. Hartline recorded the electrical activity of individual nerve fibres and found that the response to the same intensity of light was modulated by activity in neighbouring ommatidia: "The

[36] Adrian (1937).

occurrence of a purely inhibitory action in a relatively simple eye is of general interest; the role that may be played by inhibitory interaction in enhancing contrast gives it an importance to visual physiology".[37] In 1967, Hartline was awarded the Nobel Prize in Physiology or Medicine, together with Ragnar Granit (1900–1991) and George Wald (1906–1997).

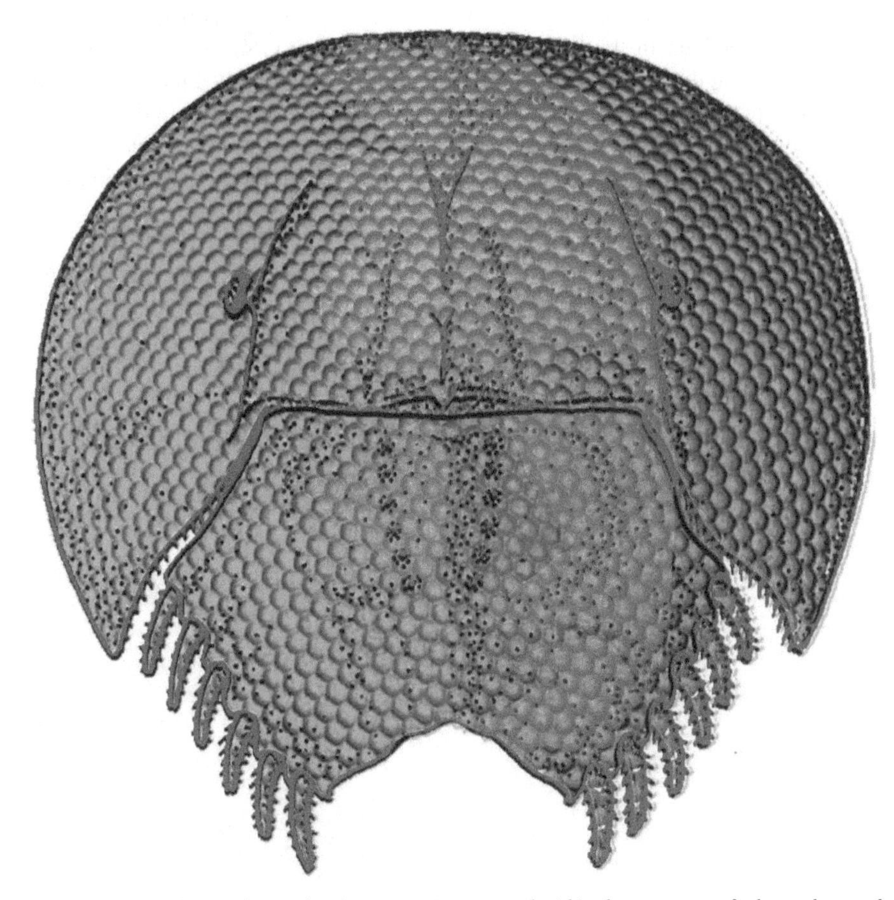

Hartline's horseshoe crab. Hartline's portrait is presented within the carapace of a horseshoe crab and combined with a matrix of ommatidia

Prior to the use of microelectrode recording it was thought that the cells in the visual pathway were excited simply by the presence of light. Now it is known that it is the pattern of light that is of importance, not solely its presence. Research on patterned stimulation at the receptor level had proceeded throughout the first half of the twentieth century, but its pace quickened thereafter. The glimmerings of pattern processing beyond the receptors emerged in the 1950s, and were amplified in the 1960s. It came as something of a surprise that retinal ganglion cells of frog responded to quite complex features of stimulation (like moving dark regions of a specific

[37] Hartline et al (1956), p. 651.

visual angle, resembling a bug), and stimulus properties that excited or inhibited neurons were referred to as 'trigger features' by Horace Barlow (1921–2020).[38] Retinal ganglion cells of cat, on the other hand, were excited by rather simpler stimulus arrangements: they are concentrically and antagonistically organized. If the centre was excited by light the surround was inhibited, and vice versa. Such an arrangement served the detection of differences in luminance well, but steady states would have little effect, since excitation nullified inhibition. This pattern of neural activity was retained in the lateral geniculate nucleus (LGN), but it underwent a radical change at the level of the visual cortex. Barlow investigated the responses of cells in the visual cortex of cats in order to determine how signals from the two eyes are combined. Barlow's research covered a broad range exploring the links between perception and physiology and he is shown in his representation of processing from the retina to the visual cortex.

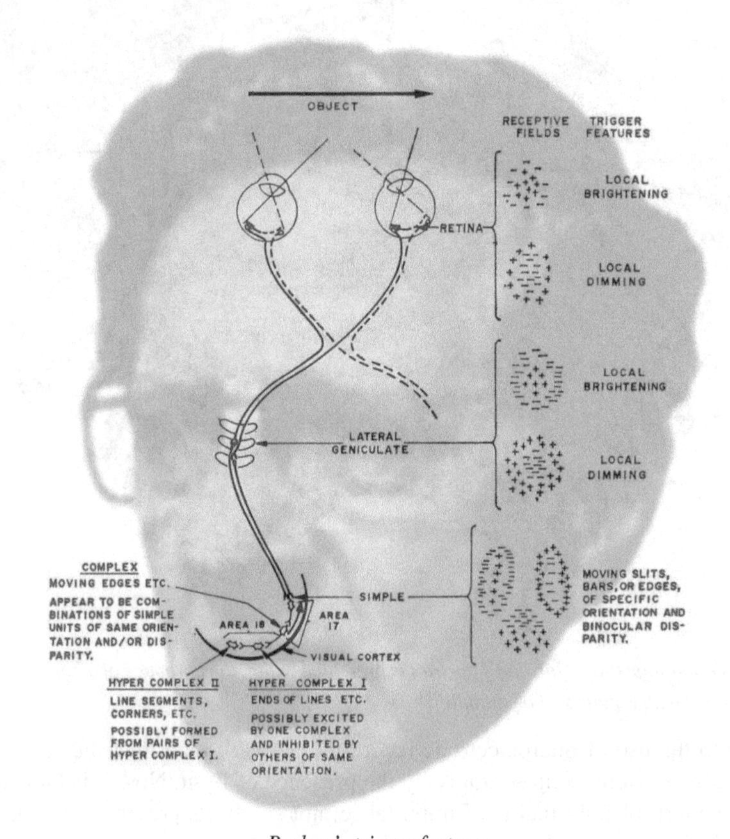

Barlow's trigger features

[38] Barlow (1953).

The activity of each retinal ganglion cell can be influenced by a particular pattern of light falling on the appropriate part of the retina. This region is called the receptive field for that cell. Most retinal ganglion cells have concentric receptive fields with central and surrounding regions that are antagonistic in their function. Thus, the processes of neural excitation and inhibition are vital in determining the ways in which cells in the visual system respond to the patterns of light falling on the retina. The retinal ganglion cells respond to changes in the pattern of illumination, rather than to steady states of uniform illumination. These changes can be spatial and temporal. The excitatory and inhibitory interconnections in the retina are the basis for the receptive field properties of the retinal ganglion cells. The antagonistic interaction between centre and surround also determines the type of stimulus that will produce a response in these cells. The maximum response is elicited by a small white or dark spot, but large responses will also be produced by grating stimuli (patterns of parallel lines). That is, the light bar of a grating will produce a large response if its projected width on the retina corresponds to the dimensions of the on-centre receptive field; conversely, a dark bar of the grating would produce a large response in an off-centre retinal ganglion cell with a receptive field of the appropriate dimensions.

As is shown in Barlow's diagram, nerve fibres from the retina project to the LGN and then to visual area 1 (V1) of the visual cortex. The neurophysiologists David Hubel (1926–2013) and Torsten Wiesel (b. 1924) found that a cortical cell will respond to a line at a specific orientation no matter what its wavelength or intensity, but it will not respond to lines that are inclined away from the preferred orientation. These orientation selective neurons were once considered to make up the majority of cells in the visual cortex. Indeed, such was the interest elicited after their discovery, by Hubel and Wiesel in 1959, that they became called feature detectors.[39] This was because the cortical neurons were selectively tuned to extract certain features contained in the pattern of retinal stimulation—in this case orientation. Direction of edge motion is another feature extracted, as are binocular disparity and colour.

The fibres in the optic tract are from similar halves of both eyes, but they project to different layers in the LGN. The first stage at which binocular integration of the neural signals occurs is in V1. The cortical neurons can be excited by appropriate stimulation of either eye, although one eye will generally have a greater influence than the other. However, at intervals of about 0.5 mm the eye preference changes abruptly to the other eye, and the sequence of orientation selective columns is repeated, rather like the stripes on a zebra. In 1981 Hubel and Wiesel were awarded the Nobel Prize for Physiology or Medicine for their research on visual neurophysiology; they are shown in the eye preference patterns they found in monkey visual cortex.

---

[39] See Hubel and Wiesel (2005).

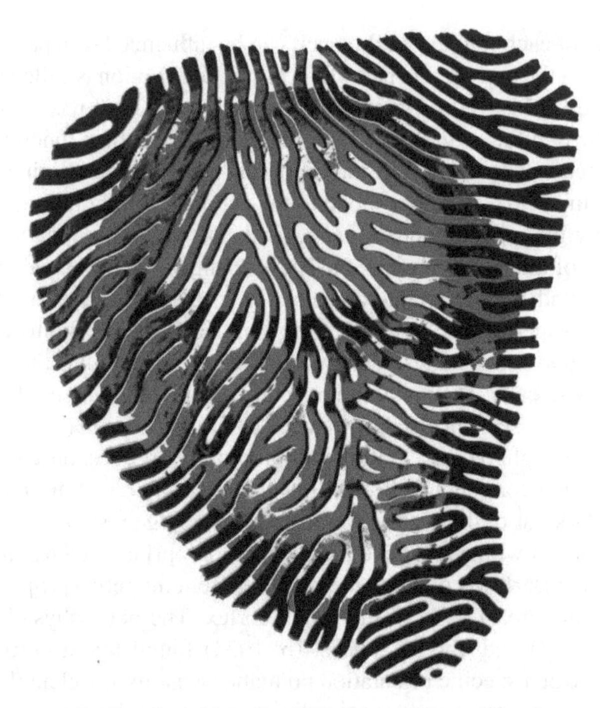

Feature detectives—David Hubel and Torsten Wiesel

The visual cortex is, therefore, a highly organized structure. It appears to break down the pattern of light falling on the eye into discrete features, like retinal location, orientation, movement and wavelength, as well as maintaining a difference between the signals from the two eyes. The binocular combination in V1 is primarily for inputs from corresponding areas of each eye. That is, most binocular V1 neurons are excited by cells from the two eyes with equivalent receptive field characteristics and corresponding retinal locations. In the thick stripes of V2 there are binocular cells that respond most strongly when the receptive field characteristics are slightly different. The inputs from each eye would have the same orientation selectivity, but the retinal locations associated with the receptive fields would be slightly different. These have been called disparity detectors because they appear to be responding to specific disparities (or horizontal retinal separations) between the stimulating edges in each eye. In addition to the analysis of retinal disparity further specializations can be found in V2.

The growing complexities of the 'wiring diagrams' of the visual system has resulted in the application of computational approaches to vision. These are epitomised in the work of David Marr (1945–1980) who sought to integrate the computational, physiological and psychological levels of understanding vision. The computational level is a theory of the task that the visual system needs to solve, as well as an understanding of the constraints that can enable solution of that task. The second level, of representation and algorithm, is a means of achieving the task, and the third level describes how the brain, or a computer, actually implements these algorithms in neural tissue or silicon: "For the subject of vision, there *is* no single equation or view that explains everything. Each problem has to be addressed from several points of view—as a problem in representing information, as a computation capable of deriving that representation, and as a problem in the architecture of a computer capable of carrying out both things quickly and reliably".[40] Marr distinguished a stage which made explicit the three-dimensional layout of the world with respect to the viewer (which he called a 2.5D sketch), potentially useful for action in the world from the more abstract 3D models which allowed object recognition. An anaglyphic sketch of Marr is enclosed within a representation of a cube.

A sketch of David Marr

[40] Marr (1982), p. 5.

Binocular Terminologies and Their Origins

Vision with two eyes has been studied for many centuries, as is evident from the previous sections, but the terminology that we use to describe it is more recent. For example, the term 'binocular' can be found in a book by Porta in 1593. His portrait is combined with the title page of his book below.[41] *De Refractione* was concerned with optics and the sixth book (on why we see one thing with two eyes) contains the Latin term 'binos oculis'. Porta has most frequently been accorded the distinction of equating the optics of a camera with those of the eye, not because he was the first to reach that conclusion, but because his description in *Magiae Naturalis* (published in 1589) was the most widely read. This book, which was an amalgam of mysticism, folklore, and science, was reprinted in many editions, and translated into several languages. Porta concluded that forming an image on a surface was an adequate account of "how vision is made", but he still considered that the image was formed on the rear surface of the lens.[42] In *Magiae Naturalis* Porta described a combination of convex lenses capable of magnifying the image of an object. His name is variously given as Giambattista della Porta, Giovanni Battista della Porta and Johannes Baptista Porta. In his book on optics, Porta also described binocular rivalry, as will be discussed later.

[41] Porta (1593).

[42] Porta (1589, 1658).

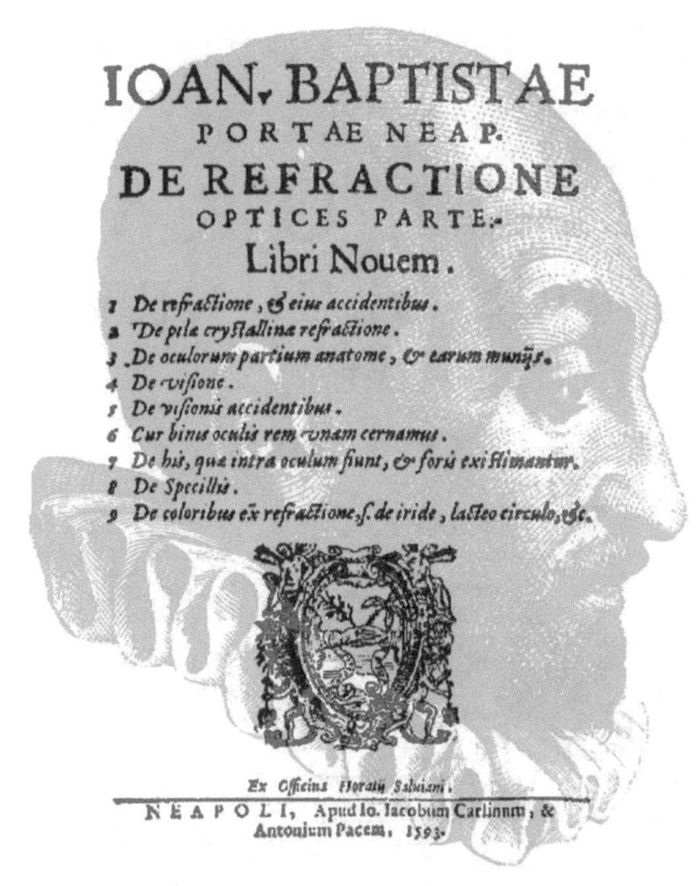

IOAN. BAPTISTAE
PORTAE NEAP.
DE REFRACTIONE
OPTICES PARTE:
Libri Nouem.

1 De refractione, & eius accidentibus.
2 De pila crystallina refractione.
3 De oculorum partium anatome, & earum munijs.
4 De visione.
5 De visionis accidentibus.
6 Cur binis oculis rem unam cernamus.
7 De his, quae intra oculum fiunt, & foris existimantur.
8 De Specillis.
9 De coloribus ex refractione, s. de iride, lacteo circulo, &c.

Ex Officina Horatij Saluiani.
NEAPOLI, Apud Io. Iacobum Carlinum, & Antonium Pacem, 1593.

Giovanni Battista della Porta and the title page of his book on refraction

Porta advanced a theory of binocular single vision that maintained we only see with one eye at one time. Two decades latter Aguilonius suggested an alternative interpretation—the images in the eyes are fused or combined.[43] He introduced the word 'horopter' and also used the Latin term 'stereographice' (stereographic). Stereographic projection was applied to geometry in the context of representing a sphere on a flat surface. Thomas Young (1773–1829) used it in this way in his lectures to the Royal Institution: "The stereographic projection of any circle of a sphere, seen from a point on its surface, on a plane perpendicular to the diameter passing through that point, is a circle".[44]

[43] Aguilonius (1613).

[44] Young (1807), p. 22.

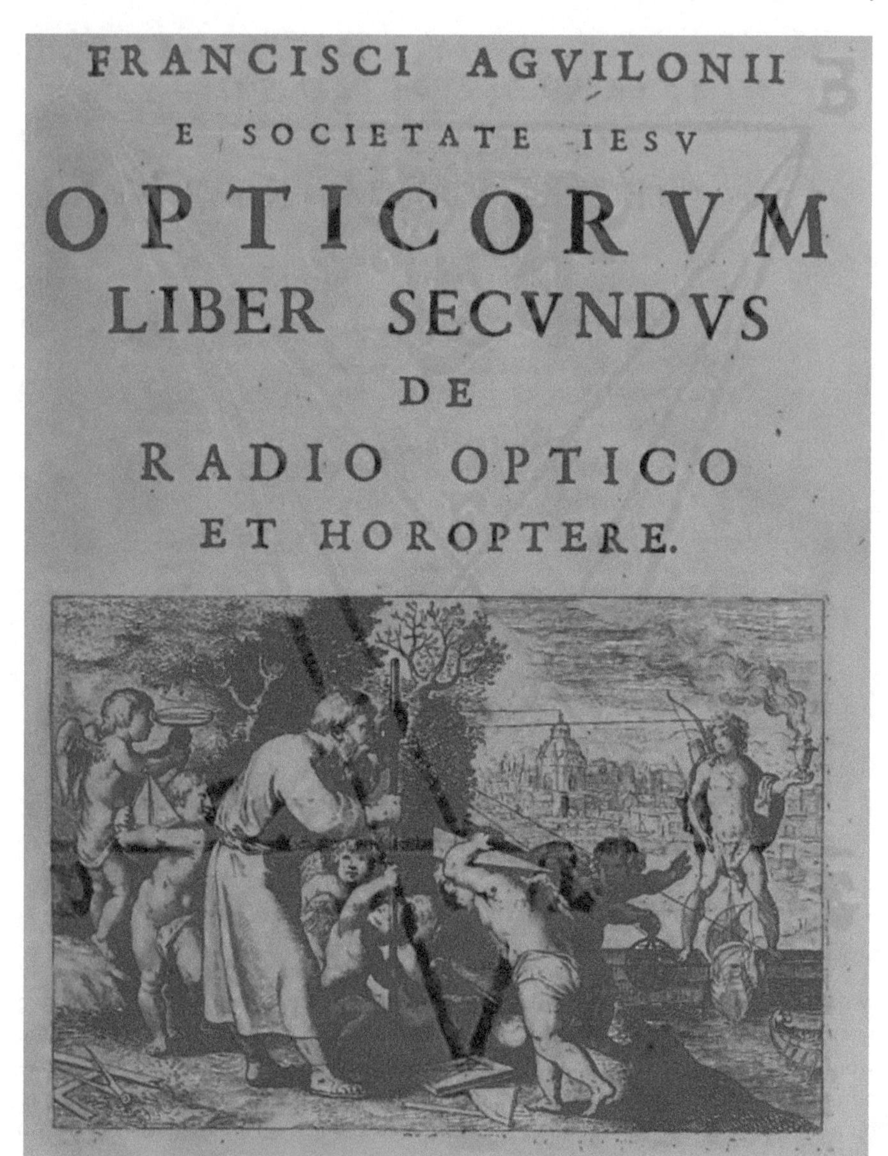
FRANCISCI AGVILONII
E SOCIETATE IESV
OPTICORVM
LIBER SECVNDVS
DE
RADIO OPTICO
ET HOROPTERE.

Title page of Book II of Aguilonius's Optics together with his diagram of the horopter showing fixation on the horopter plane (C) as well as in front of it (F) and beyond it (I)

In 1812 Jean Gabriel Augustin Chevallier (1778–1848), a Parisian optical instrument maker, described a 'stéréoscope' but it was made for projecting scenes rather than for viewing.[45] Chevallier was capitalising on the fashion for phantasmagoria that was sweeping Europe: magic lantern slides of dramatic scenes were projected

[45] The term is described in the second edition (Chevallier 1812) but not the first (Chevallier 1810). See also Wade (2021).

in all manner of locations and with special effect (like smoke) in order to create the feelings of fear and wonder in the spectators.[46]

LE

CONSERVATEUR

de la Vue,

2.me Édition

Considérablement augmentée

DÉDIÉ

à S.M. le Roi de Westphalie,

Par J. G. A. CHEVALLIER,

Ingénieur Opticien de Sa Majesté.

Membre de plusieurs académies.

1812.

Jean Chevallier and the title page of his book in which a magic lantern device called a stéréoscope is described

What we now know as a 'stereoscope' was invented by Wheatstone in the early 1830s and he named it as such when his account of the instrument and his experiments with it were published. Following his discussion of the disadvantages of

[46] Phantasmagoria are delightfully described by Mannoni (2000). The title of his book *The great art of light and shadow* is based on that of Kircher (1646) who gave an account of early magic lanterns.

previous methods for combining images in two eyes he stated "The frequent reference I shall have occasion to make to this instrument, will render it convenient to give it a specific name, I therefore propose that it be called a Stereoscope, to indicate its property of representing solid figures".[47] In his second memoir on binocular vision published in 1852 Wheatstone referred to "new experiments relating to stereoscopic appearances".[48]

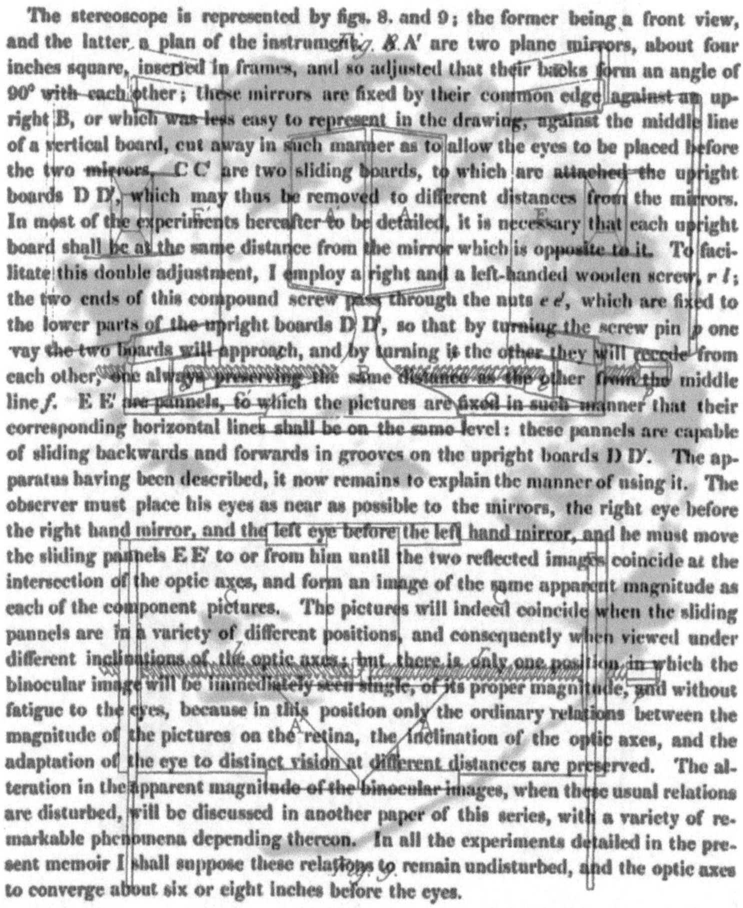

The stereoscope is represented by figs. 8. and 9; the former being a front view, and the latter a plan of the instrument. A A′ are two plane mirrors, about four inches square, inserted in frames, and so adjusted that their backs form an angle of 90° with each other; these mirrors are fixed by their common edge against an upright B, or which was less easy to represent in the drawing, against the middle line of a vertical board, cut away in such manner as to allow the eyes to be placed before the two mirrors. C C′ are two sliding boards, to which are attached the upright boards D D′, which may thus be removed to different distances from the mirrors. In most of the experiments hereafter to be detailed, it is necessary that each upright board shall be at the same distance from the mirror which is opposite to it. To facilitate this double adjustment, I employ a right and a left-handed wooden screw, *r l*; the two ends of this compound screw pass through the nuts *e e′*, which are fixed to the lower parts of the upright boards D D′, so that by turning the screw pin *p* one way the two boards will approach, and by turning it the other they will recede from each other, one always preserving the same distance as the other from the middle line *f*. E E′ are pannels, to which the pictures are fixed in such manner that their corresponding horizontal lines shall be on the same level: these pannels are capable of sliding backwards and forwards in grooves on the upright boards D D′. The apparatus having been described, it now remains to explain the manner of using it. The observer must place his eyes as near as possible to the mirrors, the right eye before the right hand mirror, and the left eye before the left hand mirror, and he must move the sliding pannels E E′ to or from him until the two reflected images coincide at the intersection of the optic axes, and form an image of the same apparent magnitude as each of the component pictures. The pictures will indeed coincide when the sliding pannels are in a variety of different positions, and consequently when viewed under different inclinations of the optic axes; but there is only one position in which the binocular image will be immediately seen single, of its proper magnitude, and without fatigue to the eyes, because in this position only the ordinary relations between the magnitude of the pictures on the retina, the inclination of the optic axes, and the adaptation of the eye to distinct vision at different distances are preserved. The alteration in the apparent magnitude of the binocular images, when these usual relations are disturbed, will be discussed in another paper of this series, with a variety of remarkable phenomena depending thereon. In all the experiments detailed in the present memoir I shall suppose these relations to remain undisturbed, and the optic axes to converge about six or eight inches before the eyes.

Wheatstone in his text describing the mirror stereoscope together with Figs. 8 and 9 from his memoir of 1838

William Oughter Lonie (1822–1894) was a teacher of mathematics at Madras College, St. Andrews and in 1856 he was awarded the Prize Essay on the Stereoscope. In it he referred to the camera as "an instrument now indispensably requisite for

[47] Wheatstone (1838), p. 374.

[48] Wheatstone (1852), p. 2.

stereoscopy".[49] The essay is fulsome in its praise of David Brewster (1781–1868) and his stereoscope and it is hard to avoid the conclusion that this influenced Brewster in awarding the prize to Lonie. Nonetheless, Lonie does appear to have introduced the term 'stereoscopy'.

The stereoscopic pictures themselves were not given a specific name by Wheatstone and this want was supplied by Oliver Wendell Holmes (1809–1894) in an article in *The Atlantic Monthly*. He wrote: "We have now obtained the double-eyed or twin pictures, or STEREOGRAPH, if we may coin a name".[50] The term is not now used as frequently as in the past and it has tended to be replaced by 'stereogram'. The translation of Helmholtz's *Handbuch* by Southall does refer to a 'stereogram'[51] but the original German is 'stereoscopic drawing' (stereoskopische Zeichnung).[52] Similarly, in the French translation it is referred to as a 'dessin stéréoscopique'.[53]

---

[49] Lonie (1856), p. 13. See also O'Shea (2017).

[50] Holmes (1859), p. 743.

[51] Helmholtz (1925), p. 440.

[52] Helmholtz (1867a), p. 728.

[53] Helmholtz (1867b), p. 920.

THE STEREOSCOPE AND THE STEREOGRAPH.

DEMOCRITUS of Abdera, commonly known as the Laughing Philosopher, probably because he did not consider the study of truth inconsistent with a cheerful countenance, believed and taught that all bodies were continually throwing off certain images like themselves, which subtile emanations, striking on our bodily organs, gave rise to our sensations. Epicurus borrowed the idea from him, and incorporated it into the famous system, of which Lucretius has given us the most popular version. Those who are curious on the matter will find the poet's description at the beginning of his fourth book. Forms, effigies, membranes, or *films*, are the nearest representatives of the terms applied to these effluences. They are perpetually shed from the surfaces of solids, as bark is shed by trees. *Cortex* is, indeed, one of the names applied to them by Lucretius.

These evanescent films may be seen in one of their aspects in any clear, calm sheet of water, in a mirror, in the eye of an animal by one who looks at it in front, but better still by the consciousness behind the eye in the ordinary act of vision... of man he was. These visible films or membranous *exuviæ* of objects, which the old philosophers talked about, have no real existence, separable from their illuminated source, and perish instantly when it is withdrawn.

If a man had handed a metallic speculum to Democritus of Abdera, and told him to look at his face in it while his heart was beating thirty or forty times, promising that one of the films his face was shedding should stick there, so that neither he, nor it, nor anybody should forget what manner of man he was, the Laughing Philosopher would probably have vindicated his claim to his title by an explosion that would have astonished the speaker.

This is just what the Daguerreotype has done. It has fixed the most fleeting of our illusions, that which the apostle and the philosopher and the poet have alike used as the type of instability and unreality. The photograph has completed the triumph, by making a sheet of paper reflect images like a mirror and hold them as a picture.

Oliver Wendell Holmes and the first page of his article coining the term stereograph

'Stereopsis' as a shorthand for stereoscopic depth perception was used by the American ophthalmologist Alexander Duane (1858–1926) in 1917. Duane's research interests were in accommodation and squint and it is the latter that is relevant to stereoscopic depth perception. He translated Ernst Fuchs's textbook of ophthalmology from German into English and it is in the fifth edition that the term stereopsis is introduced: "The stereoscope and especially the amblyoscope will show both the patient's ability to perform fusion and to secure stereoscopic vision (stereopsis)".[54] The text was added by Duane and it does not appear in earlier English editions of the book. Stereopsis was the term adopted by ophthalmologists associated with the Medical Research Laboratory at Mineoloa, New York, like Howard, Dolman, Wilmer and Verhoeff and was widely used in America thereafter.[55]

[54] Duane (1917), p. 773.

[55] See Howard (1919), Wells (1920) and O'Shea (2017).

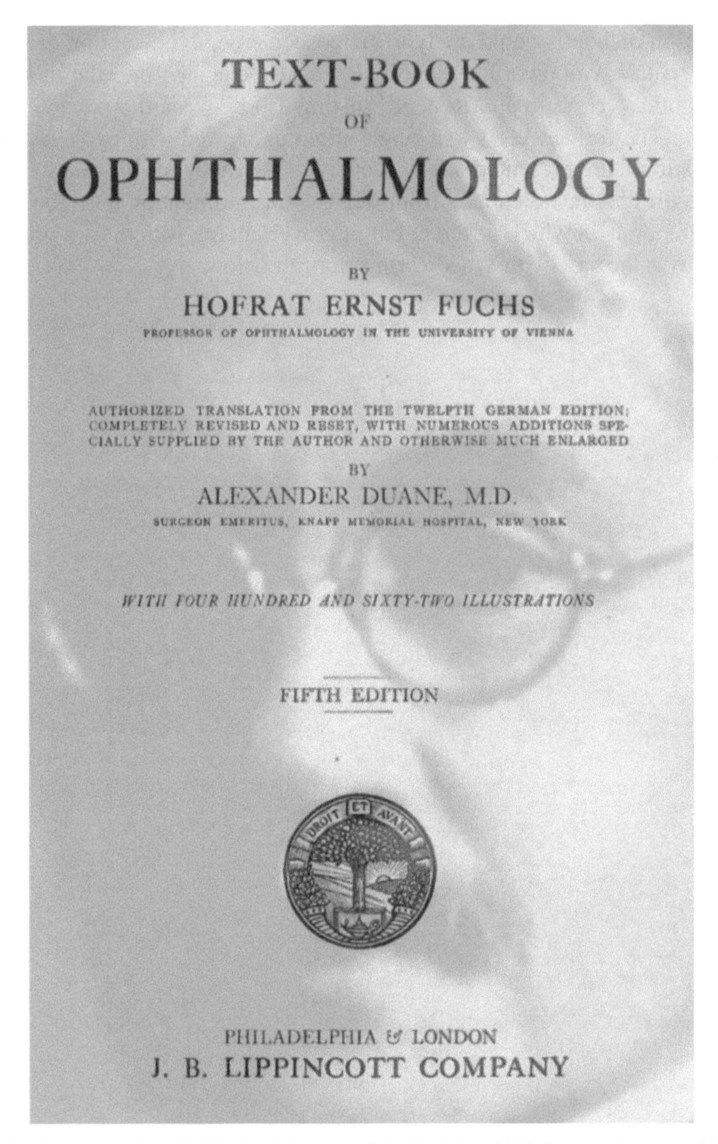

TEXT-BOOK

OF

OPHTHALMOLOGY

BY

HOFRAT ERNST FUCHS

PROFESSOR OF OPHTHALMOLOGY IN THE UNIVERSITY OF VIENNA

AUTHORIZED TRANSLATION FROM THE TWELFTH GERMAN EDITION; COMPLETELY REVISED AND RESET, WITH NUMEROUS ADDITIONS SPECIALLY SUPPLIED BY THE AUTHOR AND OTHERWISE MUCH ENLARGED

BY

ALEXANDER DUANE, M.D.

SURGEON EMERITUS, KNAPP MEMORIAL HOSPITAL, NEW YORK

WITH FOUR HUNDRED AND SIXTY-TWO ILLUSTRATIONS

FIFTH EDITION

PHILADELPHIA & LONDON

J. B. LIPPINCOTT COMPANY

Alexander Duane combined with the title page of the book in which the term 'stereopsis' is used

The term 'dichoptic' has not been used consistently since it was introduced by Robert Sessions Woodworth (1869–1962) in 1938 (although he used the word 'dichopic'). When Woodworth was treating binocular vision in his *Experimental Psychology* he wrote: "In a type of experiment which might be called *dichopic* (by analogy with dichotic and dichorhinic experiments in hearing and smell) discrepant stimulation is applied to corresponding parts of the two retinas".[56] According to

[56] Woodworth (1938), p. 572.

this definition dichoptic could apply to stereoscopic depth perception as well as to binocular rivalry. Woodworth's dichopic gradually changed into the more widely used dichoptic.[57] Howard and Rogers[58] distinguish between dioptic and dichoptic stimulation. The former refers to a single stimulus viewed with two eyes, whereas dichoptic stimuli are presented one to each eye, usually by means of a stereoscope. In order to differentiate dichoptic from stereoscopic stimulation it can be restricted to those situations in which different but non-overlapping patterns are presented to each eye, as is the case with Woodworth's portrait below.

[57] See Wade and Ono (2005).

[58] Howard and Rogers (1995).

Robert Woodworth and the title page of the book in which the term 'dichopic' was introduced

'Binocular lustre' refers to the metallic impression created by combining positive and negative images (particularly when they are black on white and white on black) in the two eyes. The phenomenon was described by Heinrich Wilhelm Dove (1803–1879) and he called it gloss (Glanz).[59] It was referred to as lustre in the English

[59] Dove (1851).

translation of Dove's article.[60] Brewster followed up Dove's observations and called the phenomenon binocular lustre.[61] Helmholtz referred to it as stereoscopic lustre.[62]

XII. *Ueber die Ursachen des Glanzes und der Irradiation, abgeleitet aus chromatischen Versuchen mit dem Stereoskop; von H. W. Dove.*

Die Beschreibung dessen, was man sieht, wenn man dem rechten Auge eine andere Farbe darbietet als dem linken, fällt bei verschiedenen Beobachtern sehr verschieden aus. Einige sehen abwechselnd eine Farbe nach der andern, einige farbige Flecke der einen neben farbigen Flecken der andern, endlich einige die aus beiden Farben entstehende Mischungsfarbe. Streng genommen liegt in dieser Beschreibung das Gemeinsame, das alle zugeben, dafs unter gewissen Bedingungen eine Combination beider Farben möglich sey, denn das Nacheinander mufs einen Durchgangspunkt haben, wo die abklingende Farbe eben so stark wird als die in das Bewufstseyn tretende, das Nebeneinander mufs Stellen des Ueberganges haben, da die Flecke neben einander sich nicht scharf gegen einander abgränzen.

Heinrich Dove and his text describing lustre

The concept of the 'cyclopean eye' is probably embodied in the mythological cyclops who forged thunderbolts for Zeus; in the Homeric *Odyssey* cyclops was a one-eyed giant. The location of the single eye was central in the forehead, and the locus of binocular visual direction is now referred to as the cyclopean eye. The illustration by Rubens for Book I of Aguilonius's book on optics,[63] shows the putti performing an operation beyond the scope of modern medicine—the cyclopean eye was being dissected; the putti are overseen by the watchful eye of the cosmic observer as well as the cyclops Polyphemus in the illustration below.

[60] Dove (1852).

[61] Brewster (1861).

[62] Helmholtz (1867a).

[63] Aguilonius (1613).

The one-eyed giant Polyphemus watching over putti dissecting the cyclopean eye

Nonetheless, with both eyes fixating on an object it will be seen as single and in a direction corresponding to an origin between the eyes. That this general idea is not new is evident from a diagram of Ptolemy in the next chapter. The concept was given empirical support by William Charles Wells (1757–1817) in his book on single vision with two eyes.[64] Wells did not provide any illustrations in his book but the essential features of his statements regarding visual direction with two eyes are shown together with the title page of his book. However, Wells did not give the origin of the common axis a name; this was supplied about 70 years later. Joseph Towne (1806–1879) was one of the few who cited Wells and conducted many experiments in stereoscopic vision.[65] He wrote: "that we see in the direction of the median plane of the head as from one central eye".[66]

[64] Wells (1792) reprinted in Wade (2003).

[65] Towne published nine papers on binocular vision in *Guy's Hospital Reports* between 1862 and 1870 (see Wade et al. 2006).

[66] Towne (1866), p. 301.

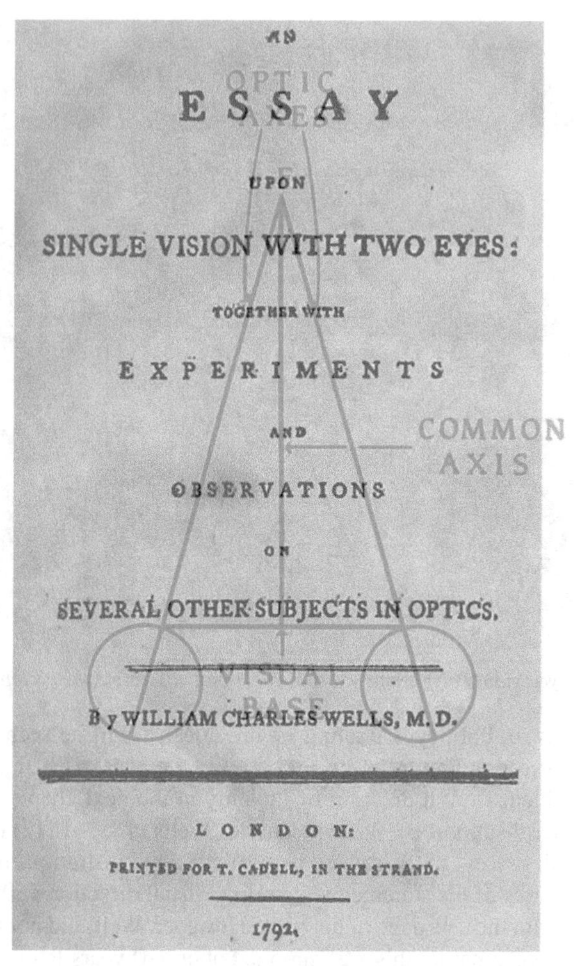
AN

ESSAY

UPON

SINGLE VISION WITH TWO EYES:

TOGETHER WITH

EXPERIMENTS

AND

OBSERVATIONS

ON

SEVERAL OTHER SUBJECTS IN OPTICS.

By WILLIAM CHARLES WELLS, M. D.

LONDON:

PRINTED FOR T. CADELL, IN THE STRAND.

1792.

The title page of the book by William Charles Wells together with a diagram of his concepts of visual direction with two eyes

Ewald Hering (1834–1916) provided the experimental confirmation of Wells's research and he referred to an imaginary single eye between the two anatomical eyes[67]:

> The direction in which the illusional images appear, is unalterably determined by the law of identical visual directions. If we conceive of the two retinal images as transferred to the retina of the imaginary single eye (cyclopean), in a manner to make all cover points coincident, and let the lines of direction of the single eye pass for the visual lines of direction: each illusional image will have its own visual direction. One must imagine the single eye, or center of visual direction, as lying in the median plane of the head.[68]

[67] Hering (1879).

[68] Hering (1942), p. 74.

Hering essentially rediscovered the principles of visual direction described by Wells, although no reference was made to Wells's earlier enquiries. The word 'cyclopean' was added in Radde's translation, as it was not present in Hering's original text which is shown below together with a portrait of Hering; he referred to it as 'an imaginary single eye'.

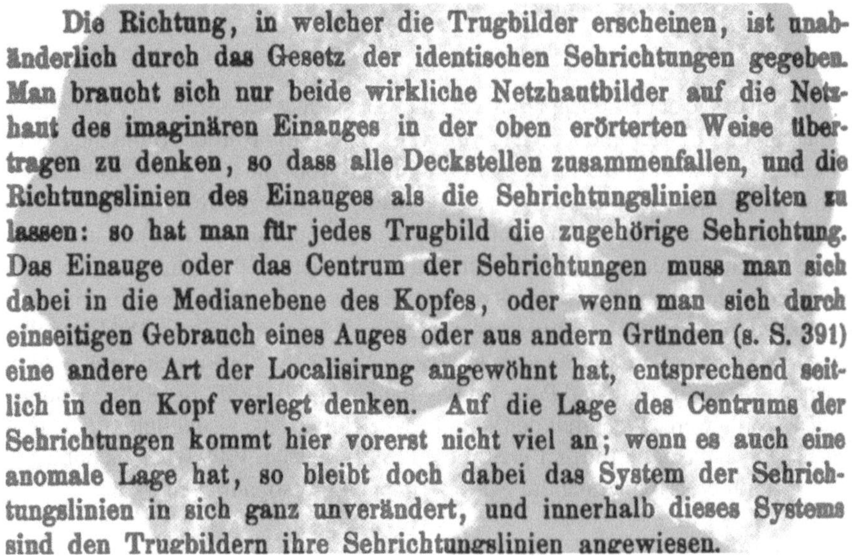

Die Richtung, in welcher die Trugbilder erscheinen, ist unabänderlich durch das Gesetz der identischen Sehrichtungen gegeben. Man braucht sich nur beide wirkliche Netzhautbilder auf die Netzhaut des imaginären Einauges in der oben erörterten Weise übertragen zu denken, so dass alle Deckstellen zusammenfallen, und die Richtungslinien des Einauges als die Sehrichtungslinien gelten zu lassen: so hat man für jedes Trugbild die zugehörige Sehrichtung. Das Einauge oder das Centrum der Sehrichtungen muss man sich dabei in die Medianebene des Kopfes, oder wenn man sich durch einseitigen Gebrauch eines Auges oder aus andern Gründen (s. S. 391) eine andere Art der Localisirung angewöhnt hat, entsprechend seitlich in den Kopf verlegt denken. Auf die Lage des Centrums der Sehrichtungen kommt hier vorerst nicht viel an; wenn es auch eine anomale Lage hat, so bleibt doch dabei das System der Sehrichtungslinien in sich ganz unverändert, und innerhalb dieses Systems sind den Trugbildern ihre Sehrichtungslinien angewiesen.

Ewald Hering and his 1879 text describing an imaginary single eye

The term is now in common usage, and it was introduced in this context by Helmholtz; the German 'Cyclopenauge' becomes the English 'cyclopean eye':

> Midway between the two eyes suppose there were an imaginary cyclopean eye which was directed to the common point of fixation of the two eyes, and that it rolled according to the law governing the rolling of the two real eyes. Imagine the retinal images transferred from one of the real eyes to this imaginary eye, so that the point of fixation of the imaginary eye is the same as that of the real eye. *Then the points of the retinal image will be projected out along the line of direction of the imaginary cyclopean eye.*[69]

[69] Helmholtz (1925), p. 258.

Man denke sich in der Mitte zwischen beiden Augen ein imaginäres mittleres Cyclopenauge, welches auf den gemeinsamen Fixationspunkt beider Augen gerichtet ist, und dessen Raddrehungen nach demselben Gesetze erfolgen, wie die der beiden wirklichen Augen. Man denke sich die Netzhautbilder aus einem der wirklichen Augen in dieses imaginäre Auge übertragen, so dass Blickpunkt auf Blickpunkt und Netzhauthorizont auf Netzhauthorizont fällt. Dann werden die Punkte des Netzhautbildes nach aussen projicirt, in der Richtungslinie des imaginären Cyclopenauges[9].

The eyes of Hermann Helmholtz and his 1867 text describing a cyclopean eye

Modern usage of the concept of the cyclopean eye does not necessarily correspond to that applied by Helmholtz. The term cyclopean vision is now synonymous with some representation in the brain that is combined from the two eyes.[70] Alternative terms for the conceptual central eye are binoculus and egocentre.

The term 'rivalry' also entered into use in the mid-nineteenth century. In his classic article Wheatstone described the fluctuating appearances of the letters A and S when presented to corresponding regions of the two eyes but he did not assign a name to the ensuing perception. When Brewster addressed the same issue in 1844 he called it 'ocular equivocation'.[71] The term rivalry (or 'Wettstreit' in German) was used by Hermann Meyer in 1856 and Peter Ludvig Panum (1820–1885) in 1858[72]; Panum's precise words were 'Wettstreit der Sehfelder' (rivalry of the visual fields). The term was used in English by Burtis Burr Breese (1868–1939) in 1899 when he published a monograph describing quantitative experimental investigations of rivalry.[73]

[70] As used by Julesz (1971) for example.

[71] Brewster (1844). See also Wade (2019).

[72] Meyer (1856), Panum (1858).

[73] Breese (1899).

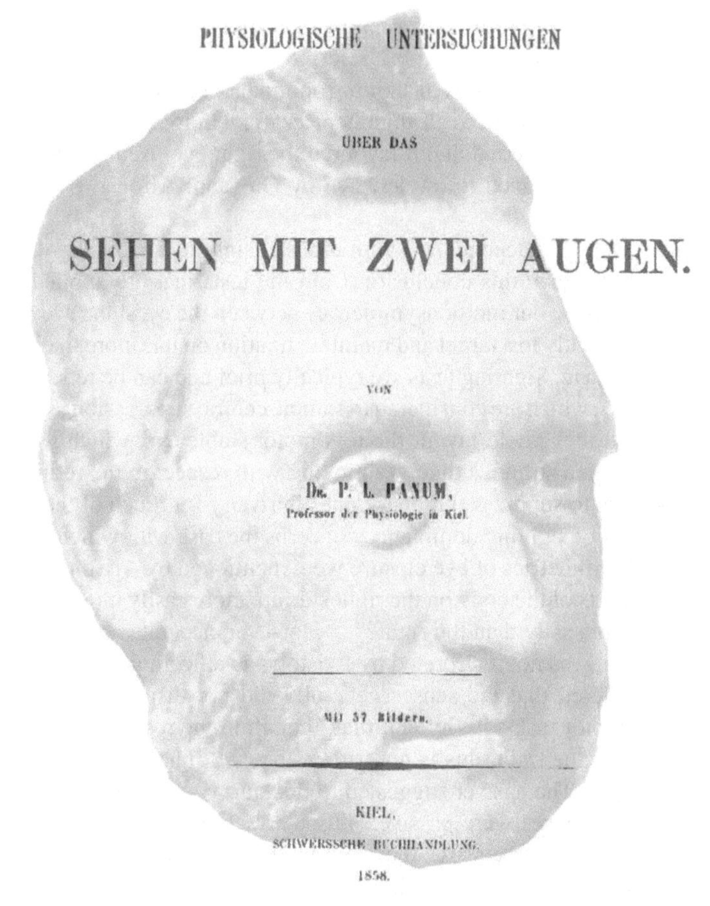

PHYSIOLOGISCHE UNTERSUCHUNGEN

ÜBER DAS

SEHEN MIT ZWEI AUGEN.

VON

Dr. P. L. PANUM,

Professor der Physiologie in Kiel.

Mit 37 Bildern.

KIEL.

SCHWERSSCHE BUCHHANDLUNG.

1858.

Panum and the title page of his book on vision with two eyes

Eye Dominances

The issue of eye dominance was raised in Chap. 1 but the situation is rather more complex than was indicated there. The singular use of the term hardly reflects the empirical confusion with which the concept is beset. Which eye is designated as dominant depends on the test that is used to assess it.[74] Tests can determine whether one eye moves more efficiently than the other, whether one eye has superior visual acuity or whether one eye suppresses the other for longer in binocular rivalry.[75] The results of these tests do not assign dominance to the same eye. The origins of the

[74] See Wade (1998a, b).

[75] Porac and Coren (1976).

tests have a long history but a realization of the different types of eye dominance is much more recent and they are now referred to as sighting, acuity and rivalry dominance, respectively. An index of sighting dominance (eye closure) was described by Aristotle, and Porta introduced an alignment test for it in 1593. Porta also described rivalry dominance and asserted that it favoured the right eye. Acuity dominance, on the other hand, was assigned to the left eye by Giovanni Alfonso Borelli (1608–1679).[76]

There might well be independent asymmetries of sighting and rivalry dominance, but it is difficult to reach this conclusion from the tests that are applied to assess them. Sighting tasks favour motor asymmetries between the eyes: the eye which can both move more quickly to a target and maintain fixation on it is more likely to be the sighting dominant one. Sighting tests are typically brief and can be repeated several times. Rivalry tests, which are considered to sample central suppression asymmetries, are longer lasting; they could favour the less motor stable eye, which would result in that eye receiving a stimulus that was moving with respect to the retina. Moving patterns are known to suppress stationary ones in rivalry for much of the time.[77]

One simple test of sighting dominance concerns the ease with which a single eye can be closed. Asymmetries of eye closure were mentioned by Aristotle: "Why is it that, though the parts of the body on the right side are more easily moved, the left eye can be closed more easily than the right?".[78] However, accurate as this observation might have been, it was not interpreted by Aristotle as reflecting on eye dominance. In fact, he considered that the senses generally did not display the asymmetries associated with motor tasks: "Why is it that, though in the rest of the body the left side is weaker than the right, this is not true of the eyes, but the sight of both eyes is equally acute?".[79] The answer suggested to this query was that the muscles were active whereas the senses were passive; muscular activities could be modified by practice, unlike the passive senses. Accordingly, right side superiority was assigned to habit, because it could be changed, as in the case of the ambidextrous. Porta is usually acknowledged as describing the first test of sighting dominance: "If someone places a staff in front of himself, and sets it against some obvious crack in the wall opposite, and notices the place, then when he shuts the left eye he will not see the staff to have moved from the crack opposite. The reason is that one sees with the right eye, just as one uses the right hand and foot, and someone using the left eye or hand or foot is considered a monster. But if the observer closes the right eye, the staff immediately shifts to the right side".[80] It is possible that an illustration of Porta's sighting test was made by Rubens, as a frontispiece to Book III of the book on optics by Aguilonius: an observer, with his left eye closed, points to a stick held at some distance from him. Aristotle's portrait is combined with the illustration.

---

[76] Borelli (1673).

[77] Wade and de Weert (1986).

[78] Ross (1927), p. 959b.

[79] Ross (1927), p. 959b.

[80] Porta (1593), p. 143.

A portrait of Aristotle together with Rubens' illustration of the cosmic observer with one eye closed pointing at a stick held by a putto

Porta is better known for describing the conditions under which binocular rivalry can be observed: viewing different books with the two eyes results in visibility of only one of them.[81] The relevant text is on the same page of his book as that relating to a sighting test. A translation of the text into English is shown together with his portrait. His first argument to support right eye superiority involved the phenomenon of binocular rivalry, which he also demonstrated with a simple test. Porta did entertain the possibility of alternating dominance for those brief periods in which the 'visual virtue' is withdrawn from the right eye and borrowed by the left. Indeed, his statements about binocular rivalry sit uneasily with others he made, and diagrams he presented, which suggest that the signals from the two eyes are combined rather than in competition. These diagrams probably relate to the third argument Porta presents, namely, that an advantage conferred by two eyes is to increase the field of vision. Porta used his observations on sighting and rivalry dominance to support a delightfully parsimonious theory of binocular single vision, namely that we use but one eye at once. This theory was championed by many others in the eighteenth century. Suppression theories of binocular single vision were set in conflict with fusion theories of the type proposed by Aguilonius and by Descartes.

[81] Porta (1593), pp. 142–143.

Nature has given us two eyes, one on the right and the other on the left, so that if we are to see something on the right we use the right eye, and on the left the left eye. It follows that we always see with one eye, even if we think both are open and that we see with both. We may prove it by these arguments: To separate the two eyes, let us place a book before the right eye and read it; then if someone shows another book to the left eye, it is impossible to read it or even see the pages, unless for a short moment of time the power of seeing is taken from the right eye and borrowed by the left. We observe the same thing happening in other senses: if we hear someone talking with the right ear we cannot listen to another with the left ear; and if we wish to hear both we shall hear neither, or indeed if we hear something with the right we lose the same amount from the left. Similarly, if we write with one hand we cannot play a lyre with the other, nor count out coins. There is another argument. If someone places a staff in front of himself, and sets it against some obvious crack in the wall opposite, and notices the place, then when he shuts the left eye he will not see the staff to have moved from the crack opposite. The reason is that one sees with the right eye, just as one uses the right hand and foot, and someone using the left eye or hand or foot is considered a monster. But if the observer closes the right eye, the staff immediately shifts to the right side. There is a third argument - that nature made two eyes, one beside the other, so that one may defend a man from attackers from the right and the other from the left. This is more obvious in animals, for their eyes are separated by half a foot, as is seen in cattle, horses and lions. In birds one eye is opposite the other; consequently, if things must be seen both on the right and on the left, the power of seeing must be engaged very quickly for the mind to be able to accomplish its functions. For these reasons the two eyes cannot see the same thing at the same time.

Giovanni Battista della Porta and text translated from his book providing support for right eye dominance

Acuity dominance is said to be independent of the other forms, and its early history would tend to support this. Although Aristotle stated that both eyes were equally acute, Borelli asserted that the left eye was stronger than the right. He based this claim on the greater distinctness of the view with his left eye, and on the sharper vision of small letters viewed through a tube when using his left eye. Borelli, like Porta, was a Neapolitan. He was a distinguished mathematician and he held chairs at the Universities of Pisa, Rome, and Messina. His work in physiology was concerned mainly with the application of mechanics to the circulation of blood in the body, and to the limb movements of animals. He speculated that the asymmetry might be due to the left optic nerve having a better blood supply than the right. The assertion of left eye superiority was challenged by Nicolas Le Cat (1700–1768), who noted that in addition to those who have a stronger left eye, others had a stronger right, and there were yet others (like himself) who had eyes 'perfectly equal'.[82] It is not clear what aspects of eye dominance Le Cat was considering, because he seemed to conflate distinctness of vision in an eye with its vigour.

Paired portraits of Giovanni Borelli with his left and right eye superimposed

Thus, a characteristic of sighting eye dominance (eye closure) was described by Aristotle, although he did not appreciate its significance. In 1593 Porta introduced tests for both sighting and rivalry dominance, which were assigned to the right side. Acuity dominance (favouring the left eye) was claimed by Borelli. These conclusions were usually based on individual observation, and few attempts at comparisons between individuals were undertaken. Individual differences in acuity between the eyes were appreciated in the eighteenth century, whereas the study of individual differences in eye dominances started gradually in the nineteenth and gained momentum in the twentieth century.

[82] Le Cat (1744).

References

Adrian ED (1937) Synchronized reactions in the optic ganglion of Dytiscus. J Physiol 91:66–89
Aguilonius F (1613) Opticorum Libri Sex. Philosophis juxta ac Mathematicis utiles. Moreti, Antwerp
Alberti LB (1435/1966) On painting (trans: Spencer JR). Yale University Press, New Haven, CT
Barlow HB (1953) Summation and inhibition in the frog's retina. J Physiol 119:69–88
Borelli A (1673) Observations touchant la force inégale des yeux. J Sçavans 3:291–294
Breese BB (1899) On inhibition. Psychol Monogr 3:1–65
Brewster D (1844) On the knowledge of distance given by binocular vision. Trans R Soc Edinb 15:663–674
Brewster D (1855) Memoires of the life, writings, and discoveries of Sir Isaac Newton. Constable, Edinburgh
Brewster D (1861) On binocular lustre. Report of the British Association for the Advancement of Science, pp 29–31
Cajal SRy (1893) La rétine des vertébrés. La Cellule 9:119–257 (English edn 1972) The structure of the retina (trans: Thorpe SA, Glickstein M, Thomas, CC). Springfield, IL
Chevallier JGA (1810) Le conservateur de la vue. Paris
Chevallier JGA (1812) Le conservateur de la vue, 2nd edn. Paris
Clayton M (2019) Leonardo da Vinci. A life in drawing. Royal Collection Trust, London
Daxecker F (2004) The physicist and astronomer Christoph Scheiner: biography, letters, works. Publications at Innsbruck University, Innsbruck
Descartes R (1637/1902) La Dioptrique. In: Adam C, Tannery P (eds) Oeuvres de Descartes, vol 6. Cerf, Paris, pp 81–228
Dove HW (1851) Ueber die Ursache des Glanzes und der Irradiation, abgeleitet aus chromatischen Versuchen mit der Stereoskop. Annalen der Physik und Chemie 83:169–183
Dove HW (1852) On the stereoscopic combination of colours, and on the influence of brightness on the relative intensity of different colours. Lond Edinb Dublin Philos Mag J Sci 4:241–249
Duane A (1917) Fuchs's text-book of ophthalmology, 5th edn. Lippincott, Philadelphia
Duke-Elder S (ed) (1961) System of ophthalmology. The anatomy of the visual system, vol 2. Kimpton, London
Harris J (1775) A treatise of optics: containing elements of the science; in two books. White, London
Hartline HK, Wagner HG, Ratliff F (1956) Inhibition in the eye of Limulus. J Gen Physiol 39:651–673
Helmholtz H (1867a) Handbuch der physiologischen Optik. In: Karsten G (ed) Allgemeine Encyklopädie der Physik, vol 9. Voss, Leipzig
Helmholtz H (1867b) Optique physiologique (trans: Javal É, Klein T). Masson, Paris
Helmholtz H (1925) Helmholtz's treatise on physiological optics, vol 3 (trans: Southall JPC). Optical Society of America, New York
Hering E (1879) Der Raumsinn und die Bewegungen des Auges. In: Hermann L (ed) Handbuch der physiologie, vol 3. Vogel, Leipzig, pp 341–601
Hering E (1942) Spatial sense and movements of the eye (trans: Radde A). American Academy of Optometry, Baltimore
Holmes OW (1859) The stereoscope and the stereograph. Atlantic Mon 3:738–748
Holmgren F (1866) Method att objektivera effecten af ljusintryck på retina. Upsala Läkareförenings Förhandlingar 1:177–191
Howard HJ (1919) A stereomicrometer—an instrument of precision for measuring stereopsis. Trans Am Ophthalmol Soc 17:395–400
Howard IP, Rogers BJ (1995) Binocular vision and stereopsis. Oxford University Press, Oxford
Hubel DH, Wiesel TN (2005) Brain and visual perception. The story of a 25-year collaboration. Oxford University Press, Oxford
Kantola L, Piccolino M, Wade NJ (2019) Holmgren and the electroretinogram: a translation and commentary. J Hist Neurosci 28:399–415

Julesz B (1971) Foundations of cyclopean perception. University of Chicago Press, Chicago
Kemp M (1989) Leonardo on painting. Yale University Press, New Haven, CT
Kepler J (1604) Ad Vitellionem Paralipomena. Marinium and Aubrii, Frankfurt
Kircher A (1646) Ars magna lucis et umbrae. Scheus, Rome
Le Cat N (1744) Traité des Sens. Wetstein, Amsterdam
Lonie WO (1856) Prize essay on the stereoscope. The London Stereoscopic Company, London
Mannoni L (2000) Thegreat art of light and shadow. Archaeology of cinema (trans: Crangle R). University of Exeter Press, Exeter
Marr D (1982) Vision. A computational investigation into the human representation and processing of visual information. Freeman, New York
May MT (1968) Galen. On the Useful of the parts of the body. Cornell University Press, Ithaca, NY
Mazzarello P (1996) The hidden structure. A scientific biography of Camillo Golgi. Oxford University Press, Oxford
Mazzarello P (2010) Golgi. A biography of the founder of modern neuroscience. Oxford University Press, Oxford
McMurrich JP (1930) Leonardo da Vinci the anatomist (1452–1519). Williams & Wilkins, Baltimore, MD
Meyer H (1856) Ueber den Einfluss der Aufmerksamkeitauf die Bildung des Gesichtsfeldes überhaupt und die Bildung des gemeinschaftlichen Gesichtsfeldes beider Augen im besondern. Archiv Für Ophthalmol 2:77–92
Munk H (1879) Physiologie der Sehsphäre der Grosshirnrinde. Archiv Für Physiol 3:255
Nakayama K, Shimojo S (1990) Da Vinci stereopsis: depth and subjective occluding contours from unpaired image points. Vision Res 30:1811–1825
Newton I (1704) Opticks: or, a treatise of the reflections, refractions, inflections and colours of light. Smith and Walford, London
O'Shea RP (2017) On "stereoscopy" and "stereopsis" [blog post]. Retrieved from https://robertposhea.blogspot.com/2017/08/origins-of-stereoscopy-and-stereopsis.html
Panum PL (1858) Physiologische Untersuchungen über das Sehen mit zwei Augen. Schwerssche Buchhandlung, Kiel
Piccolino M (1988) Cajal and the retina: a 100-year retrospective. Trends Neurosci 11:521–525
Plater F (1583) De corporis humani structura et usu. König, Basel
Polyak S (1957) The vertebrate visual system. University of Chicago Press, Chicago
Porac C, Coren S (1976) The dominant eye. Psychol Bull 83:880–897
Porta JB (1589) Magiae naturalis. Libri XX. Silviani, Naples
Porta JB (1593) De Refractione. Optices Parte. Libri Novem. Carlinum and Pacem, Naples
Porta JB (1658) Natural magick. Young and Speed, London
Ross WD (ed) (1927) The works of Aristotle, vol 7. Clarendon, Oxford
Scheiner C (1619) Oculus, hoc est fundamentum opticum. Agricola, Innsbruck
Schickore J (2007) The microscope and the eye. A history of reflections, 1740–1870. University of Chicago Press, Chicago
Schultze M (1866) Zur Anatomie und Physiologie der Retina. Arch Mikrosk Anat 2:175–286
Strong DS (1979) Leonardo on the eye. Garland, New York
Taylor J (1738) Le mechanisme ou le nouveau traité de l'anatomie du globe de l'oeil, avec l'usage de ses différentes parties, & de celles qui lui sont contigues. David, Paris
Towne J (1866) Contributions to the physiology of binocular vision, section VII. Guy's Hosp Rep 12:285–301
Treviranus GR (1837) Beiträge zur Aufklärung der Erscheinungen und Gesetze des organischen Lebens, vol 1, issue 3. Resultate neuer Untersuchungen über die Theorie des Sehens und über den innern Bau der Netzhaut des Auges. Heyse, Bremen
Vesalius A (1543) De humani corporis fabrica. Libri septem. Oporini, Basel
Vishwanath D (2014) Toward a new theory of stereopsis. Psychol Rev 121:151–178
Wade NJ (1998a) A natural history of vision. MIT Press, Cambridge, MA
Wade NJ (1998b) Early studies of eye dominances. Laterality 3:97–108

Wade NJ (2003) Destined for distinguished oblivion. The scientific vision of William Charles Wells (1757–1817). Kluwer/Plenum, New York
Wade NJ (2008) Chevalier John Taylor, ophthalmiater. Perception 37:969–972
Wade NJ (2018) The disparate histories of binocular vision and binaural hearing. J Hist Neurosci 27:10–35
Wade NJ (2019) Microscopic anatomy of sensory receptors. J Hist Neurosci 28:285–306
Wade NJ (2021) On the origin of terms in binocular vision. i-Perception 12(1):1–19. https://doi.org/10.1177/2041669521992381
Wade NJ, de Weert CMM (1986) Binocular rivalry with rotary and radial motions. Perception 15:435–442
Wade NJ, Ono H (2005) From dichoptic to dichotic: historical constrasts between binocular vision and binaural hearing. Perception 34:645–668
Wade NJ, Ono H, Lillakas L (2001) Leonardo da Vinci's struggles with representations of reality. Leonardo 34:231–235
Wade NJ, Ono H, Mapp AP (2006) The lost direction in binocular vision: the neglected signs posted by Wells, Towne, and LeConte. J Hist Behav Sci 42:61–86
Wade NJ, Piccolino M (2006) Nobel stains. Perception 35:1–8
Wells DW (1920) Stereoscopic vision. In: Wood CA (ed) The American encyclopedia and dictionary of ophthalmology, vol 16. Cleveland Press, Chicago, pp 12243–12248
Wells WC (1792) An essay upon single vision with two eyes: together with experiments and observations on several other subjects in optics. Cadell, London
Wheatstone C (1838) Contributions to the physiology of vision—Part the first. On some remarkable, and hitherto unobserved, phenomena of binocular vision. Philos Trans R Soc 128:371–394
Wheatstone C (1852) Contributions to the physiology of vision—Part the second. On some remarkable, and hitherto unobserved, phenomena of binocular vision. Philos Trans R Soc 142:1–17
Woodworth RS (1938) Experimental psychology. Holt, New York
Young T (1807) A course of lectures on natural philosophy and the mechanical arts, vol 2. Johnson, London
Ziggelaar A (1983) François de Aguilón S. J. (1567–1617) scientist and architect. Institutum Historicum S I, Rome

Chapter 3
Binocular Vision

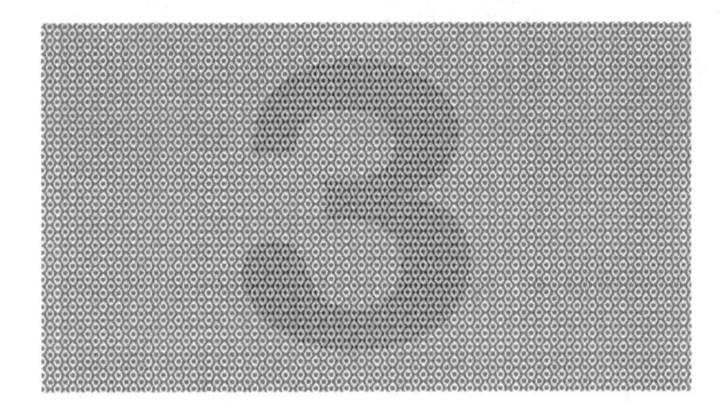

As with many aspects of vision, the optics of Euclid provided a starting point for the analysis of seeing with two eyes. Lines could be drawn from eye to object and spatial vision could be analysed in geometrical terms. According to Euclid visual angles defined visual size and visual direction followed the visual lines. Vision with two eyes presented a difficulty for this approach since the same object could have different directions for each eye. A resolution was proposed by Ptolemy who conducted experiments on binocular vision using a simple board from which he specified the conditions for singleness of vision, the distinction between crossed and uncrossed disparities, and the direction in which objects are seen with two eyes. The ways the two eyes move was an issue of constant interest. From the time of Aristotle it was appreciated that both eyes move in a similar way so that they seemed to function as a single organ. Moreover, objects appeared in a direction from a point between the eyes rather than from either eye alone, leading to the concept of a central or cyclopean eye. Most attention was directed to eye rotations in the horizontal or vertical planes and rotations around the visual axis were neglected until the nineteenth century. Concern with abnormal binocular eye movements, as in strabismus, was concentrated on the procedures for correcting it rather that investigating the visual consequences. However, the visual problems caused by loss of vision in one eye were remarked upon. Many novel techniques were devised for viewing objects with two eyes but they focused on competition between the eyes rather than their cooperation. The same applied to the early binocular viewing devices that were invented before the stereoscope; although they were binocular they were not stereoscopic. Singleness of vision was considered to follow from stimulation of corresponding points on the two eyes and these could be defined geometrically as lying on a circle passing through the rotation centres of each eye and the point of fixation. The elegance of this analysis was undermined by Wheatstone's experiments with the instrument he invented—the stereoscope.

N. Wade, *Vision and Art with Two Eyes*, Vision, Illusion and Perception 3,
https://doi.org/10.1007/978-3-030-77995-5_3

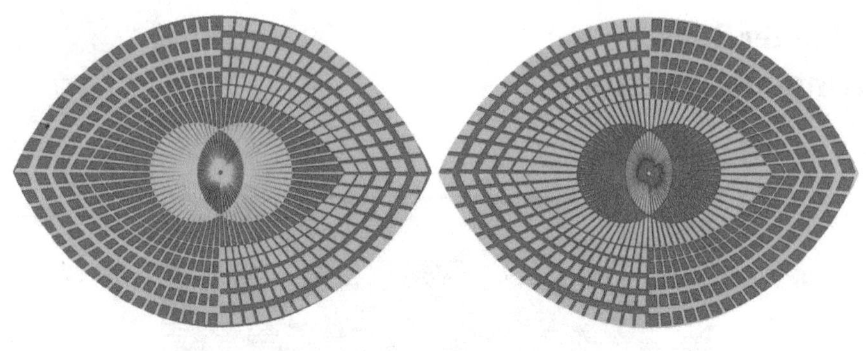

Binocular visions

In binocularity we find one of the supremely psychological phenomena of vision, hence the constant interest that has been given to it over many centuries. As with many other aspects of visual perception, binocularity has been analysed in terms of physical optics. In his book on *Optics* Euclid (ca. 325–265 BC) examined binocular vision with the consistency that he had adopted for other aspects of spatial vision—it could be reduced to optical projections and visual angles.[1] In fact, his discussion of the projections from two eyes was rather cursory, being restricted to three different sizes of sphere with respect to the interocular distance. Euclid was aware of the gross differences in the images in each eye, on the basis that light travels in straight lines and the analysis of the visual cone. He described the projections of a sphere to each eye when its diameter was equal to the interocular separation, and when it was greater and smaller. For the three diagrams shown below the sphere is represented by A. In the uppermost diagram the eyes are represented by E and Δ: "When a sphere is seen by both eyes, if the diameter of the sphere is equal to the straight line marking the distance of the eyes from each other, the whole hemisphere will be seen".[2] In the middle figure the eyes are represented by E and Δ: "If the distance between the eyes is greater than the diameter of a sphere, more than the hemisphere will be seen".[3] In the lowermost figure the eyes are at B and Γ: "If the distance between the eyes is less than the diameter of the sphere, less than a hemisphere will be seen".[4]

---

[1] Many versions of Euclid's Optics have been produced usually in Greek or Latin. The line figures in the illustration are from Heiberg (1895).

[2] Burton (1945), p. 362.

[3] Burton (1945), p. 362.

[4] Burton (1945), p. 363.

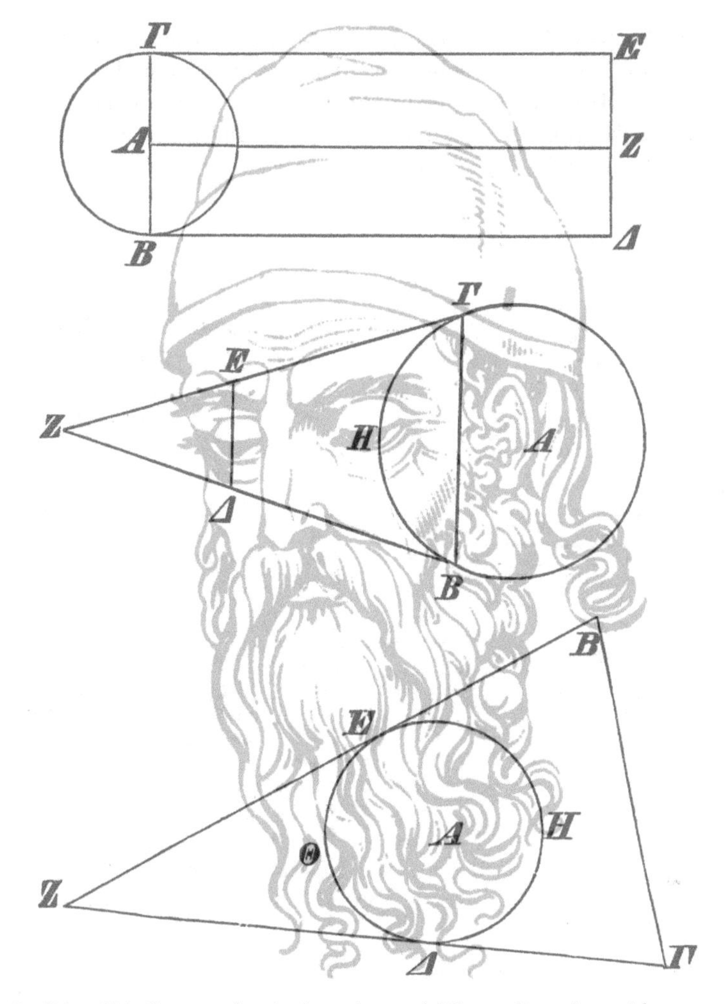

Euclid and his diagrams for viewing spheres of different dimensions with two eyes

During the second century AD the situation was transformed by Ptolemy in optics,[5] and by Galen in ophthalmology. The difference between Euclid and Ptolemy is marked, and the latter can be considered as paving the way for modern approaches to binocular vision. Ptolemy carried out controlled observations of the perceived locations of vertical cylinders; from these he specified the conditions for singleness of vision, the distinction between crossed and uncrossed disparities, and the direction in which objects are seen with two eyes. In the eleventh century Ibn al-Haytham (ca. 965–1039, known also by his Latinized name of Alhazen) extended Ptolemy's work somewhat and constructed a more elaborate board for examining binocular vision.[6]

---

[5] Ptolemy's optics, like Euclid's, underwent many transitions before reaching us today—see Lejeune (1989), Smith (1996).

[6] See Sabra (1989), Smith (2001).

Remarkably little was added to Ptolemy's and Alhazen's analyses until the sixteenth century when Leonardo, like Euclid, used a sphere as the stimulus for examining binocular vision. He represented the differences between monocular and binocular observation and described the amount of the background which was obscured from each eye by the sphere.

Leonardo's diagrams of viewing a small sphere with one or two eyes

Wheatstone lamented Leonardo's use of a sphere to examine binocular vision and remarked "Had Leonardo da Vinci taken, instead of a sphere, a less simple figure for the purpose of his illustration, a cube for instance, he would not only have observed that the object obscured from each eye a different part of the more distant field of view, but the fact would also perhaps have forced itself upon his attention, that this object itself presented a different appearance to each eye. He failed to do this, and no subsequent writer within my knowledge has supplied the omission".[7] That is, a sphere will project a circle to each eye, whereas any other object would project different aspects to each eye.

[7] Wheatstone (1838), p. 372.

Binocular Single Vision

Throughout most of the recorded history of research on binocular vision the overriding interest has been in why the world looks single with two eyes rather than why objects appear in depth.[8] Moreover, much of the concern was whether vision with two eyes was superior to that with one alone. The distinction between monocular and binocular vision derived from the occasional experience of double vision when using two eyes and from the mistakes made by those with only one eye. Binocular single vision has been discussed at least since the time of Aristotle, and it has been examined experimentally since Ptolemy, who defined lines of visual correspondence for the two eyes. Aristotle did discuss binocular single vision, but it tended to be in the context of its breakdown (double vision) either by distorting one eye or in strabismus (squint): "When, therefore, being so placed that they are in a similar position to one another and midway between an upward and a downward and an oblique movement, the two eyeballs catch the visual ray on corresponding points of themselves.... If the vision of both eyes does not rest on the same point, they must be distorted".[9] Observations with two eyes were commented upon by Galen in the second century: "the axes of the visual cones must be situated in one and the same plane if single objects are not to appear double. These axes of ours have as their beginning the channels from the encephalon.... It is the bringing together of the channels [in the optic chiasm]. For if two straight lines meet at a certain, common point in their apex, they are evidently in one plane, even if they happen to be produced from that point an infinite distance in different directions".[10] Aristotle and Galen are shown below together with diagrams of eyes reconstructed from written descriptions by Magnus.[11]

[8] See Wade and Ono (2012).

[9] Ross (1927), p. 958a.

[10] May (1968), p. 498.

[11] Magnus (1901).

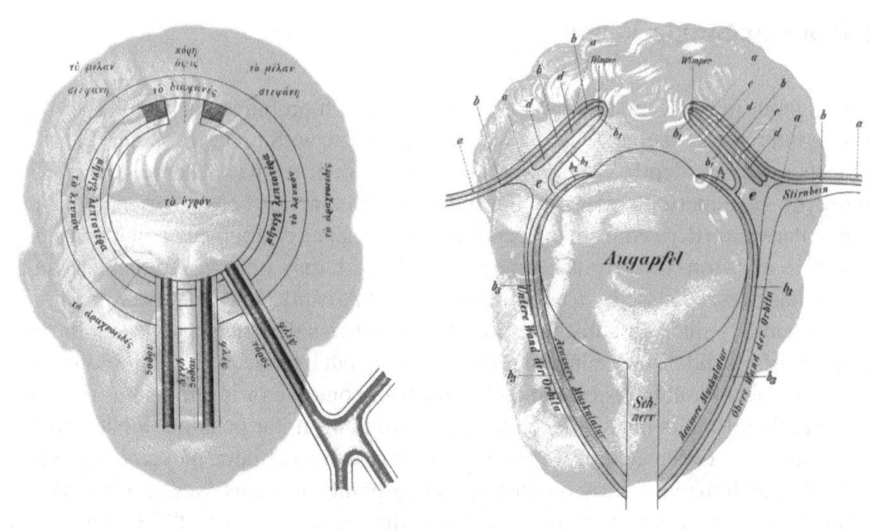

The eyes of Aristotle and Galen

Galen's approach to vision was that the eyes were nurtured by a visual spirit, the origin of which was in the ventricles of the brain. The visual spirit travelled along the optic nerves to interact with images of external objects carried in the air to the lens. Since their origin was in the brain the visual spirit separated at the optic chiasm to pass the each eye. For centuries this resulted in the view that vision with one eye alone was superior to that with two because the visual spirit was not split at the chiasm. The Galenic visual spirit pervaded Descartes' mechanistic analysis of binocular single vision, and he speculated that union of the fibres from the two eyes occurred in the pineal body. Descartes considered, on the then available anatomical evidence, that the two optic nerves remained separate at the chiasm (see Chap. 2). Thus, union was achieved in the only unpaired body in the brain as shown in the diagram below:

> the filaments 1-2, 3-4, 5-6, and the like that compose the optic nerve and extend to the back of the eye (1, 3, 5) to the internal surface of the brain (2, 4, 6). Now assume that these threads are so arranged that if the rays that come, for example, from point A of the object happen to exert pressure on the back of the eye at point 1, they in this way pull the whole of thread 1-2 and enlarge the opening of the tubule marked 2. And similarly, the rays that come from the point B enlarge the opening of tubule 4, and so with the others. Whence, just as the different ways in which these rays exert pressure on points 1, 3, 5 trace a figure at the back of the eye corresponding to that of object AB.C., so evidently, the different ways in which tubules 2, 4, 6 and the like are opened by filaments 1-2, 3-4, and 5-6 must trace [a corresponding figure] on the internal surface of the brain.[12]

[12] Hall (1972), p 84.

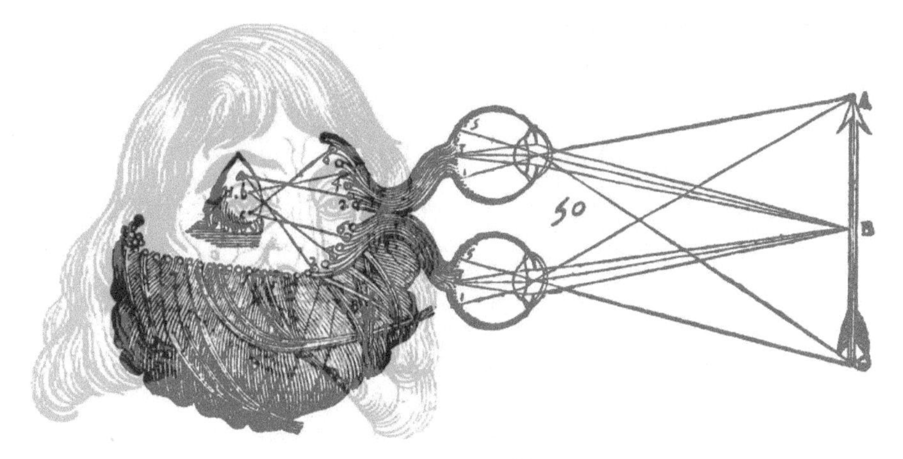

Descartes' eyes and brain

The correspondence to which Descartes was referring was probably that between objects in space and the retinal image, following the analogy of eye and camera. In this way, the same text could have been illustrated with either one or two eyes, and the contrast with the description of binocular vision via the analogy to the blind is not so stark. Descartes was able to link optics with anatomy and this has had a lasting impact. From the seventeenth to the early nineteenth century binocular vision had been interpreted in terms of convergence of the eyes and the stimulation of corresponding points on the two retinas. The two eyes were considered to operate like a range finding device, and it is clearly the case that convergence varies with distance.

The situation remained much the same until the beginning of the nineteenth century. The horopter was shown to be a circle rather than a plane. This was formalised in 1818 by Gerhard Ulrich Anton Vieth (1763–1836) and verified eight years later by Johannes Peter Müller (1801–1858). It has become known as the Vieth-Müller circle.[13] Vieth and Müller gave a geometrical analysis of binocular single vision, defining a circle passing through the point of fixation and the centres of rotation of each eye: points on the circumference of this binocular circle stimulated corresponding points and were seen as single. With regard to the figure below, Vieth wrote:

> The images *M* and *N* lie in *that* case equally far from *A* and *B*, if the angles *p* and *x* are equal, in which case this applies; if *X* lies *on the circumference of a circle*, which passes through *O* and *U* and *P*, then every angle on this circumference will have the same subtense with respect to *O U*. Therefore if by the expression *corresponding points* one understands such points which lie in the same direction in both eyes and are *equidistant* from *A* and *B*, which appears to me to be the correct meaning, and one asserts one sees things singly whose images fall on such corresponding points, then, according to this rule, one sees that thing singly which is situated within the boundary of a sphere which passes through *O*, *U*, *P*, and does not lie in the plane, which one names the *horopter*.[14]

[13] Vieth (1818), Müller (1826).

[14] Vieth (1818), p. 238.

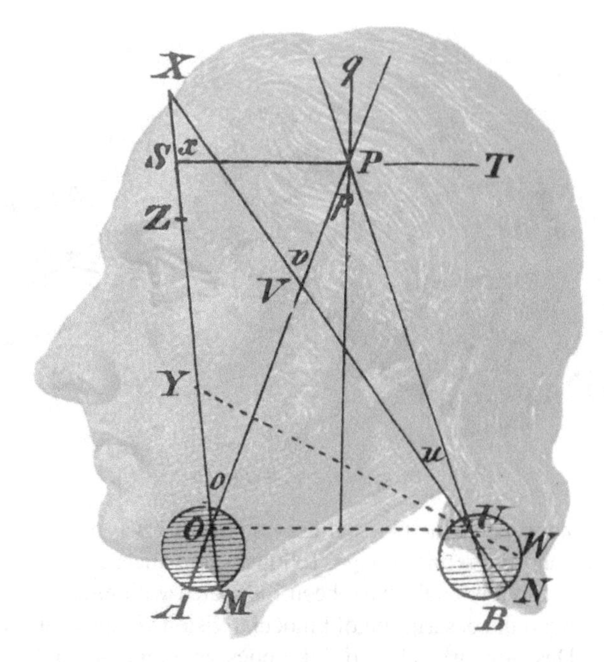

Gerhard Vieth and his diagram of stimulating corresponding points

Both Vieth and Müller adopted the term 'horopter' as introduced by Aguilonius two centuries earlier. Müller later augmented his geometrical description of the circle of single vision by linking it with identical retinal points: "The horopter is therefore always a circle, of which the chord is formed by the distance between the eyes, or, more correctly, between the points of decussation of the rays of light in the eyes, and of which the size is determined by three points, – namely, by the two eyes, and the point to which their axes converge".[15] In this way, there were only two possible states of perception—single vision when objects fell on the circumference of the circle and double vision otherwise, and singleness was served by a fixed organic relation between nerve fibers. Thus, in the year that saw publication of Wheatstone's article on stereoscopic depth perception we find a statement denying its possibility.

[15] Müller (1838/1843), p. 1196.

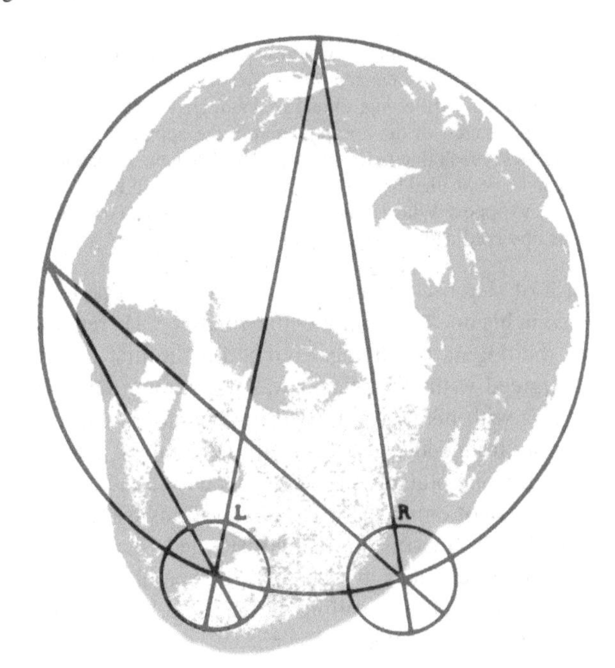

Johannes Müller and a drawing of the binocular circle taken from Wheatstone's article

It is this elegant edifice that Wheatstone undermined with his demonstration of singleness and depth from images with slight retinal disparities: he remarked that "objects whose pictures do not fall on corresponding points of the two retinæ may still appear single".[16] Wheatstone was well aware of the originality of both his observations and his interpretation of them, hence the meticulousness of his experiments in their support. Not only did he argue that singleness and depth could be observed with stimulation of non-identical retinal points but also that the stimulation of identical points could result in double vision.[17] Moreover, the stereoscope, perhaps more than any other instrument, ushered in the era of experimentation for spatial vision. It fulfilled the scientific desire to examine binocular vision by observation and experiment. As Towne put it: "The introduction of the stereoscope inaugurated a new epoch in the physiology of vision, opened a wide field for further inquiry, and suggested additional methods of investigation, while the theory of binocular vision has been greatly modified by results which have been obtained through the medium of the instrument".[18]

The impact of Wheatstone's experiments was felt acutely by sensory physiologists in Germany and it was expressed in articles and books.[19] As an example of the opposition to Wheatstone's views Alfred Volkmann (1801–1877) wrote:

---

[16] Wheatstone (1838), p. 384.

[17] See Ono and Wade (1985), Wade and Ono (1985).

[18] Towne (1862), p. 70.

[19] See Ono and Wade (1985).

> If Wheatstone were right, the teaching of vision would be in danger of being overturned; for one only needs to remember Johannes Müller's excellent investigations into the identical retinal points and their positioning, to realize that rejection of the doctrine of identity, the doctrine of double images, of the horopter, and of the apparent distance between two images in the visual field break down, i.e. precisely those teachings that are based on exact experiments that can be controlled by dimensions and numbers. It is therefore very natural for the physiologists to oppose Wheatstone's conclusions and to endeavour to reconcile the phenomena that he observed with the teaching of the identical retinal points.[20]

The significance of stereoscopic phenomena was fully appreciated by Albrecht Nagel (1833–1895) in his book on vision with two eyes.[21] He remarked that Wheatstone had shaken the dogma of identical points for the first time and that hardly any physiologists agreed with him; Nagel was not among them. Panum[22] sought to salvage the dogma with his concept of fusional areas so that single vision was expanded to a region around the geometrically constrained circumference of the Vieth-Müller circle. Despite praising the stereoscope for facilitating experiments on binocular vision, it was not enlisted for the observations Nagel made with simple line stimuli—he adopted a free-fusion method. He made many manipulations with paired vertical or near vertical lines in each eye, with combinations of inclined lines, curves and crosses are as well as rivalling vertical and horizontal lines like those examined earlier by Panum and later by Towne.

Die stereoskopischen Erscheinungen in ihrer Beziehung zu der Lehre von den identischen Netzhautpunkten.

Von

A. W. Volkmann.

Die Erscheinungen, welche das Stereoskop darbietet, veranlassen den Physiologen zu zwei Fragen, erstens nämlich: warum verschmelzen zwei Bilder, deren Conturen sich nicht decken, zu einem einzigen, und zweitens: aus welchem Grunde können einfache Linearzeichnungen den Eindruck eines körperlichen machen und demnach Raumanschauungen in der Dimension der Tiefe veranlassen? Beide Fragen sind schon oft untersucht worden, aber schwerlich jemals in dem erforderlichen Umfange. Daher sind auch die Antworten auf dieselben noch keineswegs präcis, und in manchen speciellen Fällen noch ganz unzulässig. Unter diesen Umständen glaube ich den vielbesprochenen Gegenstand noch einmal aufnehmen zu dürfen. Die gegenwärtige Abhandlung untersucht von den beiden in Anregung gebrachten Fragen zunächst nur die erste.

Bekanntlich hatte Wheatstone, der berühmte Erfinder des Stereoskops, aus seinen Versuchen den Schluss ge-

Archiv für Ophthalmologie. II. 2. 1

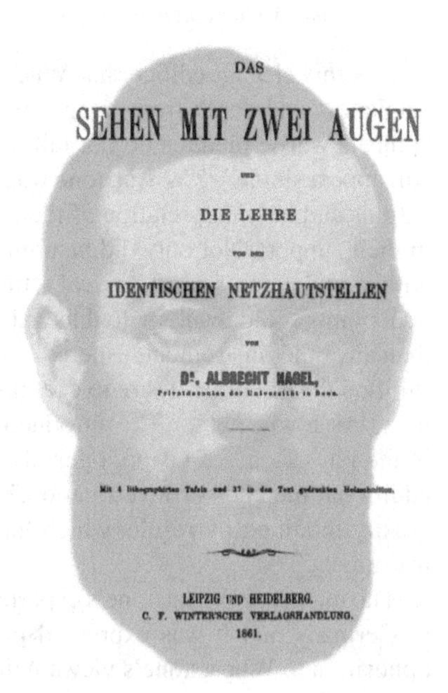

DAS

SEHEN MIT ZWEI AUGEN

DIE LEHRE

IDENTISCHEN NETZHAUTSTELLEN

Dr. ALBRECHT NAGEL,

LEIPZIG UND HEIDELBERG.

C. F. WINTER'SCHE VERLAGSHANDLUNG.

1861.

Left, Alfred Volkmann and the opening page of his article on stereoscopic vision. Right, Albrecht Nagel and the title page of his book on vision with two eyes

[20] Volkmann (1859), p. 2. The article has been translated into English by H. J. Simonsz—see Volkmann (2020, 2021a, b, c, 2022). See also Wade (2022a).

[21] Nagel (1861). Part of the book has been translated into English by H. J. Simonsz—see Nagel (2020a, b, c).

[22] Panum (1858).

Wheatstone was well aware of the impact that his observations and his interpretation of them would have. Not only did he argue that the Vieth-Müller circle does not define binocular single vision but he also rejected the link with corresponding retinal points:

> The same reasons, founded on the experiments in this memoir, which disprove the theory of Aguilonius, induce me to reject the law of corresponding points as an accurate expression of the phenomena of single vision. According to the former, objects can appear single only in the plane of the horopter; according to the latter, only when they are in the circle of single vision; both positions are inconsistent with the binocular vision of objects in relief, the points of which they consist appearing single though they are at different distances before the eyes. I have already proved that the assumption made by all the maintainers of the theory of corresponding points, namely that the two pictures projected by any object in the retinæ are exactly similar, is quite contrary to fact in every case except that in which the optic axes are parallel.[23]

Binocular Vision and Eye Movements

Perhaps the most distinctive feature of eye movements is their binocularity—the eyes tend to move together. Aristotle commented that there were certain movements that were not possible, namely, divergence beyond the straight ahead, and movements of one eye upward and the other downward. Much of the early commentary concerned the observation that when the two eyes are directed from one location to another they tend to move together. For example, Aristotle remarked:

> Why is it that we can turn the gaze of both eyes simultaneously towards the right and the left and in the direction of the nose, and that of one eye to the left or to the right, but cannot direct them simultaneously one to the right and the other to the left? Similarly, we can direct them downwards and upwards; for we can turn them simultaneously in the same direction, but not separately. Is it because the eyes, though two, are connected at one point, and under such conditions, when one extremity moves, the other must follow in the same direction, for one extremity becomes the source of movement to the other extremity.[24]

This statement is important because it is based on the assumption that the eyes move as a unit. Such conjoint motion of the eyes is not always in the same direction, as Aristotle noted in his distinction between version (movements in the same direction) and vergence eye movements, although he did not use these terms.

Thus, the initial descriptions of how the eyes move were made in the context of binocular vision. Indeed, this is one of the first areas of vision that received experimental attention. Ptolemy appreciated that monocular and binocular visual directions were not necessarily the same. In order to confirm this empirically, he constructed a board on which he could place vertical rods at different distances in the midline. There followed a description of one of the most commonly used examples of crossed and uncrossed visual directions: with fixation on the far rod, the

[23] Wheatstone (1838), p. 390.

[24] Ross (1927), pp. 957b–958a.

nearer one appeared double, and to the left with the right eye and to the right with the left eye; the reverse occurred with fixation on the nearer rod. Essentially the same demonstration is now more frequently made with two fingers, rather than rods, held at different distances from the eyes. Ptolemy stated that singleness of vision with two eyes occurred when the two visual directions corresponded, thus introducing the concept of correspondence into binocular vision.[25] He modified his board to take three rods and found that objects appeared single to two eyes when they were in the same plane as the fixation point. These facts were interpreted in terms of the visual axes and the common axis.

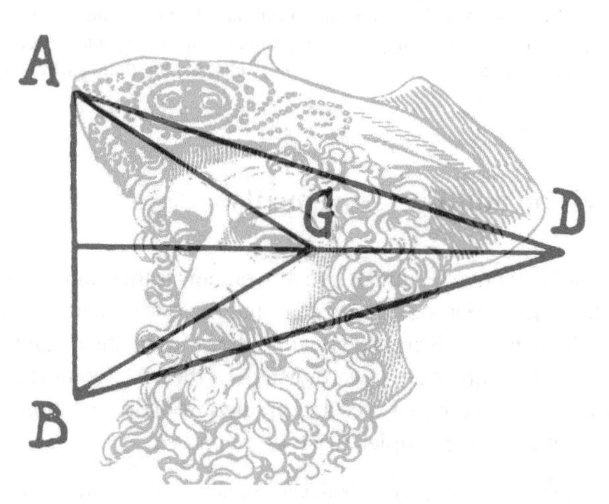

Ptolemy and his diagram of viewing objects at different distances with two eyes

Ptolemy gave a similar description of eye movements to that attributed to Aristotle, but he did appreciate the function eye movements served: they resulted in corresponding visual pyramids for the two eyes, thereby assisting binocular vision:

> These phenomena [of binocular combination] occur only by virtue of the horizontal separation of the eyes since the height and the depth of the eyes are the same. Both visual axes turn until the bases of the pyramids coincide on the object. It is possible for the eyes to turn in opposite directions left and right, but not up and down. They retain their vertical position, but can converge horizontally.[26]

Alhazen incorporated many aspects of Ptolemy's analysis of vision. The translation of Alhazen's book on optics into Latin awakened medieval Western scholars to the physics of light, its mathematical treatment, and its application to vision.[27] Alhazen integrated his analysis of binocular vision with that of the conjoint movements of the eyes:

---

[25] See Howard and Wade (1996).

[26] Lejeune (1956), p. 34.

[27] Russell (1996, 2008, 2010).

> We say, then, that when a beholder looks at a visible object, each of his eyes will regard that object; when he gazes at the object, both of his eyes will equally and similarly gaze at it; when he contemplates the object, both of his eyes will equally contemplate it; and when sight moves over the object in order to contemplate it, both eyes will move over it and contemplate it. When the beholder fixes his sight on an object, the axes of both eyes will converge on the object, meeting at a point on its surface. When he contemplates the object, the two axes will move together over the surface of the object and together pass over all its parts. And, in general, the two eyes are identical in all their conditions, and the sensitive power is the same in both of them, and their actions and affections are severally always identical. When one eye moves for the purpose of vision, the other moves for the same purpose and with the same motion; and when one of them comes to rest, the other [likewise] is at rest... When both eyes are observed as they perceive visible objects, and their actions and movements are examined, their respective actions and movements will be found to be always identical.[28]

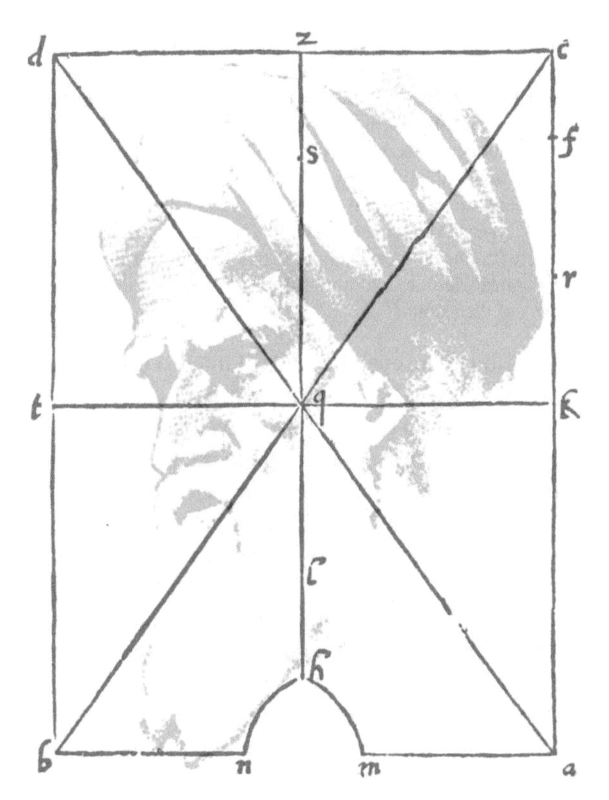

Ibn al-Haytham (Alhazen) combined with his board for viewing cylinders with both eyes. The eyes were located at a and b with fixation on a wax cylinder at q; cylinders at t and k were seen double

Thus, Alhazen proposed that the two eyes acted together in order to achieve single vision, and that the movements of the eyes were coordinated to this end. He supported the proposal, it would seem, by observing the eyes of another person while

---

[28] Sabra (1989), p. 229.

they were inspecting an object. An empirical method for confirming the coordinated movements of the eyes was described by Robert Smith (1689–1768)—placing a finger over the closed eyelid and feeling the movement when the open eye scanned a scene. He elaborated on the analysis of binocular single vision by recourse to the stimulation of corresponding points in the two eyes, and he related the misalignment of the two eyes to double vision:

> The axis of the eye is a line drawn through the middle of the pupil and of the crystalline, and consequently falls on the middle of the retina. And the axes of both eyes produced are called the optick axes. When the optick axes are parallel or meet in a point, the two middle points of the retinas, or any two points which are equally distant from them, and lye on the same sides of them either toward the right hand or left hand, or upwards or downwards, or in any oblique direction, are called corresponding points. Now we find by experience that an object or point of an object appears single when its pictures fall upon corresponding points of the retinas, and double when they do not. For when we view an object steadily, we have acquired a habit of directing the optick axes to the point of view; because its picture falling upon the middle points of the retinas are then distincter than if they fell upon any other places; and since the pictures of the whole object are equal to one another and are both inverted with respect to the optick axes, it follows that the pictures of any collateral point of the object are painted upon corresponding points of the retinas. This habit of directing the optick axes to the point in view is so strong that it is very difficult to do otherwise; insomuch as when one eye is shut and the other is in motion one may feel by ones fingers laid upon the eye-lid, that the eye which is shut, always follows the motions of the eye that is open. But if by squinting or depressing an eye with ones finger, the optick axes are not directed to the same point; in these cases objects appear double: and now it is plain that the pictures are not painted on corresponding places of the retina.[29]

Smith was bringing together several strands of interest in binocular vision. Singleness of vision with two eyes was considered to be a consequence of stimulating corresponding points on the two retinas; distinct vision was restricted to the central region of the retina; both eyes moved in unison to retain correspondence; this could be demonstrated by feeling the movements of the closed eye; double vision occurs when one eye is moved out of alignment with the other. Such squinting can be induced artificially (by displacing one open eye with the finger) and it can occur naturally. Smith also provided illustrations of crossed and uncrossed disparities, with the latter representing the disparate images in the plane of fixation but they were taken to be examples of non-corresponding stimulation and therefore of double vision. With regard to the figures below he wrote: "if while the optick axes NM, OM are directed to a mark M, we attend to an object or image q, placed anywhere within the angle NMO or its opposite, made by the optick axes produced, the object q will appear in two places, suppose at a and b, situated in the direction of the visual rays Nq, Oq. For the pictures of the object, q, which lyes between the optick axes, being both inverted with respect to the axes, must fall upon the retinas on contrary sides of the axes, and consequently upon places that do not correspond."[30]

---

[29] Smith (1738), pp. 46–47.

[30] Smith (1738), p. 47.

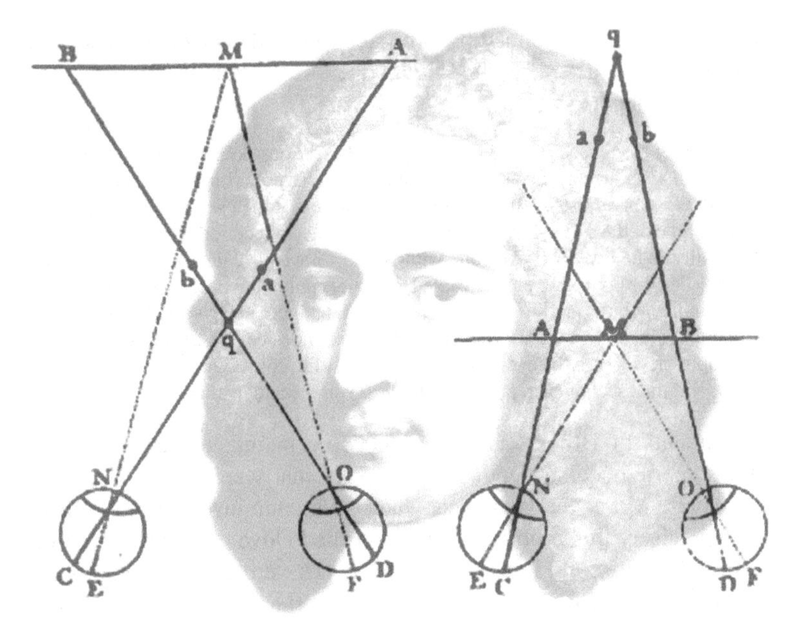

Robert Smith and diagrams of viewing objects at different distances

The study of binocular vision has been of continuing concern to students of optics since antiquity. While it has been informed by the studies of Ptolemy, Alhazen, and a host of others, the interest has been largely confined to the phenomenon of binocular single vision—the experience of a single world despite receiving two different projections from it. The situation was transformed by the invention of the stereoscope and the description of stereoscopic depth perception.[31] Wheatstone demonstrated that slightly different images presented to each eye were seen in depth rather than seen double. Such stereoscopic depth could be seen much more easily with the aid of a stereoscope and the factors involved in binocular combination could be examined with an experimental precision that had not been available previously.

Wheatstone interpreted stereoscopic depth perception in terms of psychological processes that enabled combination of the disparate images. The phenomena proved to be damaging to theories of binocular vision that were based on singleness determined by the stimulation of corresponding points. Wheatstone formed afterimages (by prolonged fixation) of differently coloured paired images and observed stereoscopic depth when they were viewed with closed eyes.

Nonetheless, alternatives, in terms of eye movements, were soon proposed. In 1841 Ernst Brücke (1819–1893) attacked Wheatstone's theory by arguing that the state of convergence changes very rapidly when viewing either solid objects or stereopairs, thereby changing the binocular visual directions of the disparate parts.[32] Much

---

[31] See Wade (1983, 1987).

[32] Brücke (1841). The article has been translated into English by H. J. Simonsz—see Brücke (2022a, b) also Wade (2022b).

the same position was maintained by Brewster who added changes in accommodation to those of convergence. With regard to the observation of a truncated cone, or stereopairs of a cone, he stated:

> Each eye, acting alone, sees the cone single, and the various points of its outline are seen more or less distinct, according as they are more or less remote from the point to which the eye is for the instant adjusted. But so rapid is the motion of the eye, and so quickly does it survey the whole of the solid, that it obtains a most distinct perception of its form, its surface, and its solidity. When we view the cone with *both eyes*, we have the same indistinctness of the outline when the optic axes are converged to a single point: but in addition to this, we have the greater indistinctness arising from every point of the figure being seen *double*, except the single point to which the axes are converged. But this imperfection, too, is scarcely visible, from the rapid view which the eyes take of the whole solid, converging their axes upon every point of it, and thus seeing each point in succession single and distinct.[33]

The arguments of Brücke and Brewster were maintained even though Dove (1841) had reported seeing depth in stereopairs that were briefly illuminated by an electric spark. Moreover, some years later Volkmann invented the tachistoscope to examine precisely this problem.[34] He confirmed Dove's observation of depth in briefly exposed stereoscopic images. Tachistoscopes were generally employed thereafter to present stimuli for periods that were too short to enable eye movements over them.

Helmholtz devoted a considerable portion of volume 3 of his *Handbuch der physiologischen Optik* to a discussion of eye movements and also stereoscopic vision.[35] His primary concern was with determining the position assumed by the eye following an eye movement and he also considered the motions of two eyes. Using afterimages projected onto a grid of vertical and horizontal lines he determined that the vertical meridia in each eye differed slightly in orientation. Helmholtz wrote: "Therefore not the *really vertical* meridians of both fields of view correspond, as has been supposed hitherto, but the *apparently* vertical meridians. On the contrary, the horizontal meridians really correspond, at least for normal eyes which are not fatigued".[36] As was the case with earlier studies of binocular eye movements, Helmholtz's interest was directed to the issue of retinal correspondence, and he was able to use the stereoscope to assist in his studies.

---

[33] Brewster (1844), p. 363.

[34] Volkmann (1859), Wade and Heller (1997).

[35] Helmholtz (1867, 1925).

[36] Helmholtz (1864), p. 196.

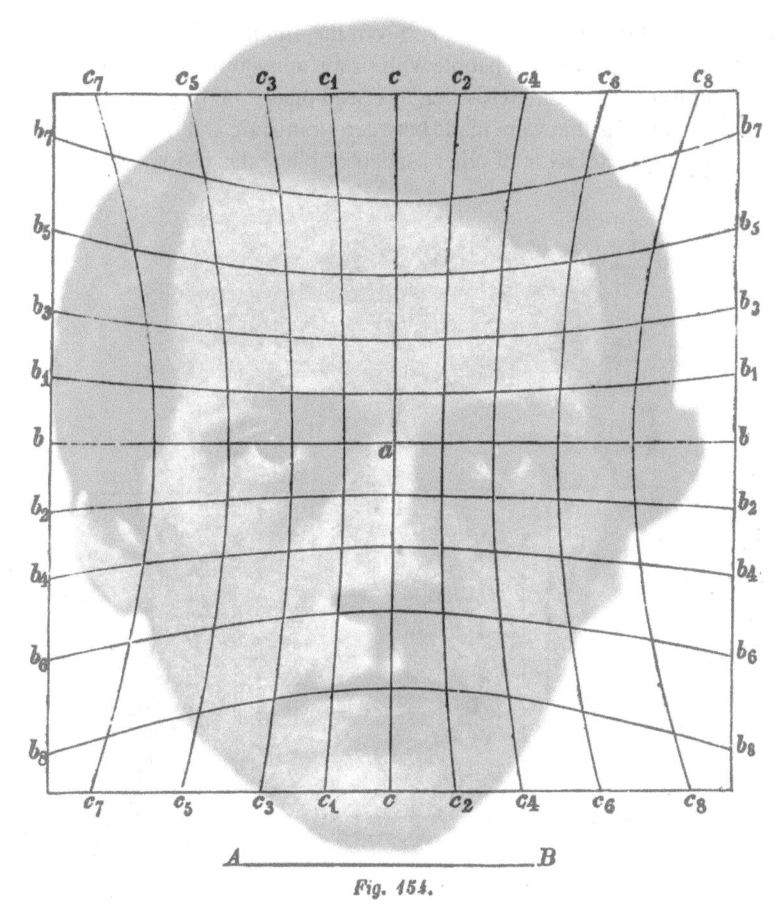

A portrait of Helmholtz as a young man is combined with his diagram of the tilted vertical meridians in different directions of gaze

Prior to the transformation generated by Wheatstone's invention of the stereoscope, binocular eye movements were considered in terms of aligning the eyes so that corresponding regions of each retina could be stimulated by the same object. When this did not occur double vision (diplopia) ensued. Strabismus or squint is a naturally occurring (and relatively common) condition involving misalignment of the eyes and uncoordinated movements of them. Usually one eye deviates laterally and it can be towards the nose (convergent) or towards the temple (divergent). Corrections for the deviations of the eyes have been advocated since antiquity and often involved putting patches over the non-deviating eye in order to encourage fixation with the deviating eye. Some of the most elaborate masks were prescribed by Georg Bartisch (1535–1606) in a book that is considered to mark the beginnings of ophthalmology[37] and they are shown below. Bartisch appreciated that strabismus was more amenable

[37] Bartisch (1583).

to correction in children than in adults. From the eighteenth century attention was directed to the manner in which people with strabismus saw objects singly. Studies of strabismus epitomize the emphasis that has been placed on binocular single vision; the many comparisons that were made between strabismic and binocular individuals dwelt on this to the exclusion of other aspects of binocular vision, like stereopsis.

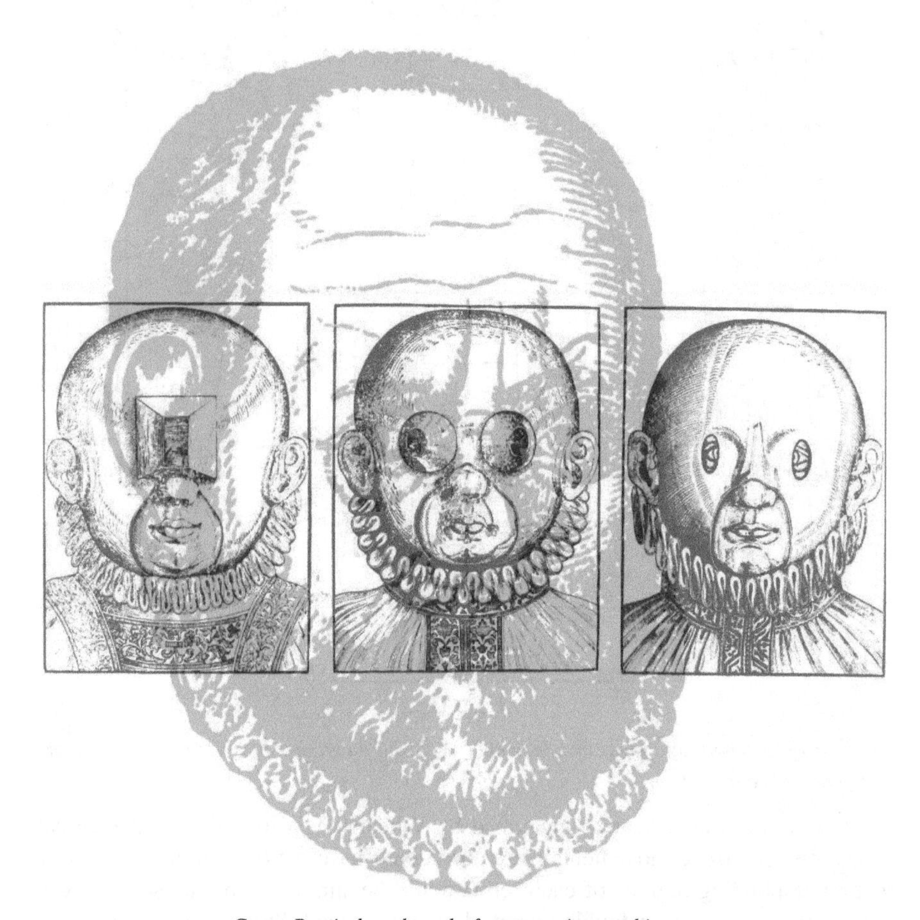

Georg Bartisch and masks for correcting strabismus

The breakdown of binocular vision in strabismus was examined in greater detail, and this resulted in further study of eye movements themselves. However, strabismus was considered largely in terms of lateral misalignment of the eyes. The question of eye rotations around the visual axis (torsion) was rarely considered and was a matter of great contention in nineteenth century visual science.[38] Speculations that the oblique muscles could rotate the eyes around the z-axis were frequently stated. For example, Charles Bell (1774–1842) wrote: "By dissection and experiment it can

[38] Simonsz and Wade (2018).

be proved, that the oblique muscles are antagonists to each other, and that they roll the eye in opposite directions, the superior oblique directing the pupil downwards and outwards, and the inferior oblique directing it upwards and inwards".[39] However, evidence supporting it in humans was hard to find and when it was presented it was usually contended.

The possibility of torsion in opposite directions seemed fanciful and yet this is what Nagel proposed in order to maintain cyclofusion for lines inclined in opposite directions relative to the horizontal.[40] Similar rotations about the vertical resulted in a depth effect with no cyclovergence. As Nagel noted, achieving cyclofusion with single lines required a lot of practice and patience. He returned to the issue later and showed that the cyclofusion could be produced more readily with arrays of lines.[41] The involvement of cyclovergence remained hotly debated until photographic recording of eye movements verified it.[42]

Cyclofusing Albrecht Nagel. Nagel's portrait and the lines are inclined two degrees clockwise and counterclockwise of the horizontal. Cyclofusion can be experienced by viewing the arrays of lines; they will appear horizontal with two eyes

[39] Bell (1803), p. 174.

[40] Nagel (1861).

[41] Nagel (1868, 1871).

[42] Crone and Everhard-Halm (1975)

Monocular and Binocular Vision

It might seem that the simplest way of determining the advantages of vision with two eyes is to compare it with using one alone. Perceptual experience is typically unified and coherent: objects are seen as having a specific size, shape, color, and a location in space. This obtains under conditions of binocular and monocular vision. Indeed, perceptual experience changes little when one eye is closed, but it does change, and the study of the differences has a long history. Contrary to the evidence accumulated before the late seventeenth century, it was long believed that vision with one eye was superior to that with two. Aristotle took this to be self-evident, interpreting it in terms of eye movement control. The source of much subsequent comparison was driven by the Galenic theory of visual spirits: it was transmitted to one or two eyes, and thus was more concentrated in monocular viewing. Despite Ptolemy's statement to the contrary, this opinion was repeated over the following centuries, and it was held as late as the seventeenth century, when Francis Bacon (1561–1626) attributed the advantages of aiming with one eye to this cause: "We see more exquisitely with one eye shut, than with both open. The cause is, for that the spirits visual unite themselves more, and so become the stronger. For you may see by looking in a glass, that when you shut one eye, the pupil of the other eye that is open dilateth".[43]

Many statements were made about tasks that were more difficult to perform with one eye rather than two, or stimuli that were more difficult to see. Robert Boyle (1627–1691) noted the problems that occurred when the function an eye was lost: "Haveing frequently occasion to pour Distill'd Waters and other Liquors out of one Vial into another, after this Accident he often Spilt his Liquors, by pouring quite Beside the necks of the Vials he thought he was pouring them directly Into".[44] Similarly, William Molyneux (1656–1698) referred to the loss of visual motor coordination when one eye was closed or covered: "the best [tennis] Player in the World Hoodwinking one Eye shall be beaten by the greatest Bungler that ever handled a Racket; unless he be used to the Trick, and then by Custom get a Habit of using one Eye only".[45]

[43] Bacon (1857), p. 628.

[44] Boyle (1688), p. 255.

[45] Molyneux (1692), p. 294.

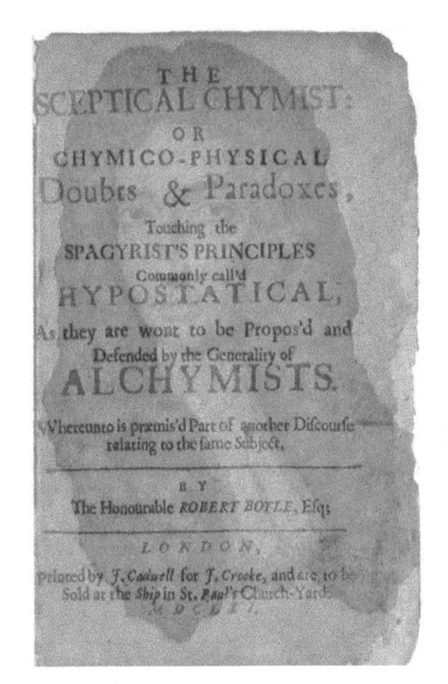

THE
SCEPTICAL CHYMIST:
OR
CHYMICO-PHYSICAL
Doubts & Paradoxes,
Touching the
SPAGYRIST'S PRINCIPLES
Commonly call'd
HYPOSTATICAL;
As they are wont to be Propos'd and
Defended by the Generality of
ALCHYMISTS.
Whereunto is præmis'd Part of another Discourse
relating to the same Subject,
BY
The Honourable ROBERT BOYLE, Esq;
LONDON,
Printed by J. Cadwell for J. Crooke, and are to be
Sold at the Ship in St. Paul's Church-Yard.
MDCLXI.

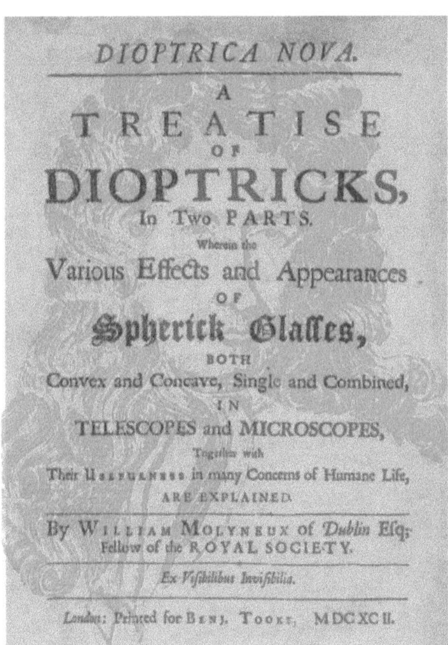

DIOPTRICA NOVA.
A
TREATISE
OF
DIOPTRICKS,
In Two PARTS.
Wherein the
Various Effects and Appearances
OF
Spherick Glasses,
BOTH
Convex and Concave, Single and Combined,
IN
TELESCOPES and MICROSCOPES,
Together with
Their USEFULNESS in many Concerns of Humane Life,
ARE EXPLAINED.
By WILLIAM MOLYNEUX of Dublin Esq;
Fellow of the ROYAL SOCIETY.
Ex Visibilibus Invisibilia.
London: Printed for BENJ. TOOKE, MDCXCII.

Boyle and Molyneux with the title pages of the books in which they described deficits in using only one eye

Wheatstone, with his knowledge of binocular depth perception, enquired why more errors are not made in monocular viewing. His answer was that many cues for depth and distance exist, and that alternatives are used when one source is not available. Most particularly, he proposed that motion of the head (motion parallax) was a potent substitute for retinal disparity.

Binocular Viewing Techniques

The simplest binocular viewing technique is the one we employ almost all the time—using both eyes to look at three-dimensional objects. In the context of stereoscopic pictures it is called 'free-viewing' and it was refined for combining two adjacent pictures by crossing the eyes (overconverging) so that the left stimulus was seen by the right eye and vice versa. The technique was refined by Sebastien Le Clerc (1637–1714) by viewing through a small hole.[46]

[46] Le Clerc (1679).

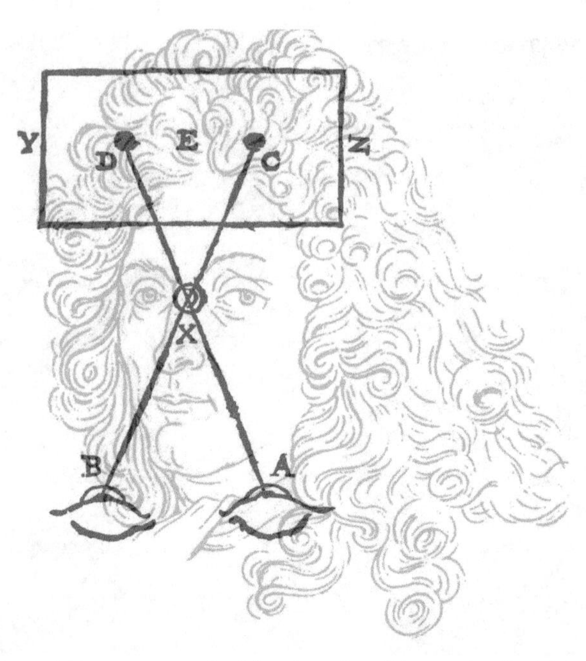

Le Clerc and his aperture method for observing different stimuli with each eye

In the context of experiments on binocular single vision, Jean Théophile Desaguliers (1683–1744) devised a method of combining different stimuli in the two eyes that was to become widely employed in other studies of binocular vision, namely, placing an aperture in such a position that two more distant, adjacent objects were in the optical axes of each eye.[47] Desaguliers used the method to examine both binocular single vision and binocular colour combination.

[47] Desaguliers (1716).

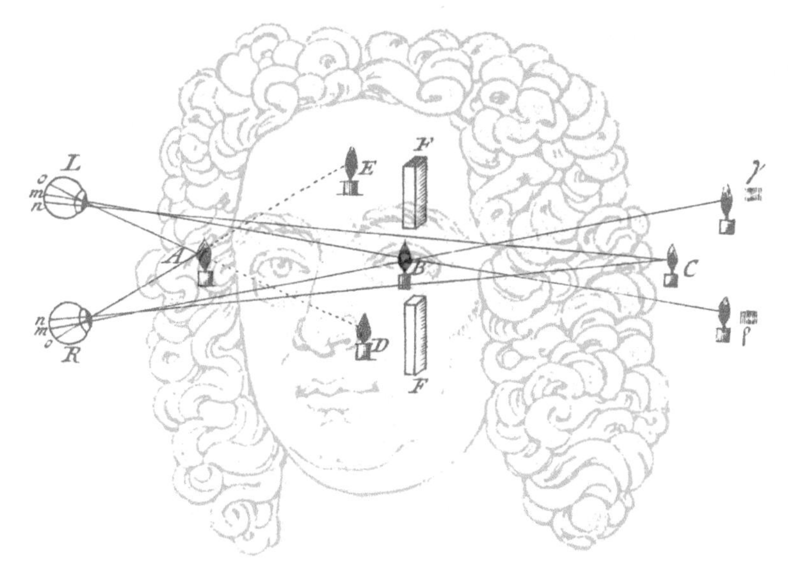

Desaguliers and his illustration of combining different patterns in each eye

Following Newton's experiments on colour mixing, the combination of different colours presented to corresponding regions of each retina became an issue of theoretical importance. Indeed, it was Desaguliers, an advocate of Newtonian optics, who was amongst the first to draw attention to the phenomenon. In particular, he showed that dichoptically presented coloured lights rival rather than combine as in Newton's experiments on colour mixing. Using the same experimental apparatus as he employed for his studies of binocular single vision, he replaced the candles with patches of different coloured silks and observed that colour mixing did not occur. Moreover, if the coloured patches were made more intense, the rivalry was more compelling. That is, no colour combination took place dichoptically, and the colour rivalry is more evident with intense stimuli. Newton, in an unpublished manuscript,[48] had stated that it was impossible for two objects to appear in the same place, because he believed that fibres from corresponding locations in each eye united at the chiasm before passing to the brain. A similar argument would apply to colour, and Desaguliers' observation was certainly in line with Newton's prediction. Nonetheless, the report set in train a series of studies that attempted to examine the phenomenon.

Desaguliers' method was applied by Taylor, who added the refinement of placing coloured glasses in front of candle flames; he found that colours combined rather than engaged in rivalry.[49] Etienne-François Du Tour (1711–1784) provided a clear description of binocular colour rivalry. He achieved dichoptic combination by another means: he placed a board between his eyes and attached blue and yellow fabric in equivalent positions on each side, or the fabric was placed in front of the fixation point. When he converged his eyes to look at them they did not mix but alternated

[48] See Wade (1987) for details about the manuscript.

[49] Taylor (1738).

in colour. Du Tour also applied the method of observing the colours through an aperture,[50] as adopted by Desaguliers, and obtained similar results.

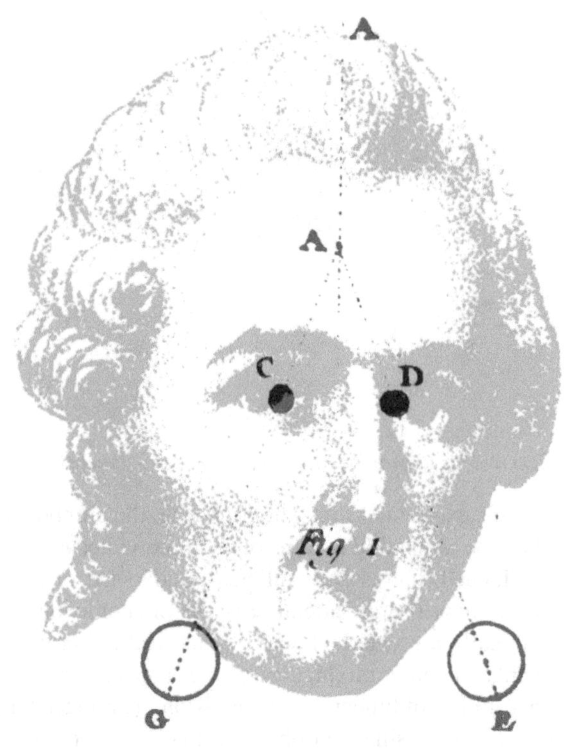

Du Tour and his method for binocular combination of different coloured patterns

Yet another technique was to view different coloured objects through two long tubes, one in each optic axis. This method was used by the philosopher Thomas Reid (1710–1796) who saw the colours combined although his description was not without its ambiguity: the colours were not only said to be combined, but also one "spread over the other, without hiding it".[51] Wheatstone tried most of these methods and illustrated some of them in his first article on the stereoscope.[52]

Dichoptic colour combination could be examined with greater ease after the stereoscope had been invented: different coloured patches could be placed on the separate arms of the stereoscope so that the ensuing experience could be reported. Wheatstone found that blue and yellow discs engaged in rivalry rather than combination. After more than one hundred fifty years of research it is evident that many factors

[50] Du Tour (1760).

[51] Reid (1764), p. 326.

[52] Wheatstone (1838).

determine whether mixture or rivalry occurs such as luminance, saturation, stimulus duration and colour difference.[53]

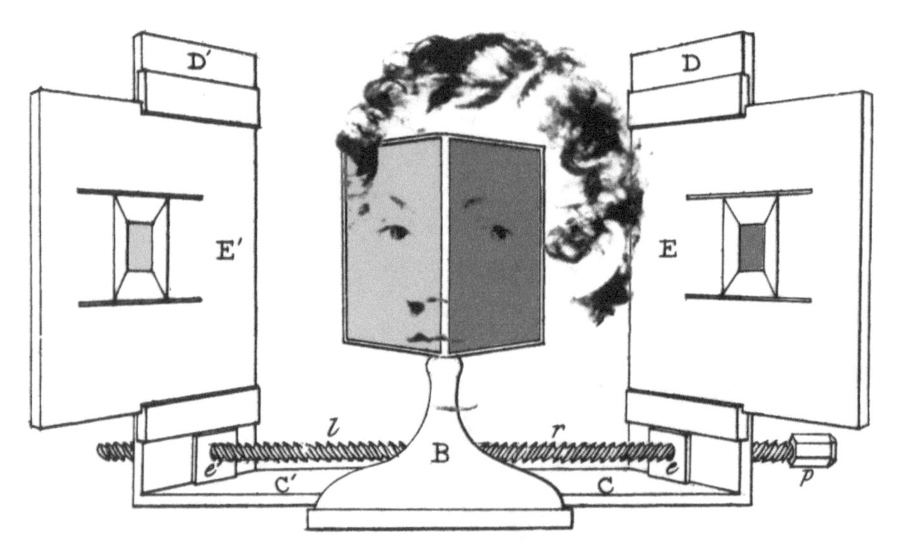

Wheatstone viewing different colour patches in his mirror stereoscope

Early Binocular Instruments

A wide variety of binocular instruments had been devised before the stereoscope was invented in the 1830s. Binocular versions of telescopes and microscopes were available in the eighteenth century, but they did not add to the understanding of binocular vision. One of the contenders as inventor of the telescope, Hans Lippershey (1570–1619), took out a patent for a binocular telescope in 1608.[54] At the beginning of the seventeenth century, the theory that vision with one eye was superior to that with two was being questioned and a range of binocular instruments was produced in that century. For example, In 1645 Schyrleus de Rheita (1604–1660) described (but did not illustrate) a binocular telescope, unlike Chérubin d'Orléans (1613–1697) and Johannes Zahn (1631–1707)—both of whom also illustrated binocular microscopes.[55] The telescopes (called binocles) consisted of paired tubes with parallel instruments and they were constructed in the belief that vision with two eyes was superior to that with one. Chérubin d'Orléans wrote: "The two eyepieces of the binocle are placed in their own tubes in such a way that one can see distinctly through

[53] See Erkelens and van Ee (2012) for a contemporary assessment of binocular colour mixing.

[54] See Schmitz (1982).

[55] Schyrleus de Rheita (1645), d'Orléans (1671, 1677), Zahn (1685, 1686).

each separately, and they can be adjusted as required for the two visual axes, so that the two eyes, which are looking at the same time, each with its own view, together see just one and the same object".[56] Chérubin d'Orléans also illustrated a binocular microscope and described it as follows: "It is known how to construct a novel type of microscope, in order to see the smallest object very agreeably and conveniently, represented entirely to the two eyes together, with a size and distinctness which surpasses all that we have seen until now with this type of microscope".[57]

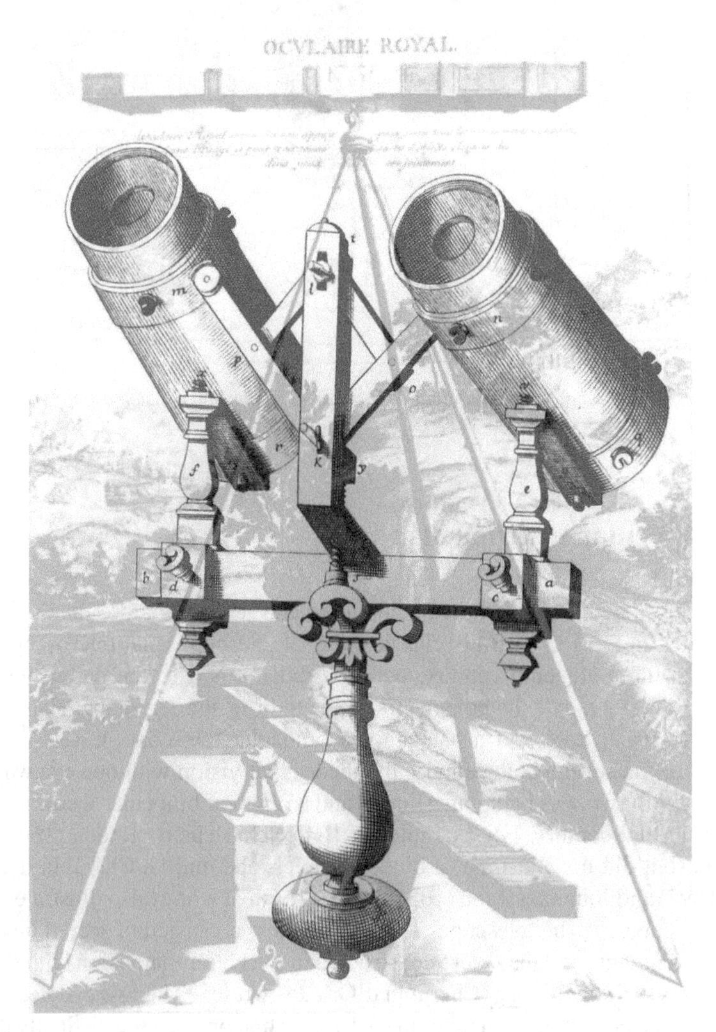

Binocular telescopes illustrated by Chérubin d'Orleans

[56] d'Orléans (1671), p. 73.
[57] d'Orléans (1671), p. 77.

The same principles were applied to telescopes and microscopes and Zahn's model is shown with a specimen on the viewing plate. Zahn's illustration is often attributed to Chérubin d'Orléans, although his prints were steel engravings and not wood-cuts. The attribution is likely to be due to Mayall[58] who represented Zahn's figure with the legend "Chérubin d'Orléans' binocular microscope (1678)" while stating in the text that the illustration was from Zahn. Carpenter[59] repeated the claim on Mayall's authority and reprinted Mayall's illustration with the same caption. The misattribution has been repeated many times since with deference to Carpenter's authority!

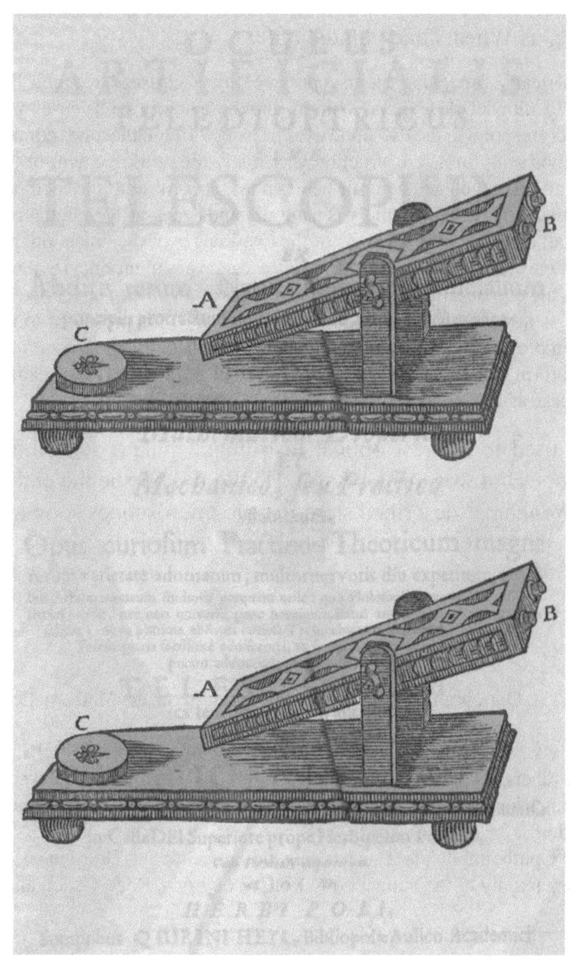

Zahn's portrait together with the title page of his book in combination with his binocular microscope

[58] Mayall (1886).

[59] Carpenter (1901).

There were questions about the advantages of a binocular microscope soon after Chérubin's book was published. In comparing monocular and binocular microscopes Robert Hooke (1635–1703) favoured the former, remarking "that with one Eye only, which is much to be preferred before that with two".[60] It is highly unlikely that these binocular microscopes would have afforded stereoscopic impressions of minute objects simply because of their construction. Wheatstone[61] noted that the arrangement of the eyepieces was such that any effects would have been pseudoscopic rather than stereoscopic. That is, the disparities were reversed so that near parts of the specimen would have had uncrossed disparity whereas that for far parts would have been crossed. The early instruments were made without an adequate understanding of binocular disparities. While these instruments were binocular, they were not stereoscopic, as Wheatstone pointed out:

> In the Père d'Orléans' binocular microscope, two object-glasses have their lateral portions cut away so as to allow of close juxta-position, and these nearly semi-lenses are so arranged, that their axes correspond to the two optic axes passing through the tubes containing the eye-pieces.The author's aim in its construction was solely the reinforcement of the impression by presenting an image to each eye, for he assumes, according to the then prevalent error, that vision by the two organs conjointly is naturally and necessarily unique, from the perfect conformity of all the homonymous parts of the two images of the object on the two retinae. The real advantage of such an instrument entirely escaped his attention; viz., that of presenting to the two eyes the two dissimilar microscopic images of an object, under precisely the same circumstances as the two unlike images of any usual object is presented to them when no instrument is employed, by which simultaneous presentment the same accurate judgment as to its real solid form, and the relative distances of all its points, can be as readily determined in the former case as in the latter.[62]

Wheatstone tried to interest optical instrument makers to manufacture stereoscopic microscopes but was unsuccessful. He did not pursue the endeavour because Riddell[63] and Wenham[64] described stereoscopic microscopes soon after.

References

Aguilonius F (1613) Opticorum Libri Sex. Philosophis juxta ac Mathematicis utiles. Moreti, Antwerp

Bacon F (1857) Sylva sylvarum: or a natural history. In: Spedding J, Ellis RL, Heath DD (eds) The works of Francis Bacon, vol 2. Longman, Simpkin, Hamilton, Wittaker, Bain, Hodgson, Wasbourne, Bohn, Richardson, Houlston, Bickers and Bush, Willis and Sotheran, Cornish, Booth, and Snow, London

Bartisch G (1583) Ophthalmoduleia. Das ist, Augendienst. Stockel, Dresden

Bell C (1803) The anatomy of the human body, vol 3. Longman, Rees, Cadell and Davies, London

[60] Hooke (1679), p. 102.

[61] Wheatstone (1853).

[62] Wheatstone (1853), p. 101.

[63] Riddell (1853).

[64] Wenham (1854).

Boyle R (1688) A disquisition about final causes of natural things. Taylor, London
Brewster D (1844) On the knowledge of distance given by binocular vision. Trans R Soc Edinb 15:663–674
Brücke E (1841) Über die stereoscopischen Erscheinungcn und Wheatstone's Angriff auf die Lehre von den identischcn Stellen der Netzhäute. Archiv für Anatomie, Physiologie und die wissenschaftliche Medicin 8:459–476
Brücke EW (2022a) About the stereoscopic phenomena and Wheatstone's attack on the theory of the identical retinal points of the retinas. Part 1. Strabismus 30:111–113
Brücke EW (2022b) About the stereoscopic phenomena and Wheatstone's attack on the theory of the identical retinal points of the retinas. Part 2. Strabismus 30:165–170
Burton HE (1945) The optics of Euclid. J Opt Soc Am 35:357–372
Carpenter WB (1901) The microscope and its revelations, 8 edn. Churchill, London
Chérubin d'Orléans (1671) La dioptrique oculaire, ou la theorique, la positive, et la mechanique, de l'oculaire dioptrique en toutes ses espèces. Jolly & Bernard, Paris
d'Orléans C (1677) La vision parfaite: ou les concours des deux axes de la vision en un seul point de l'objet. Marbre-Cramoisy, Paris
Crone RA, Everhard-Halm Y (1975) Optically induced eye torsion. I. Fusional cyclovergence. Archiv für Ophthalmologie 195:231–239
Desaguliers JT (1716) A plain and easy experiment to confirm Sir Isaac Newton's doctrine of the different refrangibility of the rays of light. Philos Trans R Soc 34:448–452
Du Tour E-F (1760) Discussion d'une question d'optique. Académie des Sciences. Mémoires de Mathématique et de Physique Présentés pars Divers Savants 3:514–530
Erkelens CJ, van Ee R (2012) Multi-coloured stereograms unveil two binocular colour mechanisms in human vision. Vision Res 42:1103–1112
Hall TS (1972) Treatise of man. René Descartes. Harvard University Press, Cambridge, MA
Heiberg IL (1895) Euclidis Optica. Teubner, Leipzig
Helmholtz H (1864) The Croonian lecture. On the normal motions of the human eye in relation to binocular vision. Proc R Soc 13:186–199
Helmholtz H (1867) Handbuch der physiologischen Optik. In: Karsten G (ed) Allgemeine Encyklopädie der Physik, vol 9. Voss, Leipzig
Helmholtz H (1925) Helmholtz's treatise on physiological optics, vol 3 (trans: Southall JPC). Optical Society of America, New York
Hooke R (1679) Lectiones Cutlerianae, or a collection of lectures physical, mechanical, geographical & astronomical, made before the Royal Society on several occasions . . . to which are added divers miscellaneous discourses. John Martyn, London
Howard IP, Wade NJ (1996) Ptolemy's contributions to the geometry of binocular vision. Perception 25:1189–1201
Le Clerc S (1679) Discours touchant de point de veue, dans lequel il es prouvé que les chose qu'on voit distinctement, ne sont veues que d'un oeil. Jolly, Paris
Lejeune A (ed) (1956) L'Optique de Claude Ptolémée dans la version Latine d'après l'Arabe de l'Émir Eugène de Sicile. University of Louvain, Louvain
Lejeune A (ed and trans) (1989) L'Optique de Claude Ptolémée dans la version Latine d'après l'Arabe de l'Émir Eugène de Sicile. Brill, Leiden
Magnus H (1901) Die Augenheilkunde der Alten. Kern, Breslau
May MT (1968) Galen. On the useful of the parts of the body. Cornell University Press, Ithaca, NY
Mayall J (1886) The microscope: lecture II. J Soc Arts 34:1007–1021
Molyneux W (1692) Dioptrica nova. A treatise of dioptricks in two parts. Tooke, London
Müller J (1826) Zur vergleichenden Physiologie des Gesichtssinnes des Menschen und der Thiere, nebst einen Versuch über die Bewegung der Augen und über den menschlichen Blick. Cnobloch, Leipzig
Müller J (1838) Handbuch der Physiologie des Menschen, vol 2. Hölscher, Coblenz
Müller J (1843) Elements of physiology, vol 2 (trans: Baly W). Taylor and Walton, London

Nagel A (1861) Das Sehen mit zwei Augen, die Lehre von den identischen Netzhautstellen. Winter, Leipzig
Nagel A (1868) Ueber das Vorkommen von wahren Rollungen des Auges um die Gesichtslinie. Archiv Für Ophthalmol 14:228–246
Nagel A (1871) Ueber das Vorkommen von wahren Rollungen des Auges um die Gesichtslinie. Zweiter Artikel 17:237–264
Nagel A (2020a) Vision with two eyes, the doctrine of the identical retina points. Part 1: the stereoscopic phenomena. Strabismus 28:55–59
Nagel A (2020b) Vision with two eyes, the doctrine of the identical retina points. Part 1: the stereoscopic phenomena. Part 2: fusion of disparate images. Strabismus 28:99–108
Nagel A (2020c) Vision with two eyes, the doctrine of the identical retina points. Part 1: the stereoscopic phenomena. Part 3: stereoscopic experiment: fusion of disparate images including cyclofusion. Strabismus 28:164–172
Ono H, Wade NJ (1985) Resolving discrepant results of the Wheatstone experiment. Psychol Res 47:135–142
Panum PL (1858) Physiologische Untersuchungen über das Sehen mit zwei Augen. Schwerssche Buchhandlung, Kiel
Reid T (1764) An inquiry into the human mind. On the principles of common sense. Millar, Kincaid & Bell, Edinburgh
Riddell JL (1853) Notice of a binocular microscope. Am J Sci Arts 15:68
Ross WD (ed) (1927) The works of Aristotle, vol 7. Clarendon, Oxford
Russell G (1996) Emergence of physiological optics. In: Rāshid R, Morelon R (eds) Encyclopedia of the history of Arabic science. Routledge, London, pp 672–716
Russell GA (2008) The origins of point to point correspondence in anatomical projection prior to Descartes: Ibn al-Haytham (d. 1040). J Hist Neurosci 17:246–247
Russell GA (2010) After Galen: late antiquity and the Islamic world. In: Finger S, Boller F, Tyler KL (eds) Handbook of clinical neurology: history of neurology, vol 95. Amsterdam, Elsevier, pp 61–77
Sabra AI (ed and trans) (1989) The optics of Ibn al-Haytham. Books I–III. On direct vision. The Warburg Institute, London
Schmitz E-H (1982) Handbuch zur Geschichte der Optik, vol 2. Wayerborgh, Bonn
Schyrleus de Rheita AM (1645) Oculus Enoch et Eliæ sive radius sideromysticus. Verdussi, Antwerp
Simonsz HJ, Wade NJ (2018) Why did Donders, after describing pseudotorsion, deny the existence of ocular counterrolling together with Ruete, Volkmann, von Graefe and von Helmholtz, until Javal reconfirmed its existence? Strabismus 26:211–222
Smith AM (1996) Ptolemy's theory of visual perception: an English translation of the optics with introduction and commentary. The American Philosophical Society, Philadelphia, PA
Smith AM (2001) Alhacen's theory of visual perception. American Philosophical Society, Philadelphia, PA
Smith R (1738) A compleat system of opticks in four books. Cambridge
Taylor J (1738) Le mechanisme ou le nouveau traité de l'anatomie du globe de l'oeil, avec l'usage de ses différentes parties, & de celles qui lui sont contigues. David, Paris
Towne J (1862) Remarks on the stereoscopic theory of vision, with observations on the experiments of Professor Wheatstone—Section I. Guy's Hosp Rep 8:70–80
Vieth GUA (1818) Ueber die Richtung der Augen. Ann Phys 28:233–253
Volkmann AW (1859) Die stereoskopischen Erscheinungen in ihrer Beziehung zu der Lehre von den identischen Netzhautpunckten. Archiv Für Ophthalmol 5:1–100
Volkmann AW (2020) The stereoscopic phenomena in relation to the doctrine of identical retinal points. Part 1: the views by Wheatstone, Brücke and Panum. Strabismus 28:223–235
Volkmann AW (2021a) The stereoscopic phenomena in relation to the doctrine of identical retinal points. Part 2: horizontal, vertical and torsional amplitudes of fusion. Strabismus 29:61–71
Volkmann AW (2021b) The stereoscopic phenomena in relation to the doctrine of identical retinal points. Part 3: observable disparities differ in parts of the visual field. 29:189–194

Volkmann AW (2021c) The stereoscopic phenomena in relation to the doctrine of identical retinal points. Part 4: Wheatstone's assertion that identical images which fall on corresponding retinal points may appear double and in different positions. Strabismus 29

Volkmann AW (2022) The stereoscopic phenomena in relation to the doctrine of identical retinal points. Part 5: Summary and presentation of some paradoxical phenomena which occur in stereoscopic experiments. Strabismus 30:49–56

Wade NJ (1983) Brewster and Wheatstone on vision. Academic Press, London

Wade NJ (1987) On the late invention of the stereoscope. Perception 16:785–818

Wade NJ (2022a) Alfred Wilhelm Volkmann on stereoscopic vision. Strabismus 30(1):1–7. https://doi.org/10.1080/09273972.2022.2022835

Wade NJ (2022b) Ernst Wilhelm Brücke on stereoscopic vision. Strabismus 30(3):159–164. https://doi.org/10.1080/0927

Wade NJ, Heller D (1997) Scopes of perception: the experimental manipulation of space and time. Psychol Res 60:227–237

Wade NJ, Ono H (1985) The stereoscopic views of Wheatstone and Brewster. Psychol Res 47:125–133

Wade NJ, Ono H (2012) Two historical strands in studying visual direction. Jpn Psychol Res 54:71–88

Wenham FH (1854) On the application of binocular vision to the microscope. Trans Microscopical Soc Lond 2:1–13

Wheatstone C (1838) Contributions to the physiology of vision—Part the first. On some remarkable, and hitherto unobserved, phenomena of binocular vision. Philos Trans R Soc 128:371–394

Wheatstone C (1853) On the binocular microscope, and on stereoscopic pictures of microscopic objects. Trans Microscopical Soc Lond 1:99–102

Zahn J (1685) Oculus artificialis teledioptricus sive telescopium. Heyl, Würzburg

Zahn J (1686) Oculo artificiali teledioptrico sive telescopio, Fundamentum III. Practico-mechanicum. Heyl, Würzburg

Chapter 4
Stereoscopes

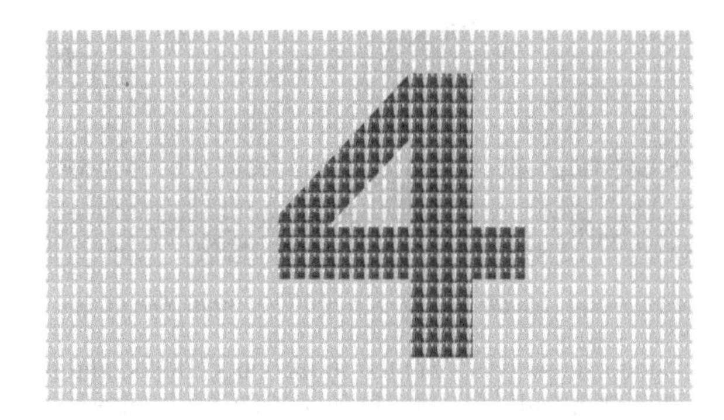

Stereoscopes transformed not only the picture of vision but also the vision of pictures. On the one hand, theories of how we see with two eyes were revolutionized by the demonstration of singleness and depth following stimulation of slightly non-corresponding points. On the other, paired pictures of distant scenes could be experienced in depth in the comfort of Victorian parlours. Wheatstone designed reflecting (mirror) and refracting (prism) stereoscopes in the early 1830s but he only described the former in his classic article published in 1838. Eleven years later Brewster announced his refracting stereoscope in which half-lenses acted as prisms and magnifiers. Thereafter numerous patents were taken out for stereoscopes predominantly based on refraction rather than reflection. Perhaps the most popular was devised by Oliver Wendell Holmes in 1861; it consisted of a pair of prisms mounted in a viewing case and an extending arm to which the card holder for photographs could be placed and adjusted for viewing distance. It became known as the American stereoscope. Alternative schemes were invented involving modifications of the stimulus as well as the viewer. Colour separations of overprinted stereoscopic images could be viewed with equivalently coloured filters in front of each eye so that only one colour could be seen by one eye. Projected light from two sources could be cross-polarized and viewed through similar polarizers. Lenticular printing techniques were developed that enabled images to be seen in stereoscopic depth without the aid of a stereoscope. The depth experienced could be exaggerated with telestereoscopes of reversed with pseudoscopes. The stimulus for the popularity of stereoscopes was their marriage with photographs. Initially stereoscopic photographs were taken by a single camera moved from one position to another but twin-lensed cameras became available from the 1850s. Wheatstone was quick to grasp the significance of photography to stereoscopy as it dispensed with the need to make two precise but systematically different drawings. He wrote "What the hand of the artist was unable to accomplish, the chemical action of light, directed by the camera, has enabled us to effect".

N. Wade, *Vision and Art with Two Eyes*, Vision, Illusion and Perception 3,
https://doi.org/10.1007/978-3-030-77995-5_4

Stereoscopic photograph of an early Wheatstone stereoscope

Wheatstone's stereoscope was invented in the early 1830s, and it opened a new world for the study of binocular vision.[1] That world was the laboratory, and with the aid of the stereoscope the methods of physics could be applied to the investigation of spatial vision. Wheatstone was able to manipulate the pictures presented to each eye and observe the depth that was produced. In so doing, he found that:

> The projection of two obviously dissimilar pictures on the two retinæ when a single object is viewed, while the optic axes converge, must therefore be regarded as a new fact in the theory of vision. It being thus established that the mind perceives an object of three dimensions by means of the two dissimilar pictures projected by it on the two retinæ, the following question occurs: What would be the visual effect of simultaneously presenting to each eye, instead of the object itself, its projection on a plane surface as it appears to that eye?[2]

The "new fact in the theory of vision" was not the projection of dissimilar pictures on the two eyes; this had long been known, as was indicated in Chap. 3. Rather, it was the perception of an object in three dimensions as a consequence of this dissimilarity.

Wheatstone's mirror stereoscope was bulky in comparison to the many varieties of lenticular and prism models that were patented in the nineteenth century.[3] 'Scott's Patent Stereoscope' shown below is an example of a refracting stereoscope with larger lenses as proposed by George Lowdon and patented by his agent, E. Ebenezer Scott, in 1856; the number of the model shown is No. 169 and it was sold in London by Negretti & Zambra.

---

[1] See Wade (1983).

[2] Wheatstone (1838), pp. 372–373.

[3] Wing (1996), Pellerin and May (2021).

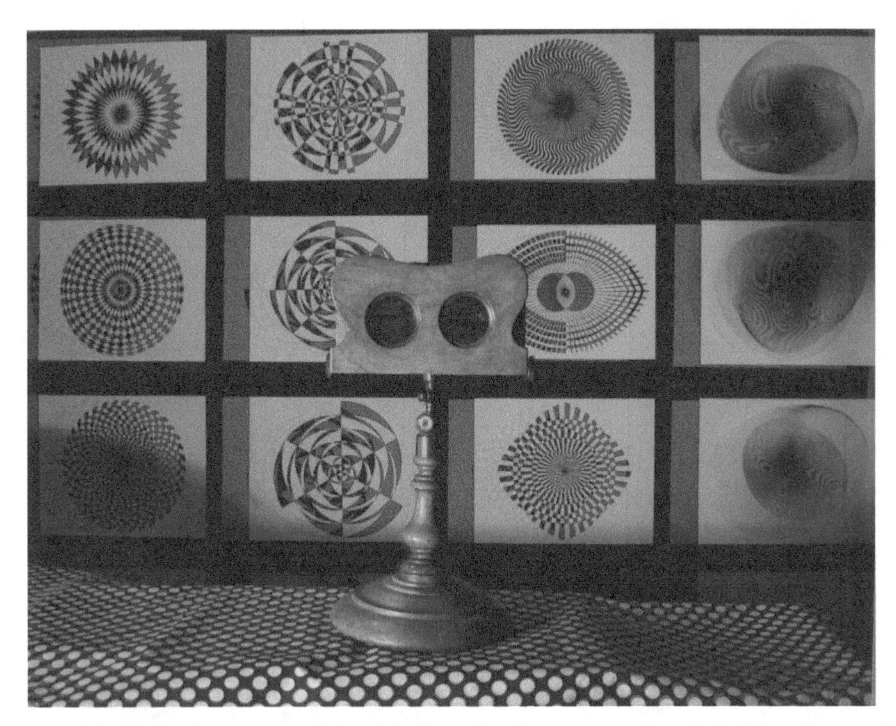

Stereoscopic stereoscope. An early model of a pedestal stereoscope is seen in front of a set of geometrical designs

Mirrors have long been adopted as aids to studying visual phenomena. Brewster enlisted them for constructing kaleidoscopes and large mirrors have been a boon to entertainment.[4] However, it was Wheatstone's use of mirrors in his stereoscope that has had, perhaps, the most profound influence on advancing an understanding of depth perception. Wheatstone is shown together with his illustrations of the mirror stereoscope taken from his article of 1838.

[4] Brewster (1819, 1832).

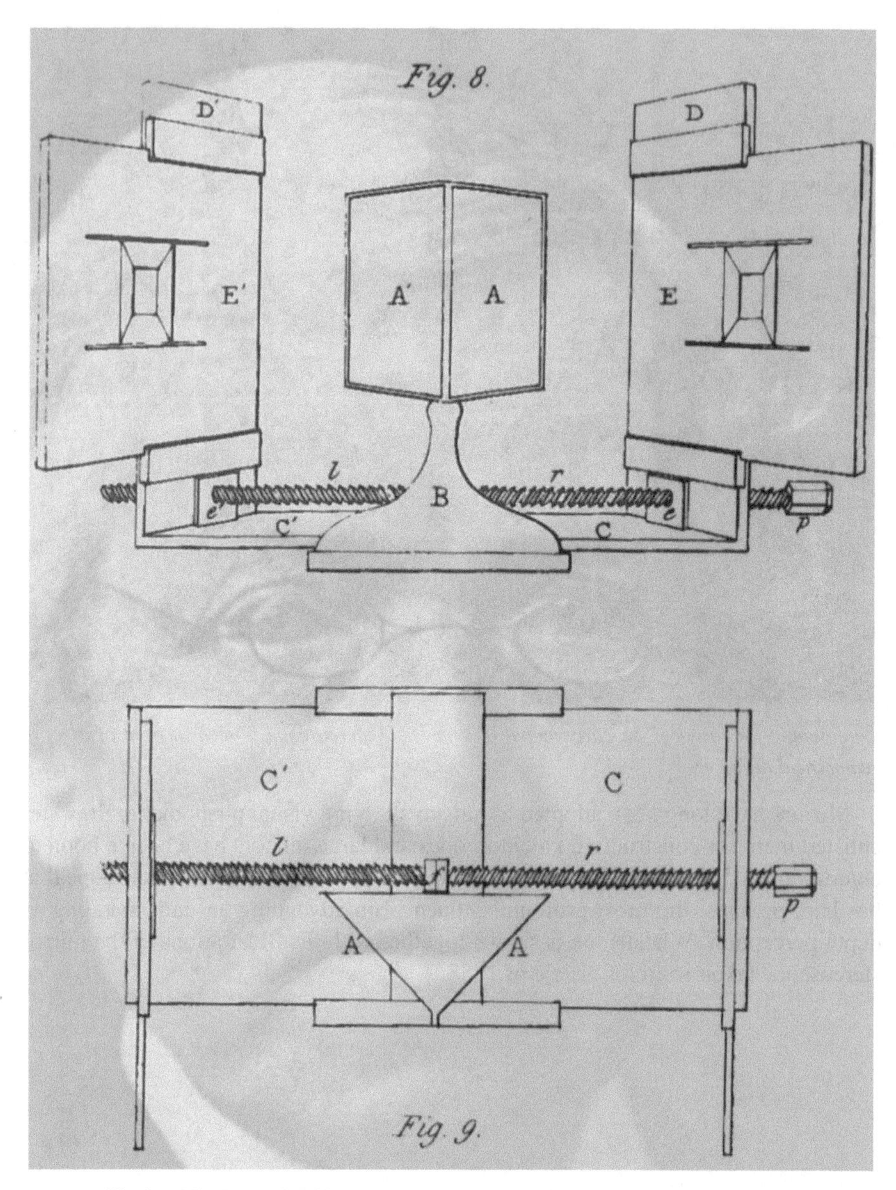

Charles Wheatstone and his mirror stereoscope as seen from in front and above

There are many methods by which different stimuli can be presented to the two eyes, as we have seen in the previous chapter. Some involved optical elements and others constrained the convergence of the eyes. After Wheatstone invented both reflecting (mirror) and refracting (prism) stereoscopes in the early 1830s some simple and novel optical techniques were introduced. Many of these involving mirrors and prisms were illustrated by Brewster in 1851 as well as by Dove in the same year.[5]

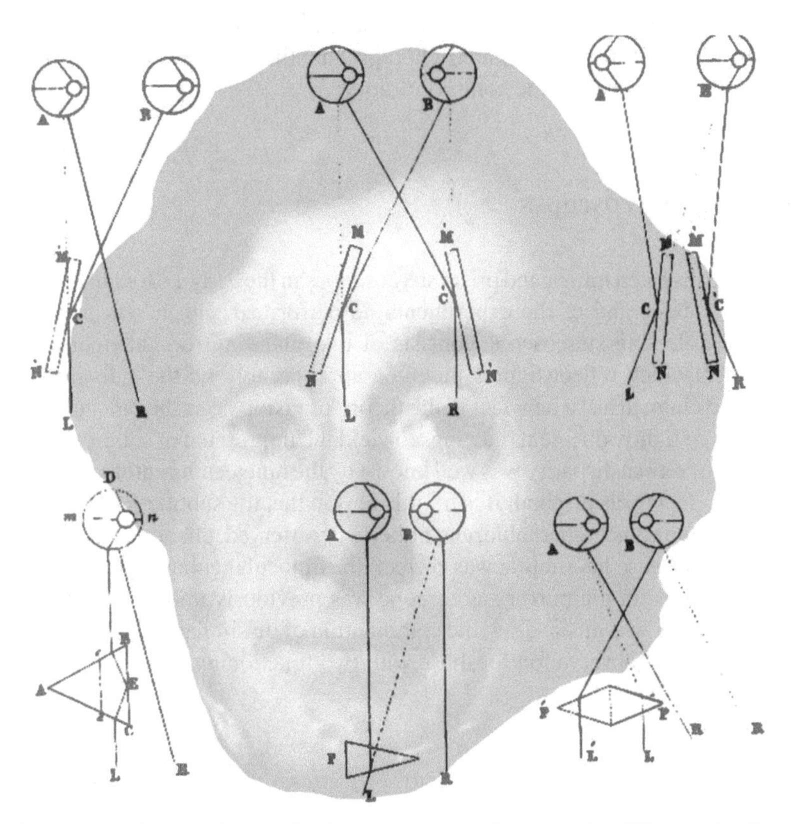

David Brewster and some mirror and prism arrangements for presenting different stimuli to each eye

Wheatstone's invention of the pseudoscope, which reversed retinal disparities, followed 14 years after the stereoscope had been made public.[6] The most popular model of stereoscope was Brewster's lenticular version.[7] The optical manipulation of disparities was also achieved with Wheatstone's pseudoscope, which reversed them, and with Helmholtz's telestereoscope, which exaggerated them.[8] The anaglyph method, enabling overprinted red and green images to be combined through similarly

[5] Brewster (1851), Dove (1851).

[6] Wheatstone (1852).

[7] Brewster (1849a, b), Morrison-Low and Bruce (2018).

[8] Wheatstone (1852), Helmholtz (1857).

coloured filters was introduced at about the same time.[9] Descriptions of more recent stereoscopic techniques can be found in Blundell[10] and Howard and Rogers.[11]

Soon after Wheatstone's invention of the stereoscope, it enjoyed a commercial success as a 'philosophical toy'. This was not only because of the instrument itself, but also its combination with paired photographs; the wonders of the world could then be seen in depth from the comfort of home.[12]

Reflecting Stereoscopes

Wheatstone invented mirror and prism stereoscopes in the early 1830s but his description of the device and of the experiments he performed with it was published in 1838. The reflecting stereoscope consists of two plane mirrors, at right angles to one another, which reflect figures mounted appropriately on the side arms. What Wheatstone claimed as "a new fact in the theory of vision" was the binocular combination of two slightly different drawings to yield the impression of solidity. While the occurrence of retinal disparity was well known by the nineteenth century, its representation in this form was original. It is for this reason that the subtitle for Wheatstone's article was: "On some remarkable, and hitherto unobserved, phenomena of binocular vision". In one sense his subtitle was correct: the binocular combination of disparate pairs of drawings in the mirror stereoscope was previously unobserved. In another sense his subtitle was misleading: the combination of dissimilar retinal images is the universal mode of observation for those with two functioning and aligned eyes.

[9] Rollmann (1853), D'Almeida (1858).

[10] Blundell (2011).

[11] Howard and Rogers (2012).

[12] See Wade (2004), Pellerin and May (2021).

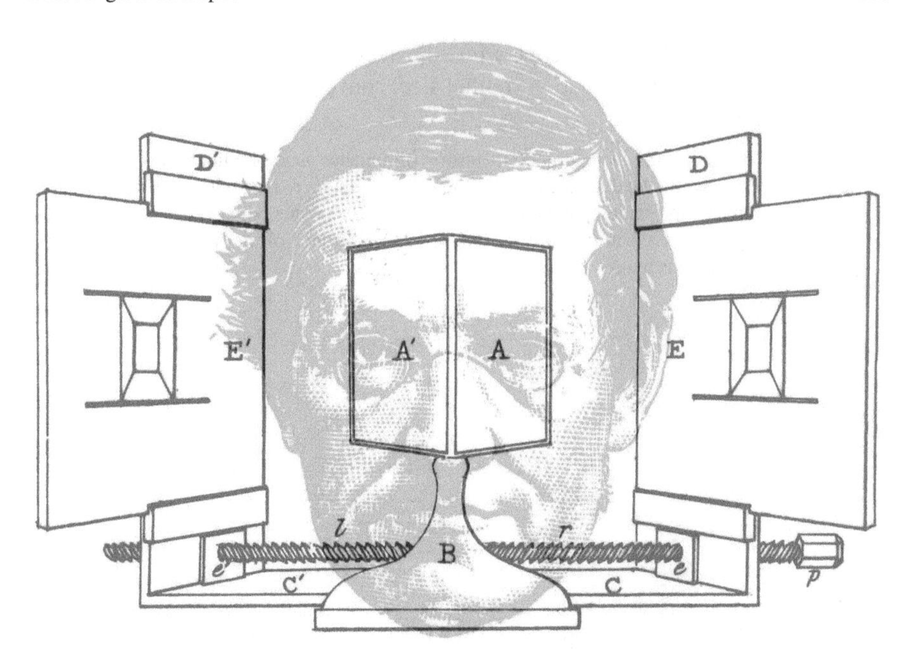

Wheatstone and his reflecting stereoscope

Wheatstone's second article on binocular vision was published fourteen years later. He described and illustrated an adjustable mirror stereoscope. The main purpose of this was to extend the range of conditions under which the two eyes could be stimulated:

> Under the ordinary conditions of vision, when an object is placed at a certain distance before the eyes, several concurring circumstances remain constant, and they always vary in the same order when the distance of the object is changed. Thus, as we approach the object, or as it is brought nearer to us, the magnitude of the picture on the retina increases; the inclination of the optic axes, required to cause the picture to fall on corresponding places on the retinæ, becomes greater; the divergence of the rays of light proceeding from each point of the object, and which determines the adaptation of the eyes to distinct vision of that point, increases; and the dissimilarity of the two pictures projected on the retinæ also becomes greater. It is important to ascertain in what manner our perception of the magnitude and distance of objects depends on these various circumstances, and to inquire which are the most, and which the least influential in the judgements we form. To advance this inquiry beyond the point to which it has hitherto been brought, it is not sufficient to content ourselves with drawing conclusions from observations on the circumstances under which vision naturally occurs, as preceding writers on this subject mostly have done, but it is necessary to have more extended recourse to the methods so successfully employed in experimental philosophy, and to endeavour, wherever it be possible, not only to analyse the elements of vision, but also to recombine them in unusual manners, so that they may be associated under circumstances that never naturally occur.[13]

[13] Wheatstone (1852), p. 2.

Wheatstone used the stereoscope with adjustable arms to vary the four circumstances mentioned in the quotation (retinal size, convergence, accommodation, and disparity). He found that:

> The perceived magnitude of an object, therefore, diminishes as the inclination of the axes becomes greater, while the distance remains the same; and it increases, when the inclination of the axes remains the same, while the distance diminishes. When both of these conditions vary inversely, as they do in ordinary vision when the distance of an object changes, the perceived magnitude remains the same.[14]

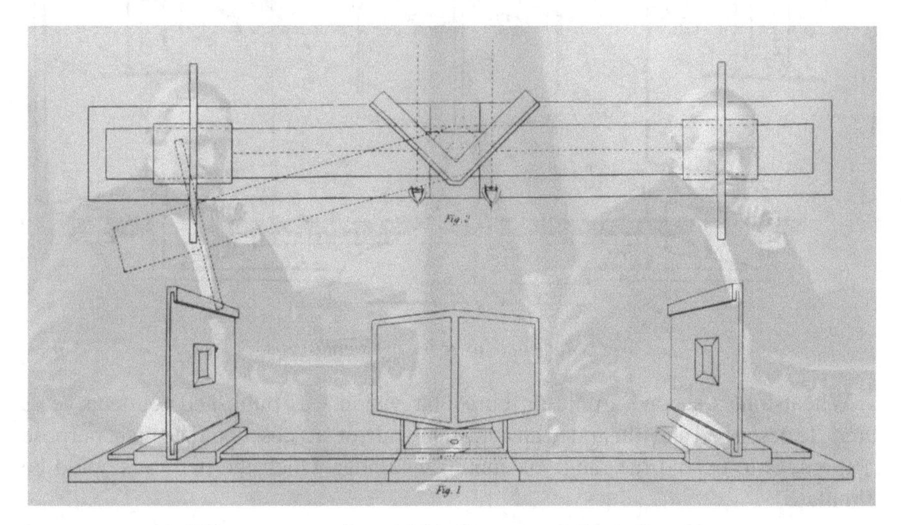

A stereo portrait of Wheatstone together with his illustrations of the adjustable mirror stereoscope seen from above and in front

Most mirror stereoscopes conformed with Wheatstone's design in placing the mirrors at right angles to one another and at 45° to the line of sight. The quality of mirrors has improved since Wheatstone's time but the orientations they occupy in the stereoscopes have remained relatively unchanged.[15]

The stereoscope made by Andy Lomas has broken away from this tradition by constructing a stereoscope in which the mirrors are separated by 60° rather than 90°. Stimuli are displayed on flat screen monitors that are dark, thereby avoiding the conflicts between the stereoscopic stimuli and the visible frames surrounding them.[16] Not only does his stereoscope deliver dynamic depth but it can be seen as an artistic structure in its own right. Initially it looks like a triptych with three representations of rotating structures but the viewing apertures attached to the central stem invite closer inspection whereupon the stereoscopic nature of the revolving and evolving forms is revealed.

---

[14] Wheatstone (1852), p. 3.

[15] Drouin (1890), Wing (1996). See also Pellerin and May (2021).

[16] Lomas A http://www.andylomas.com/constrainedForms.html.

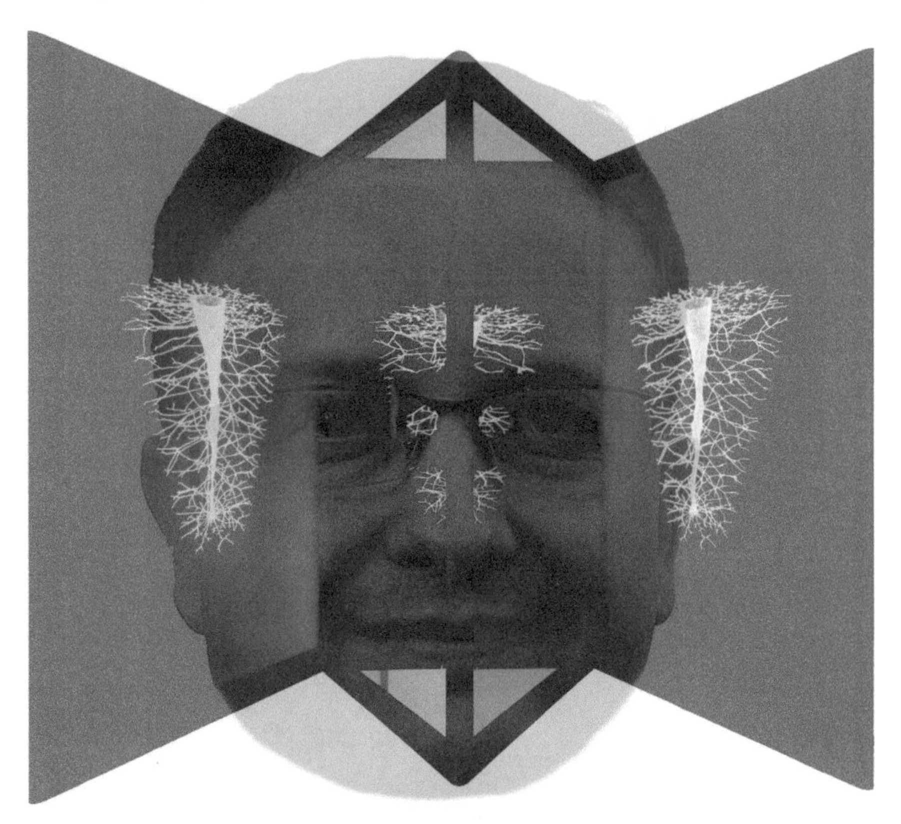

Andy Lomas and his stereoscope which displays computer-generated dynamic and evolving images

The stereoscope was very important to Helmholtz, both for the experimental world it exposed and also for his theoretical battles with Hering (see Chap. 7). In prosecuting his experimental enquiries Helmholtz developed the reflecting stereoscope so that disparities could be enhanced. This was achieved by extending the separations between the mirrors and he called the instrument a telestereoscope.[17] His diagram of the instrument is shown with his portrait.

[17] Helmholtz (1857).

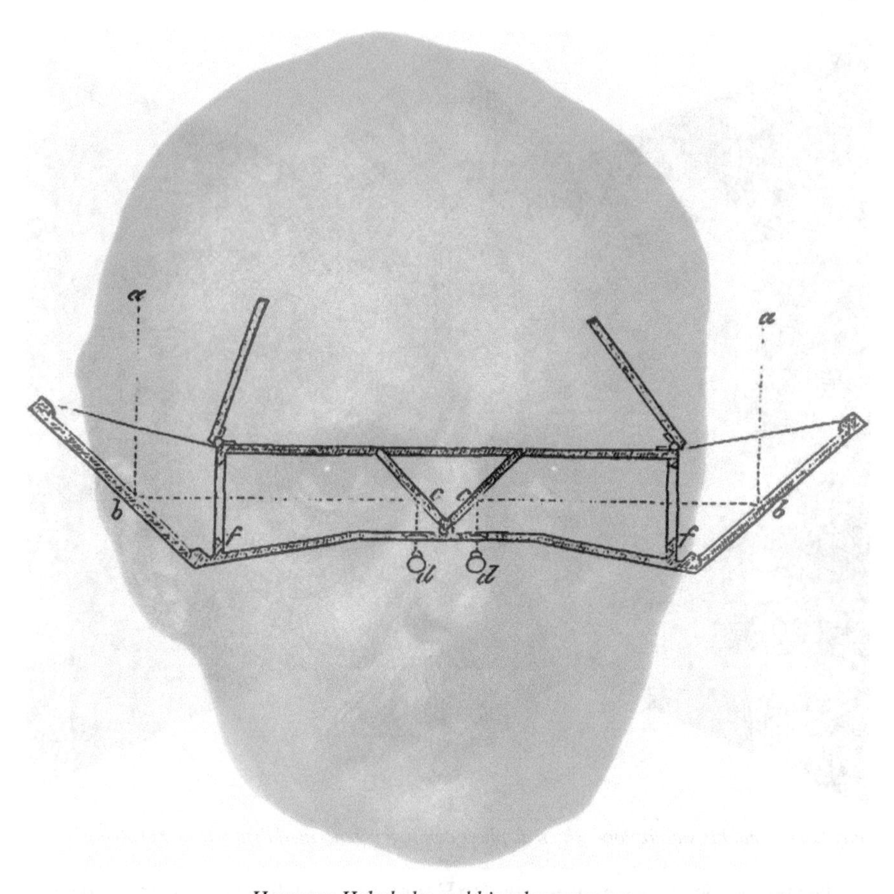

Hermann Helmholtz and his telestereoscope

Telestereoscopes have been used in aerial surveillance and as artistic devices. As an example of the latter, Terry Pope has arranged a system of mirrors to enhance disparities and he has called it a hyperscope.[18] He has also designed structures to be viewed with the hyperscope so that an enhanced impression of the depth within them is evident.

[18] Pope T. www.phantascope.co.uk.

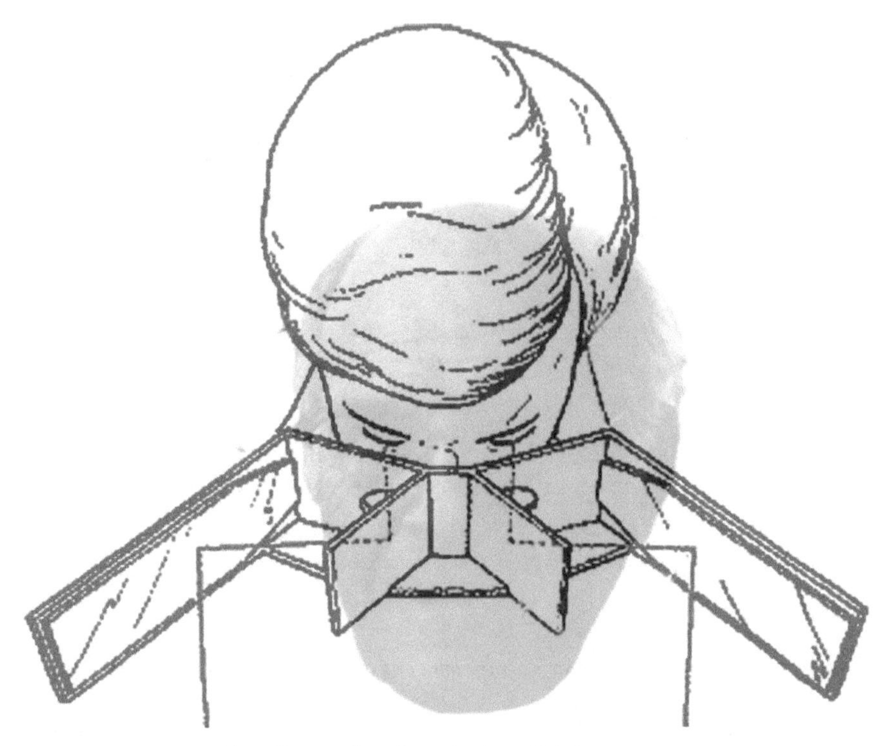

Terry Pope and a schematic view of his hyperscope

Mirrors have been employed in other binocular instruments such as pseudoscopes and creative combinations have been made by artists. They have directed attention to the instruments themselves as well as to the perceptual experiences they provide.

Refracting Stereoscopes

Wheatstone only described the reflecting stereoscope in his 1838 memoir although the instrument makers Murray and Heath constructed a refracting stereoscope for him in 1832. Wheatstone did describe and illustrate his refracting model in his second memoir of 1852.

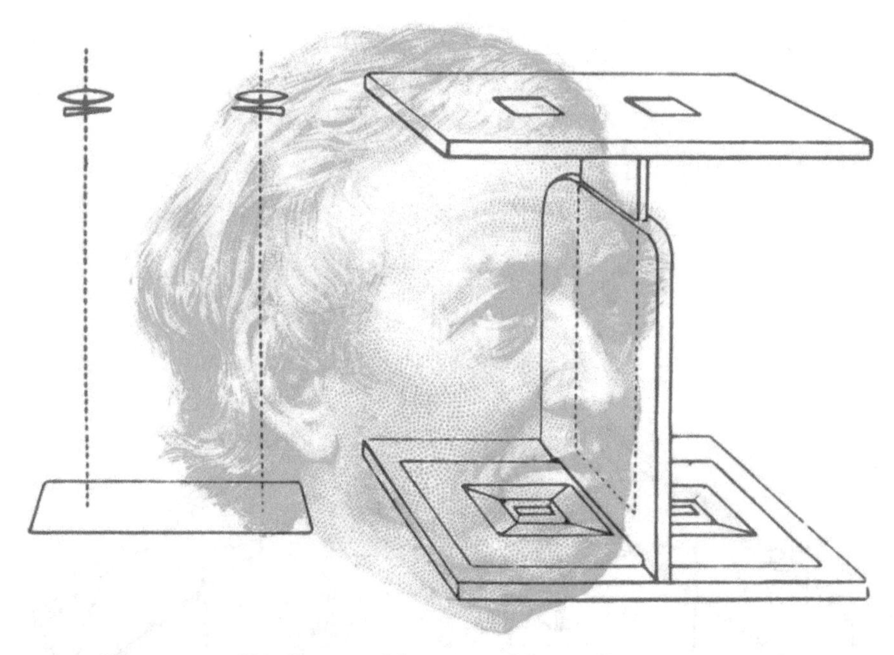

Wheatstone and his diagrams of the optics and design of his prism stereoscope

Brewster's refracting or lenticular stereoscope was presented to the meeting of the British Association for the Advancement of Science in 1849. The stereoscope was made by George Lowdon, an optical instrument maker in Dundee.

GEORGE LOWDON,

OPTICIAN AND SCIENTIST,

DUNDEE.

SKETCH OF HIS REMINISCENCES

AND CAREER.

Lowdon and the cover of a pamphlet published by the Dundee Courier and Advertiser in 1906

Lowdon had earlier made acquaintance with Brewster:

> who had at this period (1849) invented his stereoscope, and I got the making of the first one, and the sending of copies of it to many scientific men all over Europe. Later on I also improved on them, and made a great number for many years afterwards. The fault of Brewster's stereoscope was that the lenses were too small, being, in fact, only the two halves of a spectacle glass. This did not suit every eye, and in experimenting I discovered that larger lenses were an advantage. I pointed this out to Sir David, but he was wedded to his own opinion, and as I feared that the idea might be taken up by another, I took out a patent for my improvement – which experience has since amply justified – but my action was, unfortunately, resented by Sir David, and gave rise to considerable friction, for which I did not consider I was to blame, seeing I had pointed out the improvement, and he had refused it.[19]

[19] Lowdon (1906), pp. 7–8.

Brewster and his lenticular stereoscope. The images in the stereoscope are portraits of his good friend Sir Walter Scott

Unlike Wheatstone, Brewster made many diagrams of the optics associated with his refracting stereoscope, like the one below which was published in *The Stereoscope*.[20] The half-lenses acted as magnifiers as well as refractors thereby reducing the dimensions of the stereoscope.

[20] Brewster (1856).

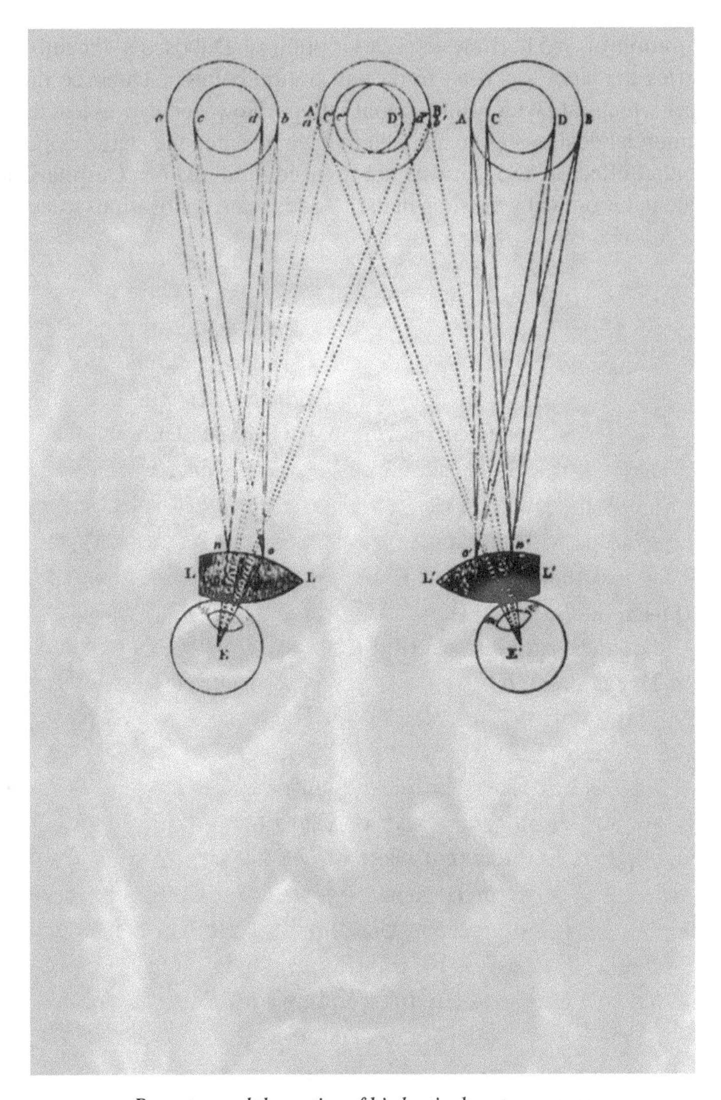

Brewster and the optics of his lenticular stereoscope

The disagreement between Lowdon and Brewster led Brewster to seek another optical instrument maker to produce it. None in Britain would accept the proposal because of Brewster's reputation. In 1850 he travelled to Paris where Abbé François Moigno (1804–1884) introduced him to the optical instrument maker Louis Jules Duboscq (1817–1886), who made the stereoscopes thereafter.[21] According to Brewster it was one of Duboscq's models that was presented to Queen Victoria at the

[21] See Wade (2012).

Great Exhibition of 1851. Brewster's description of Duboscq's "beautiful stereoscope" carries the latter's portrait in the illustration below.[22] Duboscq made many stereoscopes which sold widely throughout Europe. However, he was less than honest in his commercial dealings as he claimed to have invented the stereoscope and filed a patent to that effect in 1852; it was not revoked until 1857.[23] Despite Duboscq's dubious patent, his optical workshop in Paris added many innovations to stereoscopy.

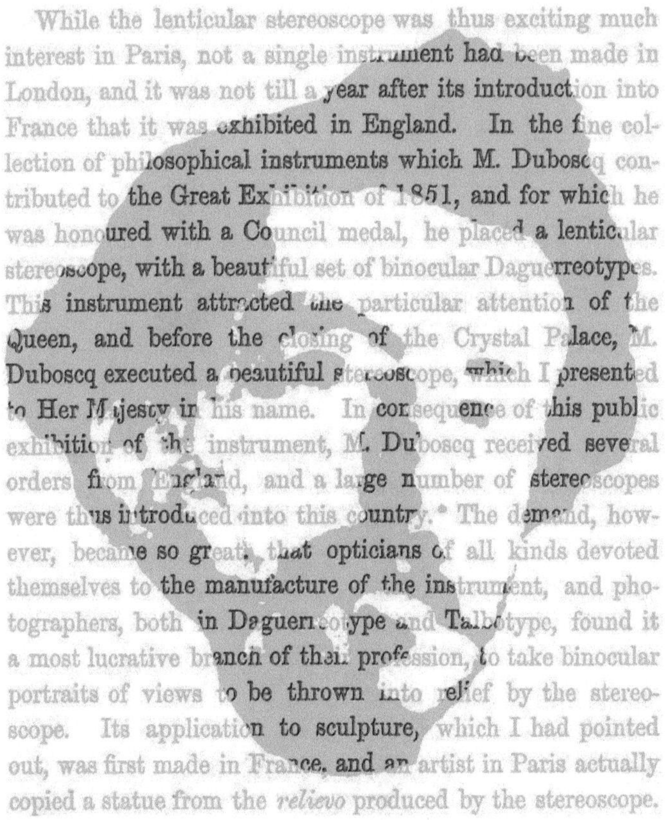

While the lenticular stereoscope was thus exciting much interest in Paris, not a single instrument had been made in London, and it was not till a year after its introduction into France that it was exhibited in England. In the fine collection of philosophical instruments which M. Duboscq contributed to the Great Exhibition of 1851, and for which he was honoured with a Council medal, he placed a lenticular stereoscope, with a beautiful set of binocular Daguerreotypes. This instrument attracted the particular attention of the Queen, and before the closing of the Crystal Palace, M. Duboscq executed a beautiful stereoscope, which I presented to Her Majesty in his name. In consequence of this public exhibition of the instrument, M. Duboscq received several orders from England, and a large number of stereoscopes were thus introduced into this country.* The demand, however, became so great, that opticians of all kinds devoted themselves to the manufacture of the instrument, and photographers, both in Daguerreotype and Talbotype, found it a most lucrative branch of their profession, to take binocular portraits of views to be thrown into relief by the stereoscope. Its application to sculpture, which I had pointed out, was first made in France, and an artist in Paris actually copied a statue from the *relievo* produced by the stereoscope.

Duboscq in text from Brewster's book The Stereoscope *describing the presentation of a Duboscq model to Queen Victoria at the Great Exhibition in 1851*

In 1859 Oliver Wendell Holmes wrote an article on stereoscopes and stereoscopic pictures which he called stereographs.[24] Two years later he described his own design of stereoscope. It consisted of a pair of prisms mounted in a viewing case and an extending arm to which the card holder for photographs could be placed and adjusted

[22] Brewster (1856), p. 31.

[23] Mannoni (2000).

[24] Holmes (1859) see also Silverman (1993).

for viewing distance.[25] It became known as the 'American stereoscope' and Holmes later described the reasons for constructing it:

> This simple stereoscope was not constructed by accident, but was the carrying out of a plan to reduce the instrument to its simplest terms. Two lenses were necessary, and a frame to hold them. I procured two of the best quality, and cut a square frame for them out of a solid piece of wood. A strip of wood at right angles to this was required to hold the pictures. I shaped one, narrow in the middle, broad at both ends; at one end to support the lenses, at the other to hold the stereographs, which were inserted in slots cut with a saw at different distances. A partition was necessary, which I made short, but wedge-shaped, widening as it receded from the eye. A handle was indispensable, and I made a small bradawl answer the purpose, taking care that it was placed so far back as to give the proper balance to the instrument, — a point which bungling imitators have often overlooked. A hood for the eyes was needed for comfort, at least, and I fitted one, cut out of pasteboard, to my own forehead. This primeval machine, parent of the multitudes I see all around me, is in my left hand as I write, and I have just tried it and found it excellent. I felt sure this was decidedly better than the boxes commonly sold, — that it was far easier to manage, especially with regard to light, and could be made much cheaper than the old-fashioned contrivances. I believed that it would add much to the comfort and pleasure of the lover of stereoscopic pictures.[26]

[25] Holmes (1861).

[26] Holmes (1869), p. 1.

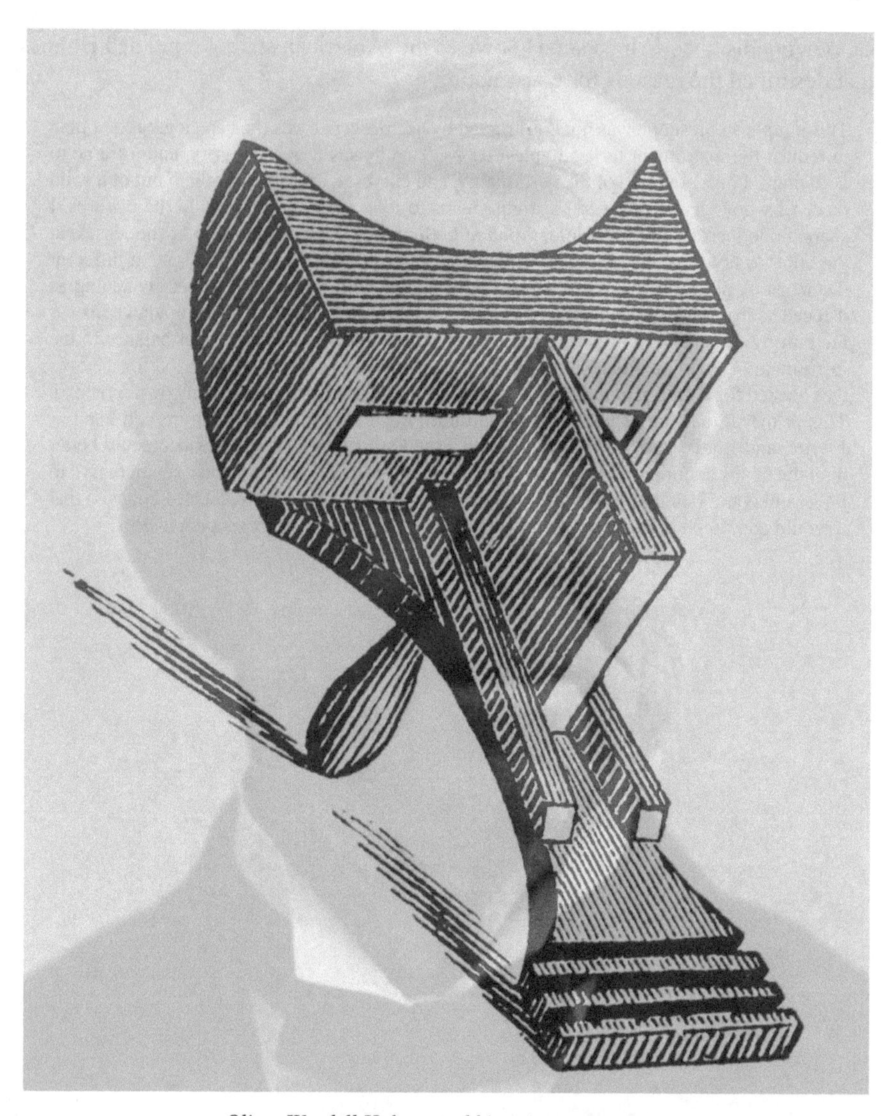

Oliver Wendell Holmes and his prism stereoscope

The Holmes stereoscope was modified further in the late nineteenth century, mainly with an extending central arm enabling fine adjustment to the optimal viewing position for an individual. A simple card-holder for the mounted stereo photographs could be moved at ease and different stereographs could be inserted into the holder. The demand for stereoscopic photographs was supplied by an increasing number of manufacturers in Europe and America, like Negretti and Zambra (founded in 1850), The London Stereoscopic Company (1856), Underwood and Underwood (1881) and the Keystone View Company (1892). The stereo photograph in the Holmes stereoscope shown below depicts a scene on the Royal Mile, Edinburgh; it was produced by

James Valentine, Dundee—Photographer by Appointment to the Queen. Valentine also produced glass transparencies of this and other pairs of Scottish Scenes for use with Brewster stereoscopes.

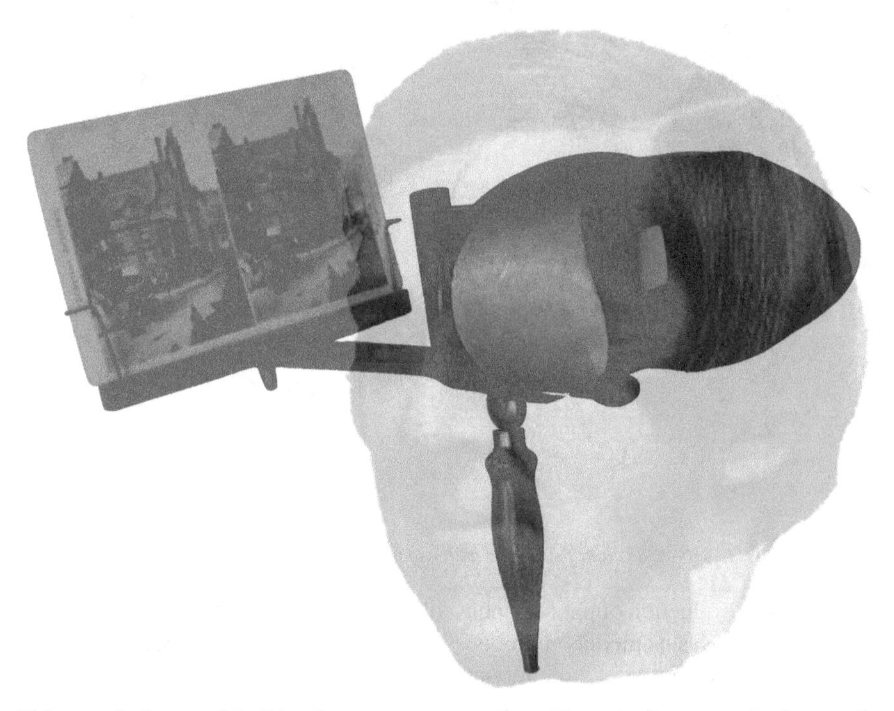

Holmes and a later model of his prism stereoscope together with a pair of stereoscopic photographs of John Knox House, Edinburgh

Many models of refracting stereoscopes were patented in the mid-nineteenth century and they were mainly distinguished by the ornate wooden containers for the simple optical parts. They are often referred to as 'wooden stereoscopes'. An extensive range of these early stereoscopes is illustrated in Paul Wing's book *Stereoscopes: The First One Hundred Years.*[27]

In addition, simple 'pocket stereoscopes' were produced so that printed stereopairs could be viewed at a specific distance, like the one shown below that was used for clinical testing. They could be used with books or printed cards and they had the advantage of compactness for transportation because the legs could fold in so that the stereoscope was almost flat. The convex lenses had a focal length equivalent to their height above the printed stereopairs.

[27] Wing (1996). See also Pellerin and May (2021).

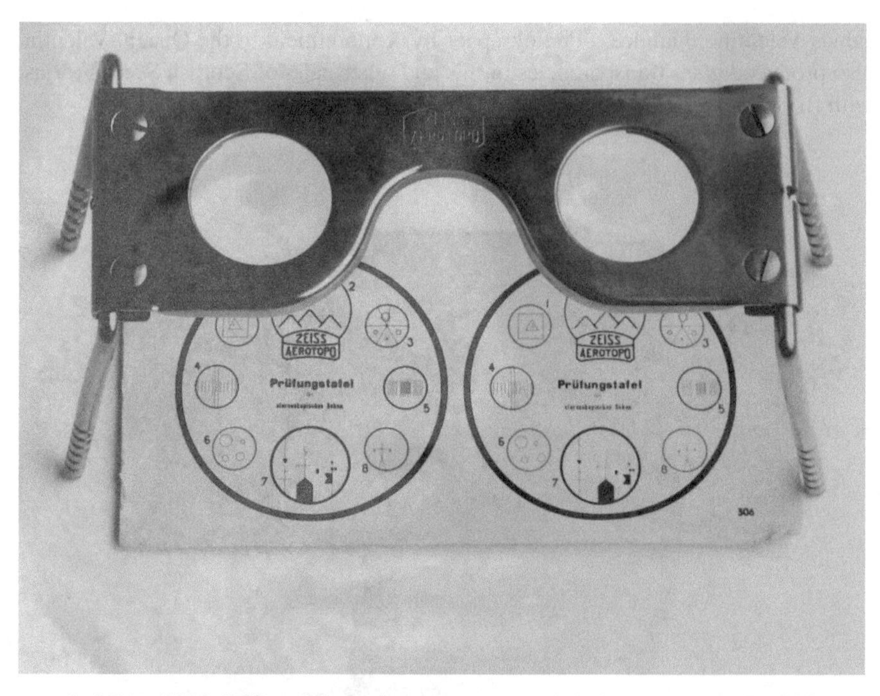

Carl Zeiss (1816–1888) and the pocket stereoscope made by his company in the 1930s

With the introduction of photographic colour transparencies a similar system was used for View-Master models, introduced in 1939 by William Gruber and Harold Graves. Stereoscopic pairs were mounted on opposite sides of a disc and viewed through paired lenses; the disc could rotate so that seven such pairs could be viewed in succession.[28]

[28] Sheppard (2016).

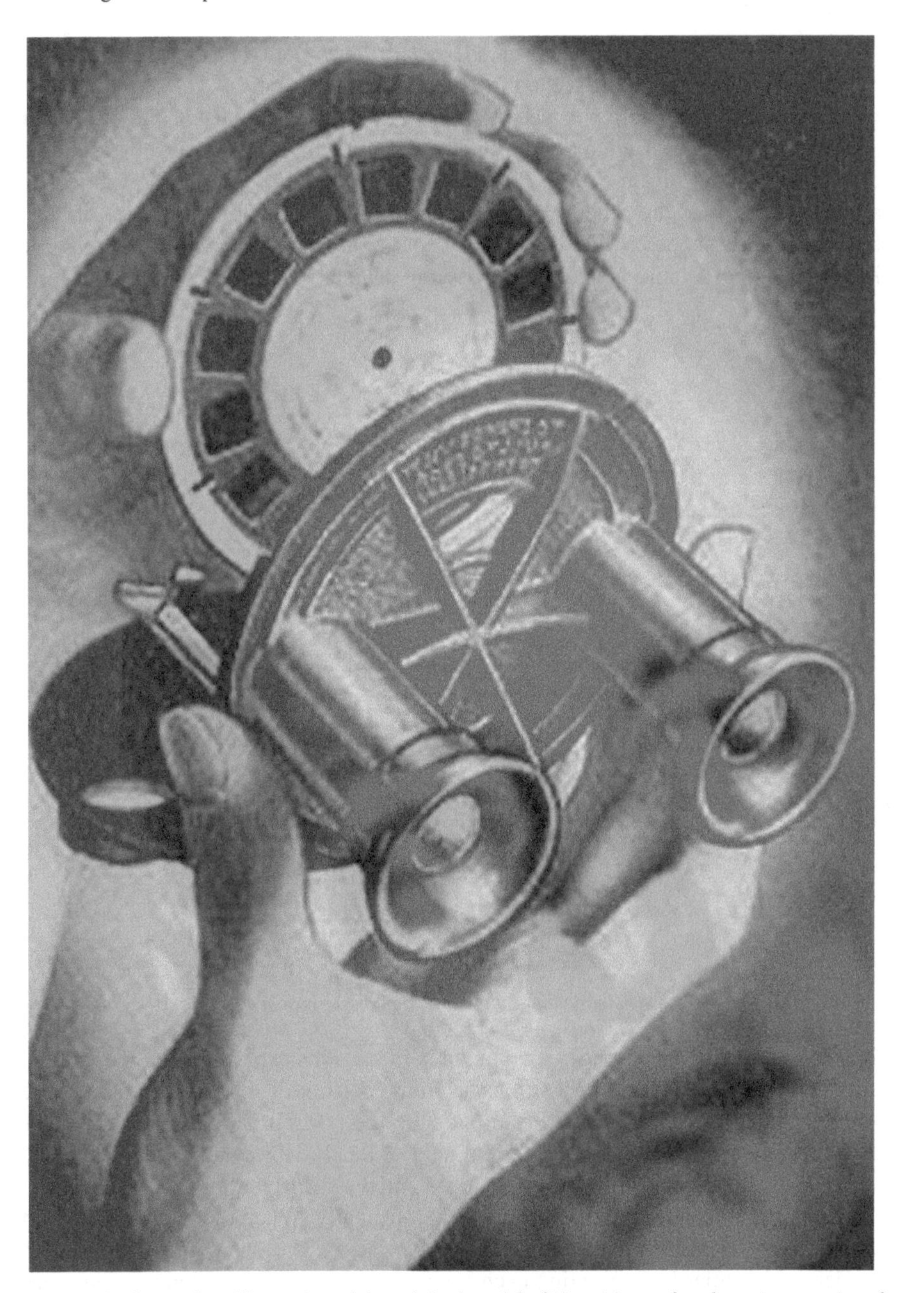

William Gruber and an illustration of the original model of View-Master for observing a series of paired photographs mounted on opposite sides of the rotatable disc

The term 'lenticular' was applied to paired lenses in the examples of refracting stereoscopes described above. Lenticular printing now refers to images in which thin cylindrical strips of images are alternatively printed beneath thin linear cylindrical lenses. Different images can be presented beneath the lenticulars so that they appear to move when viewed from different directions. They can also be used for stereoscopic pairs in which case each eye receives one of the two members of the stereopair and it is generally referred to as autostereoscopic viewing.[29] Viewed from a suitable distance the images from one set of strips are visible to one eye and those from the alternative strips to the other eye, as is shown below. Thus, the attraction of such lenticular stereoscopes is that the optical separation between the left and right eye images does not require an additional viewing device.

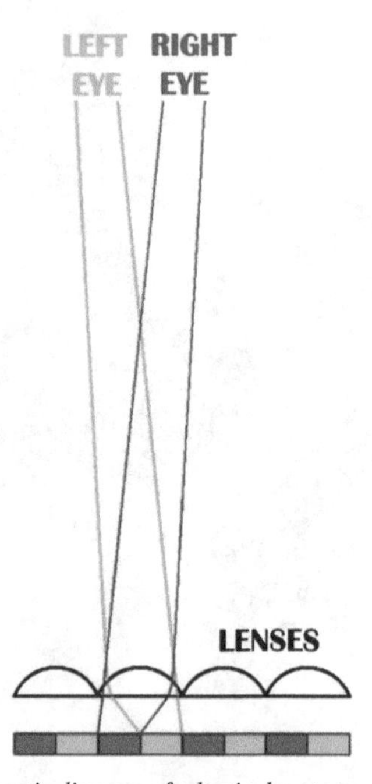

Schematic diagram of a lenticular stereoscope

The possibility of presenting two pictures in one was explored long ago and the alternatives often reflected oppositions, as in the portrait that is thought to be of the ill-fated Mary Queen of Scots. From one side the portrait is visible but the other aspect presents a *momento mori*—a reminder of death. However, it did not involve stereoscopic vision but viewing a picture painted on different sides of protruding wedges from one side or the other rather than frontally. When the picture is viewed from straight ahead neither of these is clearly visible. The two components can be rendered binocular by presenting each aspect to one eye. The technique continues to be used as in the example of the young and old representations of Bonnie Prince Charlie produced by Calum Colvin.[30]

[29] See Dodgson (2005).

[30] Colvin and Normand (2019).

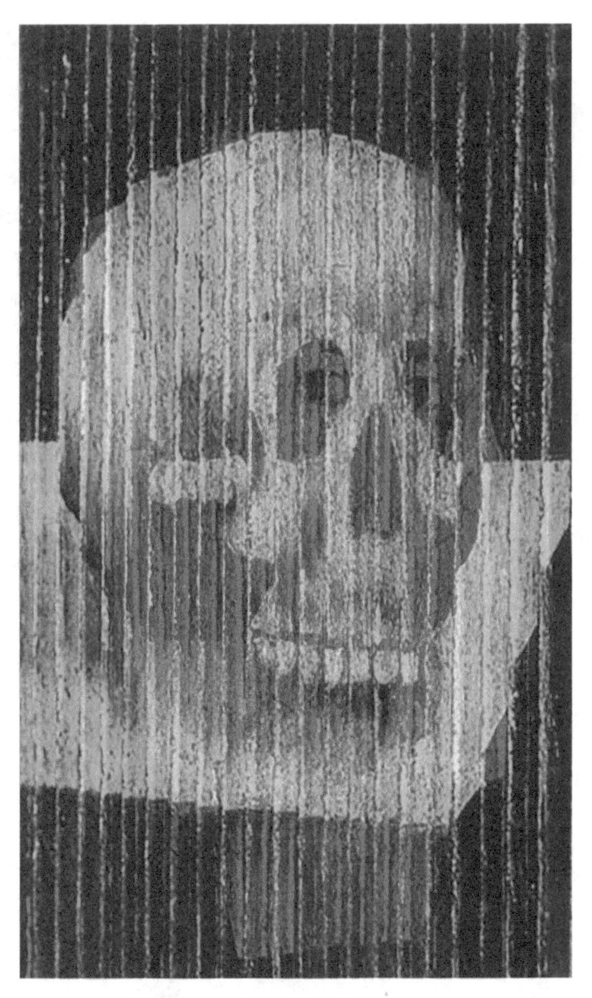

Death's head

Pretenders young and old

The possibility of lenticular stereoscopy arose at the turn of the nineteenth and twentieth centuries due largely to advances in film and printing technologies. In France, August Berthier (1865–1932) presented a barrier method in which alternating thin strips of two photographs could be seen by different eyes due to a line screen placed above them.[31] A similar system was described by Frederic Ives (1856–1937) which he called parallax stereograms. Ives developed a three-colour photographic system and also devised a novel way of presenting stereoscopic photographs.[32] This

[31] Berthier (1896) see also Timby (2001).

[32] Ives (1902).

involved interposing a set of fine parallel lines in front of a photograph so that different parts were visible to each eye.

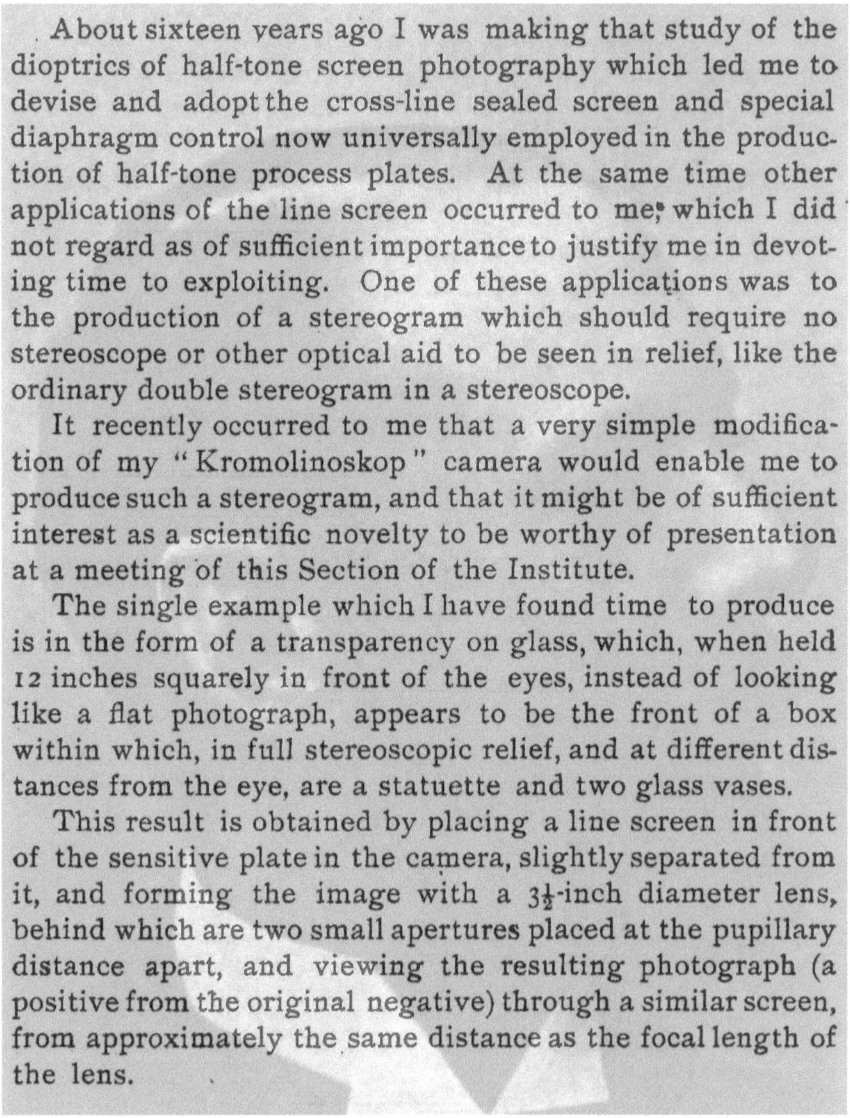

About sixteen years ago I was making that study of the dioptrics of half-tone screen photography which led me to devise and adopt the cross-line sealed screen and special diaphragm control now universally employed in the production of half-tone process plates. At the same time other applications of the line screen occurred to me, which I did not regard as of sufficient importance to justify me in devoting time to exploiting. One of these applications was to the production of a stereogram which should require no stereoscope or other optical aid to be seen in relief, like the ordinary double stereogram in a stereoscope.

It recently occurred to me that a very simple modification of my "Kromolinoskop" camera would enable me to produce such a stereogram, and that it might be of sufficient interest as a scientific novelty to be worthy of presentation at a meeting of this Section of the Institute.

The single example which I have found time to produce is in the form of a transparency on glass, which, when held 12 inches squarely in front of the eyes, instead of looking like a flat photograph, appears to be the front of a box within which, in full stereoscopic relief, and at different distances from the eye, are a statuette and two glass vases.

This result is obtained by placing a line screen in front of the sensitive plate in the camera, slightly separated from it, and forming the image with a 3½-inch diameter lens, behind which are two small apertures placed at the pupillary distance apart, and viewing the resulting photograph (a positive from the original negative) through a similar screen, from approximately the same distance as the focal length of the lens.

Frederic Ives and his description of a novel stereogram

Ives took out a patent for the system in 1903 and others were soon to follow. One of those patenting an optical lenticular picture process was the Nobel Prize winning physiologist, Walter Rudolf Hess (1881–1973). He is shown together with the diagrams from his patent application.

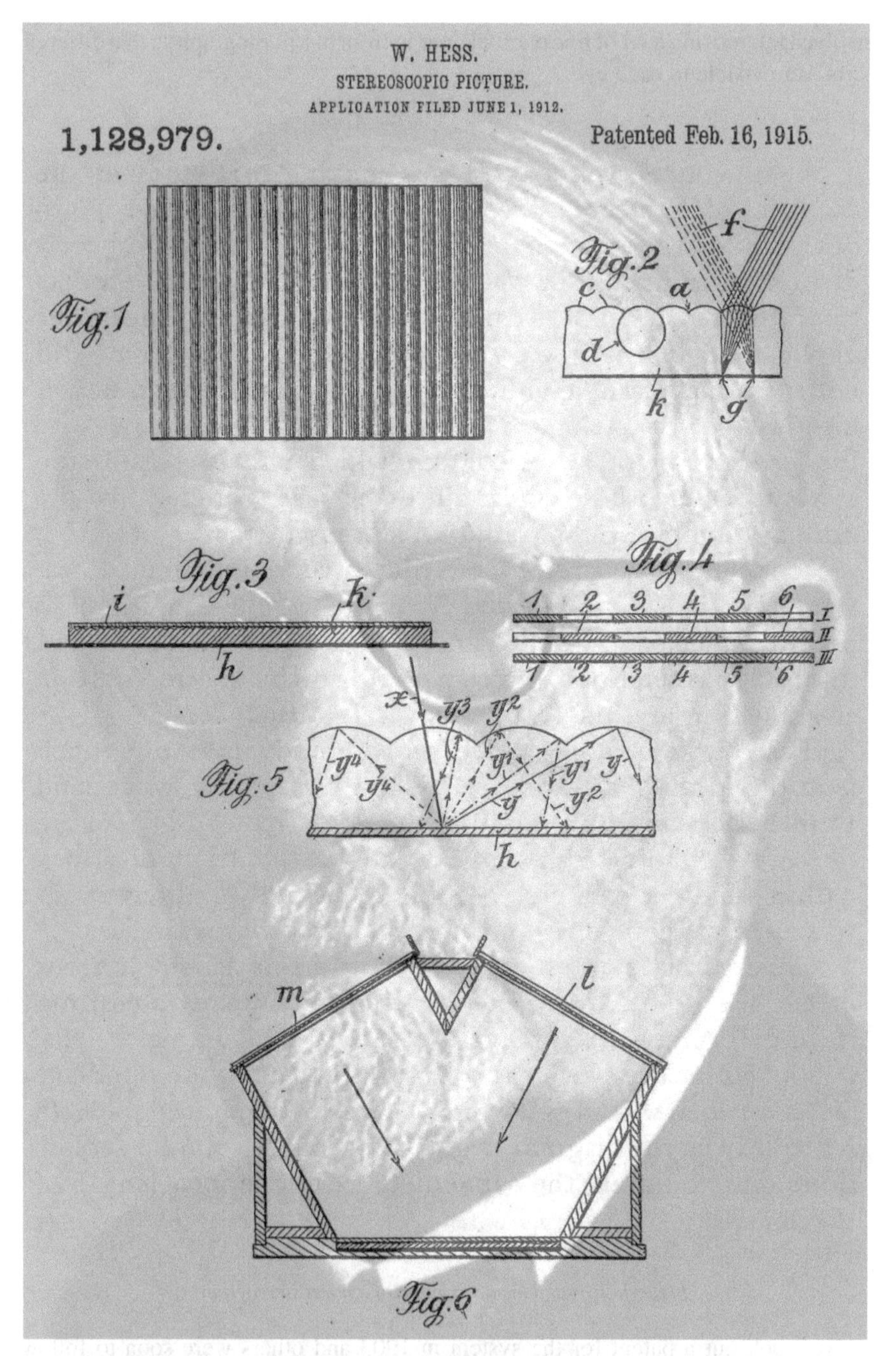

Walter Hess and his patent for a lenticular system for stereoscopic pictures

Printing technology has advanced enormously of late and increasingly intricate lenticular surfaces can be laid down. They tend to be used in advertising for animating sequences of images rather than for presenting stereoscopic pairs. Derek Michael Besant is an artist who blends both multiple views of a scene with stereoscopic effects in order to enhance the interactions between viewers and his large lenticular prints.[33] Besant creates blurred portraits which are often combined with some relevant text. Here two blurred images of his face are blended within alternate lines of gratings to give the impression of the strips of a lenticular print.

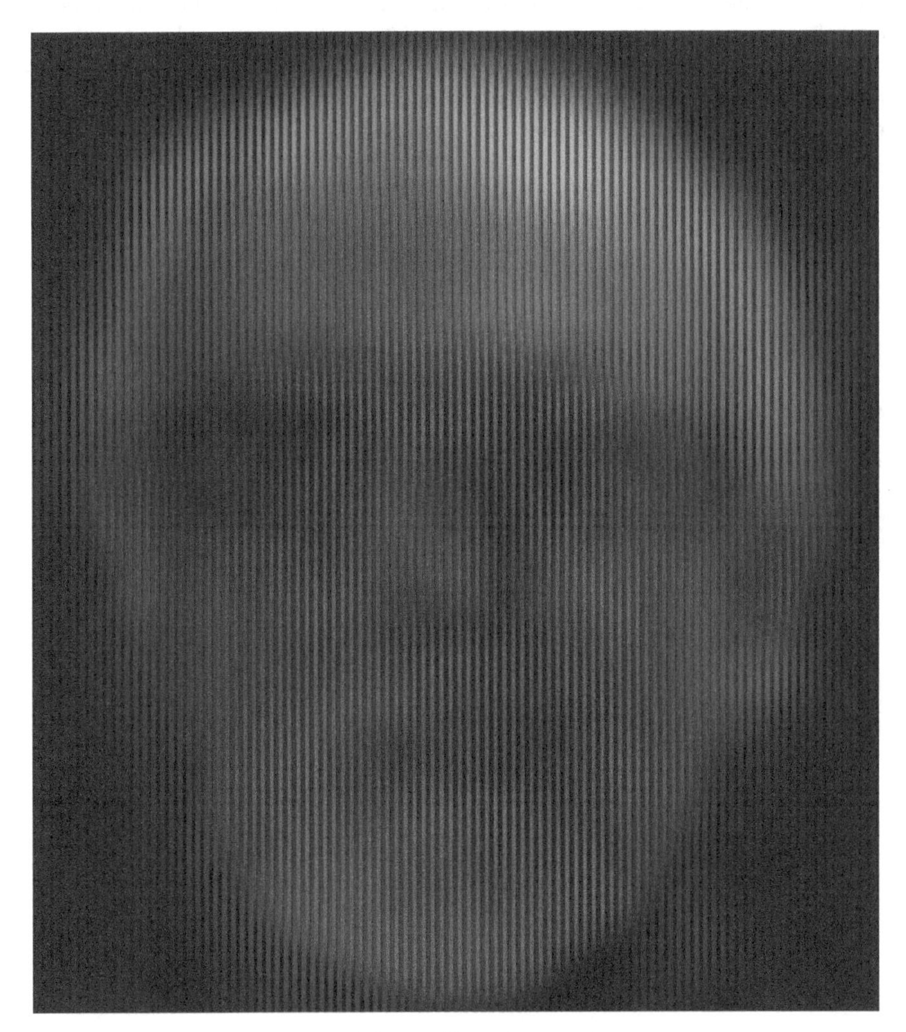

Derek Besant–Besantine portrait

[33] Besant (2018), https://www.cspaceprojects.com/meet-the-artist-derek-besant/.

Colour Separations

The realization that colours could be a source of separating the eyes was stated by Wilhelm Rollmann, a German inventor, in 1853. In an article describing two new stereoscopic techniques he wrote: "The colour stereoscope consists of coloured double drawings and two coloured glasses".[34] The colours that he found worked best were blue and yellow drawings combined with red and blue glasses. As would be expected, he reported that reversing the glasses resulted in a reversal of the relief seen. Five years later Joseph-Charles D'Almeida (1822–1880), a French physicist, described a similar system using images projected with two magic lanterns having colour filters in front of the lenses; the observer viewed the superimposed projections through similar filters. He found that combinations of red and green projections and glasses worked well. The first page of his article is shown below together with his portrait.

[34] Rollmann (1853), p. 187.

PHYSIQUE APPLIQUÉE. — *Nouvel appareil stéréoscopique;* *par* M. J.-Ch. d'Almeida.

(Commissaires, MM. Pouillet, Babinet.)

« Depuis que les expériences de M. Wheatstone ont indiqué la possibilité d'obtenir, au moyen d'images planes, les sensations que produisent les objets en relief, différents appareils ont été proposés qui permettent de réaliser les conditions requises. Ce sont les *stéréoscopes*. Au stéréoscope à réflexion de M. Wheatstone a succédé le stéréoscope à lentilles de M. Brewster, construit avec d'heureuses modifications par M. Duboscq. Dans ces derniers temps, M. Faye a fait connaître un appareil très-simple, ou plutôt une disposition qui dispense de tout appareil. Enfin, récemment, M. Claudet a découvert un moyen ingénieux d'agrandir les images et de les rendre visibles à « *deux* ou *trois* » observateurs simultanés.

» Tous ces appareils ne peuvent offrir les phénomènes qu'à un nombre très-restreint de spectateurs regardant ensemble. Dans un stéréoscope, il faut que chacun observe à son tour. Je me suis proposé d'obtenir une disposition telle, que les images fussent agrandies jusqu'à devenir visibles à plusieurs mètres de distance, et que les illusions du relief pussent être aperçues des divers points de la salle où s'exécute l'expérience. Deux procédés m'ont réussi.

» I. Au moyen de lentilles on projette sur un écran les images de deux épreuves stéréoscopiques telles que les épreuves ordinaires. Les images projetées sont amenées à se superposer, non pas trait pour trait — ce qui est impossible, car elles ne sont pas identiques — mais à peu près dans la position relative où elles se seraient présentées si les objets qu'elles reproduisent avaient été devant les yeux. Ces deux images forment sur l'écran un

Joseph-Charles D'Almeida and the text describing his system of colour separation

The proposals of Rollmann and D'Almeida had relatively little impact until the French pioneer of colour photography, Louis Ducos du Hauron (1837–1920), devised a method of over-printing red and blue or green designs in 1891; he took out a patent for "Prints, photographs and stereoscopic paintings producing their effect in broad daylight without using the stereoscope".[35] Thereafter, anaglyphs became increasingly popular as a means for printing stereoscopic drawings and photographs. They were also used for presenting sequences of images moving in depth, as D'Almeida had intimated in the conclusion to his article: "In closing this Note, I believe I should make it known that I am busy right now trying to achieve a simple combination that will allow movement to the images and reproduce in relief the effects of the phenakistiscope".[36] The phenakistiscope is a rotating device that can deliver the

[35] See Ducos du Hauron (1897), p. 483.

[36] D'Almeida (1858), p. 63.

impression of motion through a sequence of briefly presented pictures.[37] Anaglyph printing and projection became practical with Ducos du Hauron's colour separation system but the images seen were essentially black and white.

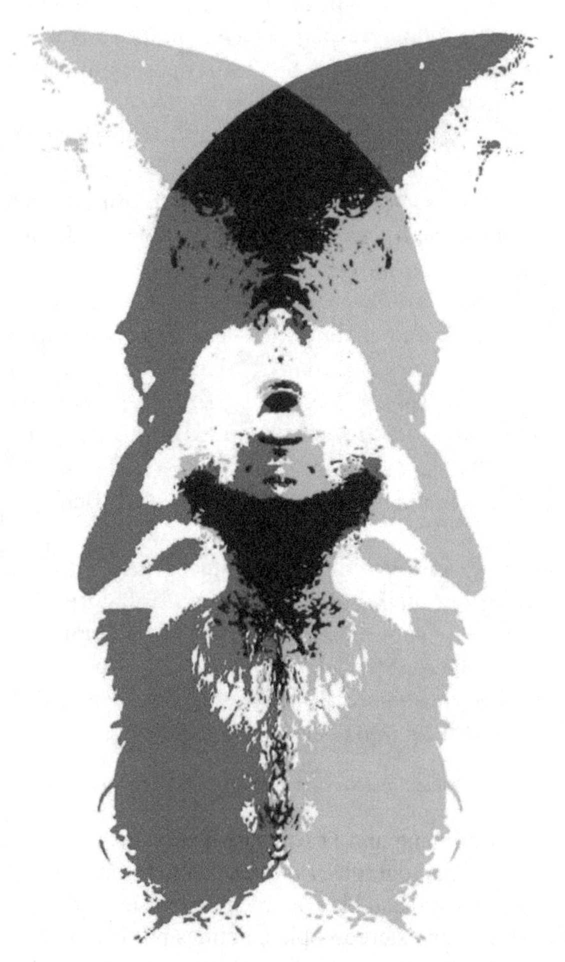

Red and cyan projected profiles of Ducos du Hauron

Light is an electromagnetic wave which oscillates in all directions. Light passing through certain types of glass becomes polarized so that the waves oscillate in one direction only. In 1828 William Nicol (1770–1851) discovered that a prism made from two parts of Iceland spar polarized light in directions at right angles to one another.[38] Between 1891 and 1895 an English scientist, John Anderton (1843–1905), mounted two Nicol prisms in the paths of light from two projectors and viewed

[37] Plateau (1833), See also Wade (2004).

[38] Nicol (1828).

the images through two Nicol prisms in opposite orientations mounted in viewing glasses; stereoscopic images were projected onto a specially silvered screen.[39] He then saw the image from one light source with one eye and the other with the other eye. He appreciated that this provided a method for presenting and viewing stereoscopic images. When sheet polarizers were manufactured in the 1930s the technique became more widely used. The advantage of this technique is that coloured images can be seen in depth or rivalry.[40] Since it is not possible to show polarized images here, Anderton is represented in contours crossing at right angles.

John Anderton introduced cross-polarized image viewing

Analglyphic images can be projected onto other scenes and then rephotographed. This was done with the pair of photographs below; they have been recombined in reverse so that opposite portraits are visible to each eye and the combined images engage in binocular rivalry.

[39] Anderton (1895) see also http://microscopist.net/FieldR.html; McBurney (2020).

[40] See Zone (2014).

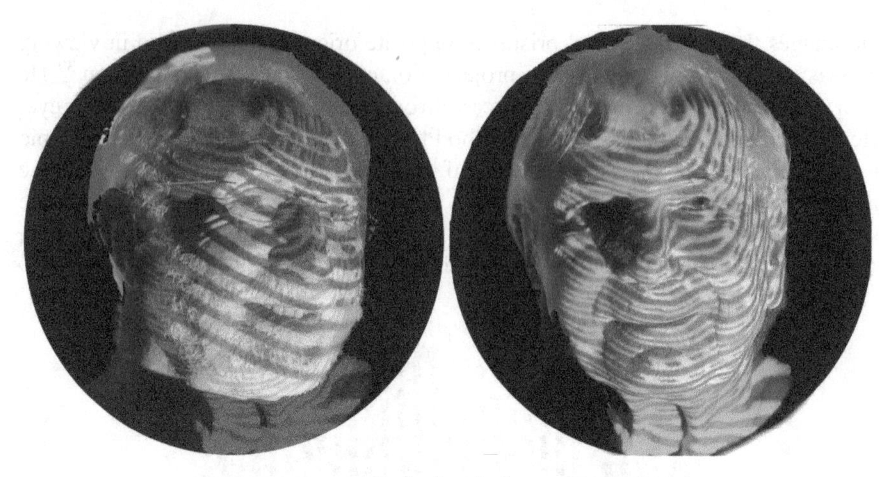

Projective rivalry

Pseudoscopes

Pseudoscopes reverse the disparities that normally exist so that concave objects appear convex or vice versa. It is as though the left and right eyes are being transposed. With flat stereoscopic pairs it is simple to reverse the stimuli presented to each eye by reversing their sides in a stereoscope. It is even easier with anaglyph glasses: the filters in front of each eye require reversing. For example, the pattern below looks like an approaching cone with the cyan filter in front of the right eye but like a receding tunnel with the filters reversed.

Cone or tunnel?

Wheatstone was well aware of the conversions of relief when reversing the stereopairs but he was intrigued by the effects of reversing disparities when viewing three-dimensional objects. To achieve this he described and named the pseudoscope in his 1852 memoir; it consisted of prisms as is shown in the illustration below. He applied it to reverse the normal relations between monocular and stereoscopic cues to depth: "With the pseudoscope we have a glance, as it were, into another visible world, in which external objects and our internal perceptions have no longer their habitual relation with each other".[41] He remarked on the difficulty of perceiving reversals of relief with the pseudoscope, and the illuminating conditions that are necessary for such reversal.

---

[41] Wheatstone (1852), p. 12.

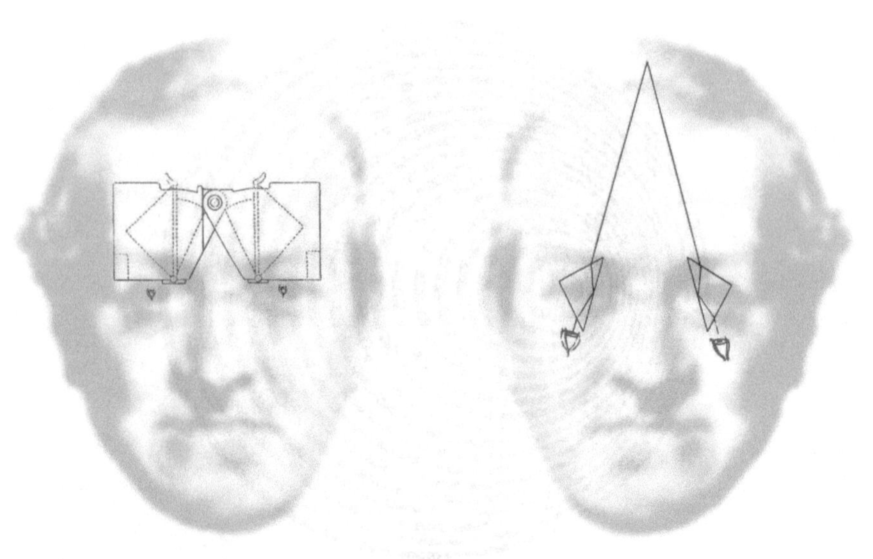

Wheatstone and two of his diagrams showing a pseudoscope for reversing the disparities in the eyes

Artists have developed pseudoscopes as a way of manipulating the perception of space in intriguing and unexpected ways but they have tended to use mirrors rather than prisms to achieve the effect. Quite elaborate constructions were made by the Swiss artist, Alfons Schilling (1934–2013). His interests in binocular vision have been expressed in random dot patterns, autostereograms, stereoscopic paintings and a variety of mirror and prism devices that confound vision.[42] These have included pseudoscopes rather similar to that illustrated by Wheatstone. Terry Pope has made models of pseudoscopes that are based on reflection rather than refraction.

[42] Schilling (1997).

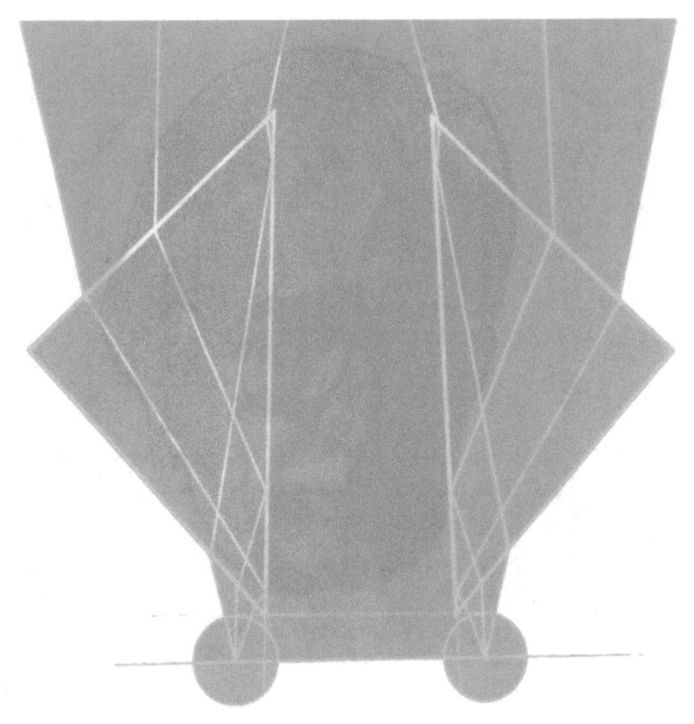

Alfons Schilling and his schematic drawing of a prism pseudoscope

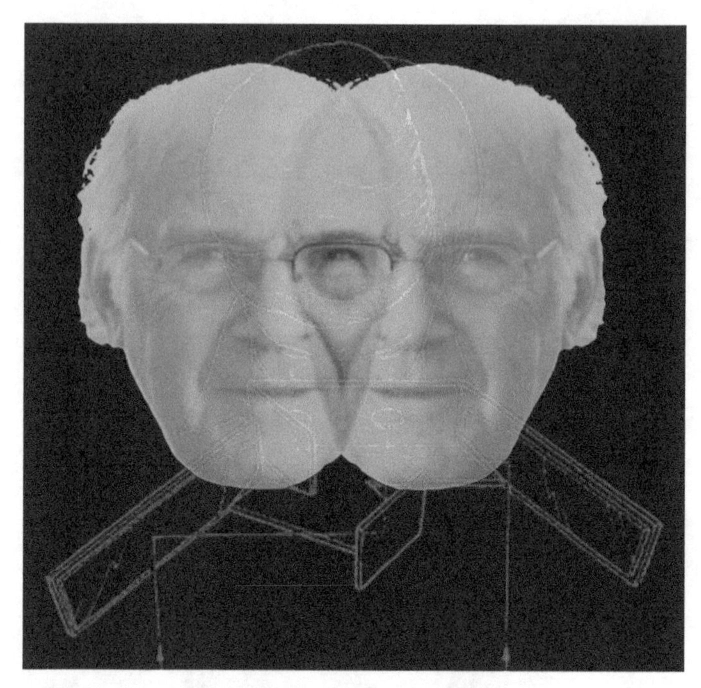

Terry Pope and his mirror pseudoscope

Binocular Cameras

The first stereoscopic photographs were made by using a single camera and moving it laterally so that two slightly different photographs of the same scene were taken in succession.[43] Various devices were introduced to avoid moving the whole camera, like sliding camera bodies and moveable lenses.[44] This was to change with the introduction of twin-lensed (binocular) cameras that could take two slightly different photographs at the same time. Brewster announced his binocular camera for taking stereoscopic photographs at the same meeting of the British Association for the Advancement of Science as the description of his lenticular stereoscope in 1849[45]; a fuller account was presented two years later and in his book on the stereoscope.[46] The camera had the lenses at a fixed separation corresponding to the interocular distance.

[43] See Klooswijk (1991).

[44] See Coe (1978), Newhall (1982).

[45] Brewster (1849a, b).

[46] Brewster (1851, 1856).

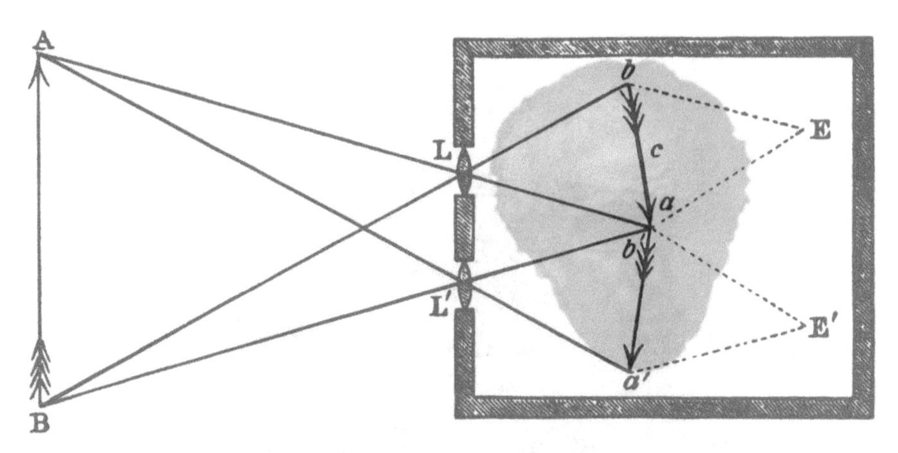

Brewster and his binocular camera

Brewster used half-lenses in his stereoscope and binocular camera because he did not consider that it was then possible to grind two lenses to have the same curvatures. John Benjamin Dancer (1812–1887) did just that with a twin-lensed binocular camera; he was an acquaintance of Brewster's also considered that the camera lenses should be separated by the distance between the eyes. Dancer made his first model in 1852 and he produced an improved, commercially available model in 1856. Dancer was an optical instrument maker in Manchester and added many refinements to binocular cameras, like variable apertures for the lenses, a ratchet system for advancing unexposed dry plates and a spirit level to assist in the appropriate alignment of the two lenses. Looking back on the introduction of photography to Liverpool and Manchester, he reminisced:

> At that time stereoscopic photographs were usually taken with one camera. After the first exposure the camera was moved to a distance on the same plane, then the second picture was taken. Photographers were not agreed as to the exact distance the camera should be moved for taking the second picture: some advocated eight inches, others several feet between the two places, I practically demonstrated that in taking the two pictures the camera should only be moved the distance of the average space between the human eyes, otherwise the pictures when viewed in the stereoscope would not present a natural appearance. I then introduced the twin lens camera; my idea was ridiculed at first by some leading photographers, and especially by Mr. Sutton, then editor of the Photographic Notes. I sent copies of my practical demonstrations to many eminent men of science, including Sir Charles Wheatstone and Sir David Brewster. The latter philosopher complimented me on my demonstration. Since that time stereoscopic cameras have been mounted with twin lenses, and both pictures taken simultaneously.[47]

[47] Dancer (1886).

John Benjamin Dancer and his binocular camera

Brewster's design was discussed in Britain and France in the early 1850s and Achille Quinet (1831–1900) made a model in 1853 and patented it two years later[48]; he called it a Quinetoscope. Binocular cameras, like those of Quinet and Dancer, removed many of the difficulties of alignment and time difference associated with using a single lens camera for stereoscopic photographs. When combined with Brewster's stereoscope, they did much to hasten the popularity of stereoscopic photography in the second half of the nineteenth century. The situation was summarised by Brewster: "The photographic camera is the only means by which living persons and statues can be represented by means of two plane pictures to be combined by means of the stereoscope; and but for the art of photography, this instrument would have had a very limited application".[49] Many new binocular cameras were made in the remainder of the century but the one that had the longest life was vérascope, created and developed by Jules Richard (1848–1930) and patented in 1893.

[48] Timby (2001).

[49] Brewster (1856), p. 135.

Jules Richard and his vérascope

There are many excellent books on the history of photography as well as cameras, most of which have sections on binocular devices sometimes with stereoviewers. The resurrection of The London Stereoscopic Company has led to the publication of several lavishly illustrated books with specially designed viewers.[50]

Stereoscopic Photography

Wheatstone and Brewster were well aware of the early researches by William Henry Fox Talbot (1800–1877) on capturing images on light-sensitive paper. In 1836, both were guests of Talbot at Laycock Abbey, prior to the Bristol meeting of the British Association for the Advancement of Science, and they corresponded about the process thereafter. Talbot's paper negative process was made public in 1839, the year after Wheatstone's first article on the stereoscope appeared. Wheatstone immediately grasped the significance of photographing scenes from two positions, so that they would be seen in depth when mounted in the stereoscope. In his second memoir on binocular vision he wrote:

> At the date of the publication of my experiments on binocular vision, the brilliant photographic discoveries of Talbot, Niepce and Daguerre, had not been announced to the world.

---

[50] May et al (2013), Pellerin and May (2014, 2021).

> To illustrate the phenomena of the stereoscope I could therefore, at that time, only employ drawings made by the hands of an artist. Mere outline figures, or even shaded perspective drawings of simple objects, do not present much difficulty; but it is evidently impossible for the most accurate and accomplished artist to delineate, by the sole aid of his eye, the two projections necessary to form the stereoscopic relief of objects as they exist in nature with their delicate differences of outline, light and shade. What the hand of the artist was unable to accomplish, the chemical action of light, directed by the camera, has enabled us to effect.[51]

The Wheatstone family (derived from a stereo-daguerreotype taken by Antoine Claudet in the mid-1850s)

Brewster was instrumental in the early development of photography in general and stereoscopic photography in particular. He championed Talbot's paper negative process against its competitor, the metal-plated Daguerreotype as well as introducing Robert Adamson to the painter David Octavius Hill. Hill and Adamson produced

[51] Wheatstone (1852), pp. 6–7.

some of the finest early photographs.[52] In St. Andrews Brewster instructed Dr. John Adamson in the art of making photographs and John passed on the skills to his younger brother, Robert. Brewster fostered the marriage between the photograph and the stereoscope and he encouraged photographers in St. Andrews to cement the relationship. Thomas Rodger set up a studio in St. Andrews where it is likely the portrait of Brewster shown below was taken.[53]

Brewster seated beside his stereoscope (derived from a stereo-calotype probably taken by Thomas Rodger)

Many stereoscopic photographs in the nineteenth century were taken with a single camera moved between the two exposures. The Scottish photographer George Washington Wilson (1823–1893) photographed landmarks throughout the country.[54] Among the many landscapes he photographed were several of the Falls of Foyers, a noted spectacle near Loch Ness. The waterfall, which is shown below, is significant

[52] Schaaf (2002), Stevenson (1999).

[53] Crawford (2011), Morrison-Low and Bruce (2018).

[54] Taylor (2018).

in the history of vision research.[55] A peripatetic lecturer, Robert Addams (1789–1875), wrote a letter to Brewster describing an illusion he had observed following observation of the descending waters.[56] Maintaining fixation on a particular part of the falls for about a minute and then looking at the rocks at the side resulted in the rocks appearing to move upwards. It is now known as the waterfall illusion. The photographs making up the stereo image below were taken from the River Foyers below the falls.

Falls of Foyers

One of the early photographers who produced stereoscopic images was Antoine François Jean Claudet (1797–1867) and he took the stereodaguerreotype portrait of Wheatstone shown above. Claudet was born in Lyon and moved to London in 1829. He was a student and then partner of Daguerre and improved the daguerreotype process. He opened the first daguerreotype studio in London and became recognised as a scientist as well as a photographer.[57] Claudet wrote about the camera separations for taking paired photographs with subjects at different distances. He photographed

[55] Swanston and Wade (1994), Wade (2018).

[56] Addams (1834).

[57] Gill (1967).

many of the contemporary luminaries of science and photography, as can be seen in the image below.

Claudet can be seen combined with photographs he took of Jacques Louis Mandé Daguerre, William Henry Fox Talbot, Charles Wheatstone and Michael Faraday (clockwise from the top left)

Like Wheatstone, Claudet sought to combine apparent depth with apparent motion.[58] In the prosecution of this research an entry in *La Lumière* of May 1852 reported that:

> M. Claudet informs us that he has constructed a stereoscope in which one can see a person moving, for example a lady working with a needle and making all the necessary movements, a smoker with his cigar moving in and out of his mouth while exhaling smoke, people who drink and toast one another in the English way, steam engines in motion, etc. M. Wheatstone, on his side, without knowing about the device proposed by M. Claudet, seeks to resolve the same problem, and in a few days the mechanisms proposed by the two physicists will be published. M. Wheatstone and the inventor of the phenakisticope (Plateau) have been struck

[58] See Wade (2012).

> for several years by the possibility of applying stereoscopic principles to the effects of the phenakisticope, but at present have not been successful.[59]

Claudet took out a patent for his instrument in 1853, but Wheatstone appears to have abandoned his attempts and did not return to them for over a decade. However, Claudet "constructed his instrument in such a manner that by means of a slide with one hole he can, by moving it rapidly in a reciprocating horizontal direction, shut one lens while the other remains open; and in continuing that motion, while one eye sees one of the two pictures, the second eye cannot see the other picture".[60] Since the motion was controlled by hand and therefore the timings will have been variable and this could have been the reason why motion was more easily seen than stereoscopic depth. Claudet's method is a precursor of the electronic shuttering systems that have been employed more recently.[61] Claudet also noted that with his alternating vision technique "Another curious phenomenon of this alternative vision is, that one cannot distinguish by which eye the object is seen by."[62]

Most stereoscopic photographs are now taken with cameras having two lenses. The aim is usually to capture the two views of a scene as they would be seen with two eyes. That is, the separation between the lenses is similar to that between the eyes. In order to capture a scene stereoscopically with a single camera then some disparity between the two exposures is necessary. This can be produced by changing the position of the camera, by movements of the objects in the scene when the camera is still, or by combinations of these. When the camera is moved between exposures then the separation will depend on the distance of the object of interest. Photographing a distant landscape with a stereo camera will result in two almost identical images. Disparities can be introduced by increasing the separation between the two exposures. A rule of thumb used by photographers is to separate the camera positions by about a thirtieth of the distance to the subject. Therefore the separations are smaller than the interocular distance for close subjects but considerably more for distant subjects. If a separation that approximates the interocular distance is used for close subjects, like the white tulips below then the disparities and depth will be amplified relative to natural viewing.

---

[59] Anon (1852), p. 88.

[60] Claudet (1865), pp. 9–10.

[61] Blundell (2011).

[62] Claudet (1865), p. 10.

White tulips

References

Addams R (1834) An account of a peculiar optical phænomenon seen after having looked at a moving body. London Edinb Philos Mag J Sci 5:373–374
Anderton J (1895) Method by which pictures projected upon screens by magic lanterns are seen in relief. https://patents.google.com/patent/US542321A/en
Anon (1852) Bulletin de la correspondence. La Lumière 288
Berthier A (1896) Images stéréoscopiques de grand format. Cosmos 34:227–233
Besant DM (2018) The dark woods (revisited). Vernon Public Art Gallery, Vernon, BC
Besant DM. https://www.cspaceprojects.com/meet-the-artist-derek-besant/
Blundell BG (2011) 3D displays and spatial interaction, vol 1. From perception to technology. Walker & Wood, Auckland, NZ
Brewster D (1819) A treatise on the kaleidoscope. Constable, Edinburgh
Brewster D (1832) Letters on natural magic addressed to Sir Walter Scott, Bart. John Murray, London
Brewster D (1849a) Description of a binocular camera. Report of the British Association. Transactions of the Sections 5
Brewster D (1849b) Account of a new stereoscope. Report of the British Association. Transactions of the Sections 6–7
Brewster D (1851) Description of several new and simple stereoscopes for exhibiting, as solids, one ormore representations of them on a plane. Trans R Scottish Soc Arts 3:247–259
Brewster D (1856) The stereoscope. Its history, theory, and construction. John Murray, London
Claudet A (1865) On moving photographic figures, illustrating some phenomena of vision connected with the combination of the stereoscope and the phenakistoscope by means of photography. Report of the British Association for the Advancement of Science. Transactions of the Sections, pp 9–10
Colvin C, Normand T (2019) The constructed world of Calum Colvin. Symbol, allegory, myth. Luath Press, Edinburgh

Coe B (1978) Cameras. From daguerreotypes to instant pictures. Marshall Cavendish, London
Crawford R (2011) The beginning and the end of the world. St Andrews, scandal, and the birth of photography. Birlinn, Edinburgh
D'Almeida J-C (1858) Nouvel appareil stéréoscopique. Comptes Rendus Hebdomadaire Des Séances De L'academie Des Sciences 47:61–63
Dancer JB (1886) Early photography in Liverpool and Manchester. Manchester City News, 22 May
Dodgson NA (2005) Autostereoscopic 3D displays. Computer 38:31–36
Dove HW (1851) Beschreibung mehrerer Prismenstereoskope und eines einfachen Spiegelstereoskops. Annalen der Physik und Chemie 83:183–189
Drouin F (1890) The stereoscope and stereoscopic photography. The Country Press, Bradford UK
Ducos du Hauron A (1897) La triplice photographique des couleures et l'imprimerie: systéme de photochromographie Louis Ducos du Hauron. Gauthier-Villars, Paris
Gill AT (1967) Antoine François Claudet (1797–1867). Photogr J 107:405–409
Helmholtz H (1857) Das Telestereoskop. Annalen der Physik und Chemie 101:494–496
Holmes OW (1859) The stereoscope and the stereograph. The Atlantic Monthly 3:738–748
Holmes OW (1861) Sun painting and sun sculpture. The Atlantic Monthly 8:13–29
Holmes OW (1869) History of the 'American stereoscope.' The Philadelphia Photographer 6:1–3
Howard IP, Rogers (2012) Perceiving in depth, vol 2. Stereoscopic vision. Oxford University Press, USA
Ives FE (1902) A novel stereogram. J Franklin Inst 153:51–52
Klooswijk AIJ (1991) The first stereo photograph. Stereo World May/June, 6–11
Lomas A. http://www.andylomas.com/constrainedForms.html
Lowdon G (1906) George Lowdon, Optician and scientist, Dundee. Sketch of his reminiscences and career. Dundee Advertiser, Dundee
Mannoni L (2000) The great art of light and shadow. Archaeology of cinema (trans: Crangle R). University of Exeter Press, Exeter
May B, Pellerin D, Fleming (2013) Diableries. Stereoscopic adventures in hell. The London Stereoscopic Company, London
McBurney S (2020) Stereoscopy on the silver screen: the analyticon and early cinema in Edinburgh, Scotland. Int J Stereo Immersive Media 4:104–119
Morrison-Low AD, Bruce D (2018) Photography and the doctor. John Adamson of St Andrews. National Museums of Scotland, Edinburgh
Newhall B (1982) The history of photography. Secker & Warburg, London
Nicol W (1828) On a method of so far increasing the divergency of the two rays in calcareous spar that only one image may be seen at a time. Edinb New Philos J 6:83–84
Pellerin D, May B (2014) The poor man's picture gallery. Stereoscopy versus paintings in the Victorian era. The London Stereoscopic Company, London
Pellerin D, May B (2021) Stereoscopy. The dawn of 3-D. The London Stereoscopic Company, London
Plateau J (1833) Des illusions sur lesquelles se fonde le petit appareil appelé récemment phénakisticope. Annales de Chimie et de Physique de Paris 53:304–308
Pope T. www.phantascope.co.uk
Rollmann W (1853) Zwei neue stereoskopische Methoden. Annalen der Physik und Chemie 166:186–187
Schaaf L (2002) St. Andrews and early Scottish photography including Hill and Adamson: Sun pictures catalogue. Kraus, New York
Schilling A (1997) Ich/Auge/Welt. The art of vision. Springer, New York
Sheppard JP (2016). file:///C:/Users/Admin/Downloads/Sheppard,%20Jaime.pdf
Silverman RJ (1993) The stereoscope and photographic depiction in the 19th century. Technol Cult 34:729–756
Stevenson S (1999) Facing the light. The photography of Hill and Adamson. National Galleries of Scotland, Edinburgh
Swanston MT, Wade NJ (1994) A peculiar optical phænomenon. Perception 23:1107–1110

Taylor R (2018) George Washington Wilson artist and photographer (1823–1893). The London Stereoscopic Company, London

Timby K (2001) Images en relief et images changeantes. Études Photographiques. http://journals.openedition.org/etudesphotographiques/246

Wade NJ (1983) Brewster and Wheatstone on vision. Academic Press, London

Wade NJ (2004) Philosophical instruments and toys: optical devices extending the art of seeing. J Hist Neurosci 13:102–124

Wade NJ (2012) Wheatstone and the origins of moving stereoscopic images. Perception 41:901–924

Wade NJ (2018) Pursuing paradoxes posed by the waterfall illusion. Perception 47:689–693

Wheatstone C (1838) Contributions to the physiology of vision—part the first. On some remarkable, and hitherto unobserved, phenomena of binocular vision. Philos Trans R Soc 128:371–394

Wheatstone C (1852) Contributions to the physiology of vision—part the second. On some remarkable, and hitherto unobserved, phenomena of binocular vision. Philos Trans R Soc 142:1–17

Wing P (1996) Stereoscopes: the first one hundred years. Transition Publishing, Nashua, NH

Zone R (2014) Stereoscopic cinema and the origins of 3-D film, 1838–1952. University Press of Kentucky, Lexington, KY

Chapter 5
Stereoscopic Vision

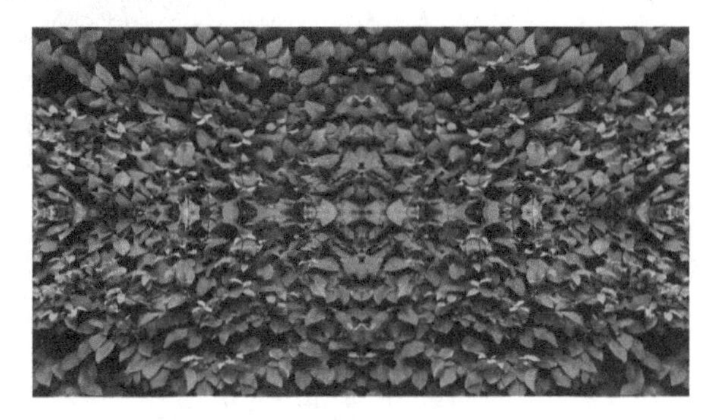

Stereoscopic vision is the normal mode of vision in the natural environment for those with two functioning eyes. However it could not be investigated easily before the invention of the stereoscope. Charles Wheatstone not only invented the instrument but he also carried out significant experiments with it. The "new fact in the theory of vision" to which Wheatstone referred was that slight and systematic dissimilarities between two pictures when viewed in a stereoscope can yield the impression of depth not seen in single pictures. The sign of perceived depth (nearer or farther) was determined by the sign of disparity (crossed or uncrossed); there was a limit to the disparities that can yield singleness and depth; and markedly different images presented to each eye will engage in rivalry. In most of his experiments Wheatstone employed simple outline stimuli so that as little pictorial depth as possible would influence what was seen. He embraced the use of stereoscopic photography for portraiture but not for scientific experiments. His desire to examine stereoscopic depth perception without object recognition eluded him. It was achieved over a century later by Béla Julesz with computer-generated random dot stereograms; no relative depth could be seen by either eye alone but only by the binocular combination of the two patterns. Julesz referred to this as cyclopean perception. Computers are not required to examine cyclopean perception as the effects can be produced with complex patterns produced from naturalistic scenes, examples of which will be illustrated in this chapter and later in the book. One binocular depth effect that was described before invention of the stereoscope is the wallpaper illusion. Repetitive patterns on wallpaper can be seen to lie in front of or behind the plane of the wall if adjacent elements are fixated. Small and systematic variations of the size and spacing between the elements can introduce relative, stereoscopic depth. This is the basis for autostereograms—computer-generated patterns that can contain features in depth if the eyes either over- or under-converge. Yet another aspect of relative depth perception depends on the colours displayed and it is called colour stereoscopy or chromostereopsis. When long- and short-wave colours (like red and blue) are placed near to one another they can appear to occupy different depths.

N. Wade, *Vision and Art with Two Eyes*, Vision, Illusion and Perception 3,
https://doi.org/10.1007/978-3-030-77995-5_5

Brewster and Wheatstone in stereoscopic depth

Wheatstone's appreciation of depth from disparity was based on a chance observation of the reflection of a candle flame in a polished metal plate:

> When a single candle flame is brought near such a [metal] plate, a line of light appears standing out from it, one half being above, and the other half below the surface; the position and inclination of this line changes with the situation of the light and of the observer, but it always passes through the centre of the plate. On closing the left eye the relief disappears, and the luminous line coincides with one of the diameters of the plate; on closing the right eye the line appears equally in the plane of the surface, but coincides with another diameter; on opening both eyes it instantly starts into relief. The case here is exactly analogous to the vision of two inclined lines when each is presented to a different eye in the stereoscope. It is curious, that an effect like this, which must have been seen thousands of times, should never have attracted sufficient attention to have been made the subject of philosophic observation. It was one of the earliest facts which drew my attention to the subject I am now treating[1]

Wheatstone illustrated this situation with a diagram representing the two reflections at different angles. The orientation difference between the two lines in Wheatstone's figure is excessive as it results in rivalry rather than depth in the median plane.

[1] Wheatstone (1838), pp. 372–373.

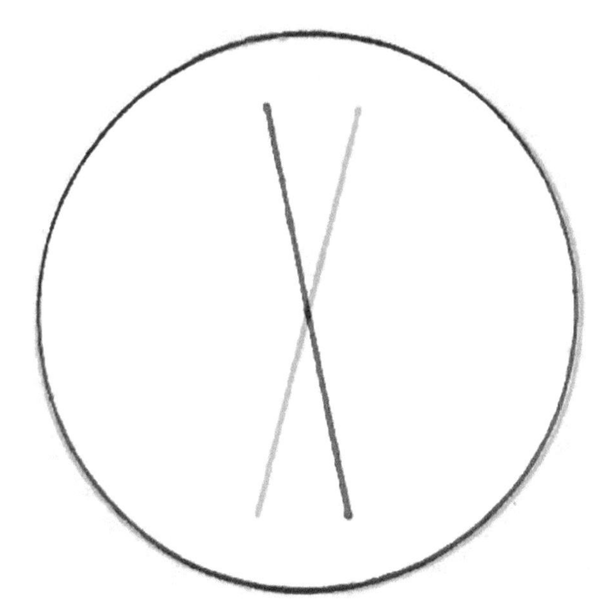

Wheatstone's illustration combining lines in different orientations

Crossed and Uncrossed Disparities

The "new fact in the theory of vision" to which Wheatstone referred was that slight and systematic dissimilarities between two pictures when viewed in a stereoscope can yield the impression of depth not seen in single pictures. Moreover, the sign of the dissimilarities (now referred to as disparities) defined the relative depth seen. Crossed disparities occurred when the common contour appeared to the right with the left eye and to the left with the right: this resulted in the contour appearing nearer than the point of fixation. Uncrossed disparities (the contour to the left with the left eye and to the right with right eye) resulted in the contour appearing farther away than the point of fixation. Thus, there is a lawful relationship between the sign of disparity (crossed or uncrossed) and the relative depth seen (nearer or farther).

Crossed (upper) and uncrossed (lower) disparities when viewed with the red/LE and cyan/RE combination. With fixation on the dot the upper line appears nearer and the lower line looks farther away

Having established that dissimilar pictures, when viewed in the stereoscope, produce the appearance of depth, Wheatstone conducted a series of systematic manipulations of the figures in order to discover the nature of the relationship. In his first article he demonstrated that the sign of disparity (crossed or uncrossed) determined the relative depth seen (nearer or farther), that there was a limit to the disparity yielding singleness of vision, that eye movements were not involved (because depth was seen in disparate afterimages), and that radically different pictures or colours resulted in rivalry. Wheatstone's second article on binocular vision was published fourteen years later in 1852. He described and illustrated an adjustable mirror stereoscope, a prism stereoscope, and a pseudoscope for reversing disparities. The main purpose of these was to extend the range of conditions under which the two eyes could be stimulated:

> Under the ordinary conditions of vision, when an object is placed at a certain distance before the eyes, several concurring circumstances remain constant, and they always vary in the same order when the distance of the object is changed. Thus, as we approach the object, or as it is brought nearer to us, the magnitude of the picture on the retina increases; the inclination of the optic axes, required to cause the picture to fall on corresponding places on the retinæ, becomes greater; the divergence of the rays of light proceeding from each point of the object, and which determines the adaptation of the eyes to distinct vision of that point, increases; and the dissimilarity of the two pictures projected on the retinæ also becomes greater. It is important to ascertain in what manner our perception of the magnitude and distance of objects depends on these various circumstances, and to inquire which are the most, and which the least influential in the judgements we form. To advance this inquiry beyond the point to which it has hitherto been brought, it is not sufficient to content ourselves with drawing conclusions from observations on the circumstances under which vision naturally occurs, as preceding writers on this subject mostly have done, but it is necessary to have more extended recourse to the methods so successfully employed in experimental philosophy, and to endeavour, wherever it be possible, not only to analyse the elements of vision, but also

> to recombine them in unusual manners, so that they may be associated under circumstances that never naturally occur.[2]

Wheatstone used a stereoscope with adjustable arms to vary the four circumstances mentioned in the quotation (retinal size, convergence, accommodation, and disparity). He found that:

> The perceived magnitude of an object, therefore, diminishes as the inclination of the axes becomes greater, while the distance remains the same; and it increases, when the inclination of the axes remains the same, while the distance diminishes. When both of these conditions vary inversely, as they do in ordinary vision when the distance of an object changes, the perceived magnitude remains the same.[3]

Crossed and uncrossed disparities yield the impression of an object being nearer or farther away, as is evident in the figures below. For *Crop circle*, a large central circle appears in depth with respect to its surround. With the eye filter combination red/LE and cyan/RE it is as though there is a circular window through which the central pattern is seen; with cyan/LE and red/RE the central circle seems to stand out in front of the surround. There is another consequence of these signed disparities. When the central circular pattern appears more distant then sideways head movement results in the impression of its movement in the opposite direction; the reverse occurs with the cyan/LE, red/RE arrangement. Moreover, the circle appears larger when apparently more distant (with uncrossed disparity). Both effects can be seen more easily with *Berry circles*: crossed and uncrossed disparities are present in the two circles left and right of the centre; lateral head movement will result in the circles moving in opposite directions. Since the circles are physically equal, the apparent difference in size is easier to see and can be verified by reversing the filters in front of the eyes. It is an instance of size constancy—the perceived size of objects is dependent on the perceived distance. If two objects subtend the same angle at the eye (as do the two circles) then the apparently more distant one will appear larger.[4]

The apparent movement is equivalent to the motion parallax that is seen with actual separations in depth between objects—objects closer than the point of fixation seem to move in the same direction relative to the fixation point whereas more distant ones have the opposite relative motion. Thus, motion parallax provides a similar source of relative distance information to retinal disparities. This is an aspect of spatial vision that has been studied in detail in recent years.[5]

---

[2] Wheatstone (1852), p. 2.

[3] Wheatstone (1852), p. 3.

[4] See Ono and Comerford (1977).

[5] Rogers and Graham (1979), Rogers (2017).

Crop circle

Berry circles

Stereoscopic Stimuli

Simple line stimuli are used in most scientific studies of stereoscopic depth perception because the separations could be accurately determined and manipulated. The line drawings are also relatively free from the monocular depth cues conveyed by pictures in perspective so that perceived depth could be assigned to disparity rather than perspective. Two examples below are taken from Wheatstone's article; they were described as "The frustum of a square pyramid" and "the outline of a frustum of a cone".

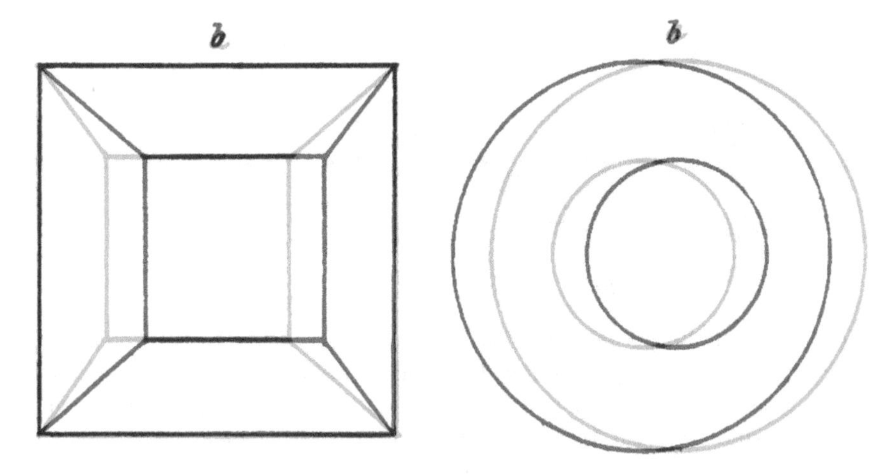

Line stereograms from Wheatstone's 1838 *article*

Wheatstone remarked that:

> For the purposes of illustration I have employed only outline figures, for had either shading or colouring been introduced it might be supposed that the effect was wholly or in part due to these circumstances, whereas by leaving them out of consideration no room is left to doubt that the entire effect of relief is owing to the simultaneous perception of the two monocular projections, one on each retina.[6]

Nonetheless, he did present some more complex line stimuli and it could be argued that he did publish the first attempt to produce stereoscopic art.[7] It represented a drawing of pair of arches that could be mounted in a stereoscope.

---

[6] Wheatstone (1838), p. 376.

[7] See Brooks (2017).

Wheatstone's stereoscopic depiction of a scene

However in 1839, the year after Wheatstone's first article on the stereoscope appeared his friend, Henry Talbot, made public his negative–positive photographic process. Wheatstone immediately grasped the significance of photographing scenes from two positions, so that they would be seen in depth when mounted in the stereoscope. He appreciated the difficulty of making two accurate perspective representations from slightly different positions and soon saw the virtues that photography offered for this endeavour: "What the hand of the artist was unable to accomplish, the chemical action of light, directed by the camera, has enabled us to effect".[8] Thereafter, photography and stereoscopy were closely entwined. Computer graphics have added further technical advantages to producing art for two eyes.

Photographs

In 1840, Wheatstone enlisted Talbot's assistance to take stereo-photographs for him; when they were sent the angular separation of the camera positions used to capture the two views was too large (47.5°) and Wheatstone suggested that 25° would be more appropriate. Klooswijk[9] has reprinted a section of Wheatstone's letter to Talbot, and has himself taken stereo-photographs of the bust Talbot probably employed from camera angles of 47.5°, 25°, and 1.75°.

[8] Wheatstone (1852), p. 7.

[9] Klooswijk (1991).

Negative and positive images of Talbot

Wheatstone showed how the photographic camera, in combination with the stereoscope, could be employed to reintroduce the dimension of depth to the perception of pictures. He achieved the result to the problem that Leonardo has struggled with so long. In Wheatstone's obituary notice in *Nature* the following comments were added by Signor Volpicelli of the Academia dei Lincei:

> Our countryman, Leonardo da Vinci, in 1500, or thereabouts, conceived and was the first to affirm, that from a picture it was not possible to obtain the effect of relief. But Wheatstone, reflecting profoundly in 1838, on the physiology of vision, invented the catoptric [reflecting] stereoscope, with which he philosophically solved the problem of the optical and virtual production of relief.[10]

---

[10] Volpicelli (1876), p. 502.

Stereoscopic photographs could even be used in portraiture. To achieve this end, Wheatstone enlisted the assistance of Henry Collen (1800–1875) to take stereoscopic photographs of Charles Babbage (1792–1871); a single camera was used to take photographs from different positions because it was difficult to find two cameras that were optically equivalent. Collen (1854) described it thus:

> In 1841, when I was one of the very few who undertook to make use of Mr. Talbot's process, Mr. Wheatstone not only had the idea of making photographic portraits for the stereoscope, but at his request, and under his direction, in August of that year, I made a pair of stereoscopic portraits of Mr. Babbage, in whose possession they still remain; and if I remember rightly, Mr. Wheatstone has previously obtained some daguerreotype portraits from Mr. Beard for the stereoscope.[11]

The reason for Collen's letter to the *Journal of the Photographic Society* was that Claudet (1853) had claimed to make the first photograph for the stereoscope. Wheatstone did not express a preference for metal-plated daguerreotypes or paper printed calotypes (or talbotypes as they were also called).

These paired photographs were taken with a single camera moved laterally between the two exposures. This was to change after Brewster announced his binocular camera and stereoscope at the 1849 meeting of the British Association for the Advancement of Science. The camera (shown in Chap. 4) took two photographs at the same time and the dimensions of Brewster's stereoscope made it more convenient for mounting stereoscopic photographs and it was this version that was to prove more popular than mirror stereoscopes.

An early model of Brewster's lenticular stereoscope

[11] Collen (1854), p. 200.

Random Dot Patterns

One stereopair presented in Wheatstone's original article in 1838 was a line drawing of a building (see above). He was reluctant to use such stimuli because of the pictorial or perspectival depth implied by them. Thereafter many attempts were made to reduce the pictorial depth in stereopairs. In addition, stereoscopic techniques were used to conceal and then reveal images. These came from an unlikely source—the microanatomist Cajal in the 1870s. His interests in photography led him to use stereophotography as a technique for transmitting secret messages.

Cajal and a diagram of his stereoscopic method of concealing and revealing messages

The technique Cajal devised is not unlike the principle of random dot stereograms developed in the twentieth century. Cajal (1901) described it thus:

> During my stereoscopic honeymoon, that is to say, long ago between the years 70 and 72, I was absorbed in imagining new fancies and recreations of this genre. My aim was to achieve a mysterious writing, which could only be deciphered with the stereoscope and usable for those people who don't want to divulge their own matters.... The game consists of making a proof [a print on glass] only with dots, lines and scribbles, or also of letters, crossed and entangles in a thousand ways. A proof in which, with the naked eye, you cannot read anything at all. And, nevertheless, as soon as you see the double image of this background in the stereoscope, a perfect legible sentence or text suddenly appears, standing out on the foreground and clearly detaching itself from the chaos of the lines or dots.[12]

[12] Bergua and Skrandies (2000), p. 71.

Cajal's technique can be illustrated with two patterns of red and cyan dots one of which carries his portrait and the other is a random array

Handmade stereoscopic dot patterns were produced by Herbert Mobbs in 1919, by Boris Kompaneysky in 1939, and more complex versions were made by Claus Aschenbrenner in 1954.[13] The principles can be described with hand drawn random patterns. Viewing the pattern below through one eye/filter and then the other will yield a flat array of black dots. This will change when both eyes are used. If the red filter is in front of the left eye and the cyan in front of the right then gradually a shape will emerge in the centre, in front of the plane of the paper. There are marked

[13] See Blundell (2011), Howard and Rogers (2012), Wade (2021).

individual differences in the time it takes for the depth to be seen. Typically a vague region initially emerges in depth and it gradually articulates to a well-defined circle. Some people see the depth immediately whereas others require several seconds or even minutes to see the depth. The stereoscopic depth can be reversed by the simple expedient of turning the colour filters round so that the red is in front of the right eye. This results in the large central circle appearing to be behind the plane of the paper, as if viewing it through a circular window. Random dot stereograms provide a very good test of stereoscopic depth perception because the planes only articulate in depth with binocular combination. Once the circle is seen in depth then it will appear to move sideways with left-to-right head movements. If the circle appears farther than the surround then it will be seen as moving in the opposite direction to the head; if it appears closer then it will seem to move in the same direction.

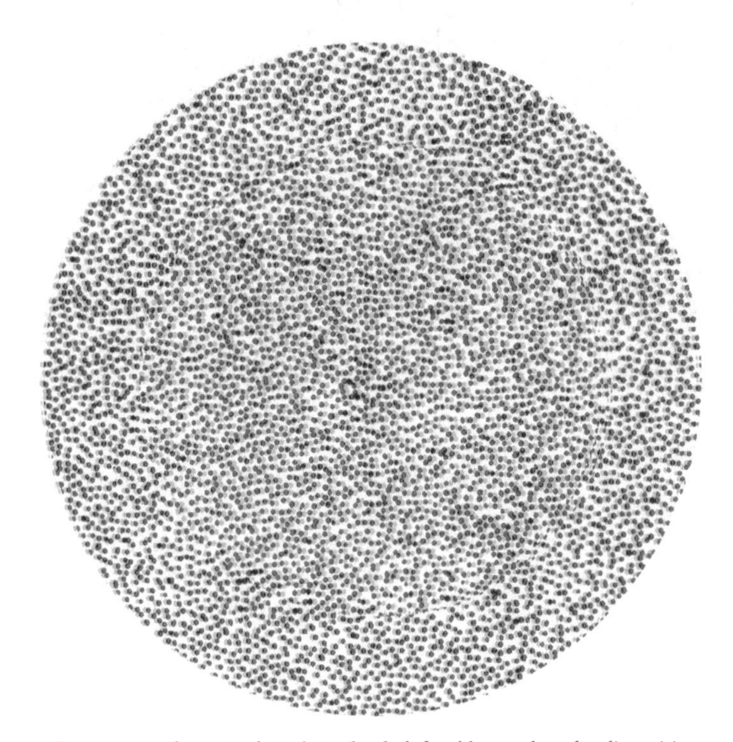

Stereogram of a central circle in depth defined by random dot disparities

A breakthrough was made around 1960 when Béla Julesz (1928–2003) enlisted the power of the computer to produce random dot stereograms.[14] Julesz wished to study stereoscopic depth perception without any knowledge of the objects that were to appear in depth. Wheatstone was well aware of the fact that object recognition could influence the depth perceived but he did not have any means of removing object properties from the stereopairs. With the advent of computer generated images, Julesz

[14] Julesz (1960).

realised Wheatstone's dream—he made random dots stereograms in which there was nothing presented to either eye alone that could indicate the depth to be seen with two eyes. Only with their combination could the depth emerge in what he called cyclopean vision. Julesz published many more complicated patterns as random dot stereograms in his book *Foundations of cyclopean perception.*[15]

Two of Julesz's colleagues, Thomas Papathomas and Kazunori Morikawa at Rutgers University, took a stereoscopic portrait of him.[16] It formed the basis for the two images shown below. The one on the right combines the stereo-portrait with a random dot pattern.

Béla Julesz in depth

The head of Julesz can be seen in silhouette in the upper half of the image below if it is viewed with red/cyan glasses, otherwise the head cannot be discerned. The silhouette appears either in front of the background or behind it dependent on the arrangement of the colour filters before the eyes. In the lower half the silhouetted area is filled with a portrait of Julesz that is itself in depth!

[15] Julesz (1971).

[16] The stereoscopic photograph can be seen in Papathomas et al. (2019).

Béla Julesz in depth in depth

Wallpaper Illusion

Another binocular depth phenomenon was described before Wheatstone invented his stereoscope but it is not dependent on retinal disparity. It was initially described early in the nineteenth century not in wallpaper but in the fluted marble of a chimneypiece.[17] With under-convergence so that adjacent elements were fused the fluting appeared to be further away and magnified relative to fixating on the same elements. However, its significance was not appreciated until after the invention of the stereoscope. In 1844 Brewster rediscovered the illusion when observing a repetitive pattern of flowers printed on wallpaper.[18] It was from such patterns that were frequently printed on wallpaper that the phenomenon derived its name. Brewster also illustrated the principles upon which the illusion depends, and his portrait is combined with it below. With under-convergence (combining an element on the left with the left eye with an adjacent one on the right with the right eye) the surface appears more distant and larger. With over-convergence the reverse occurs.

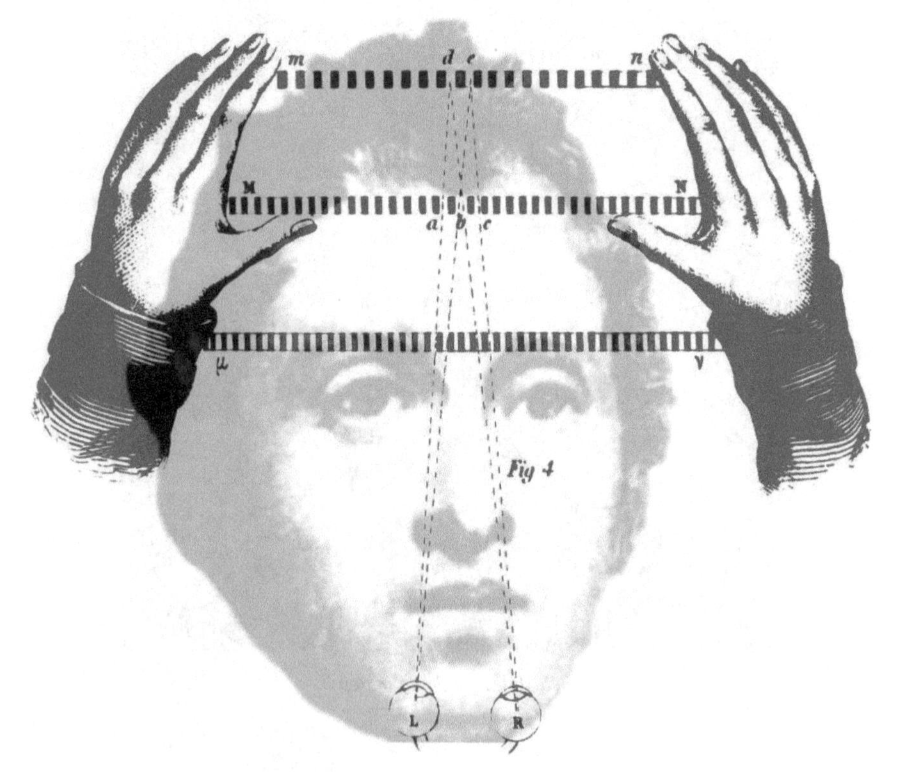

Brewster and his illustration of viewing a repetitive pattern and the planes in which it will appear with under- and over-convergence

[17] Blagden (1813).

[18] Brewster (1844a, b).

The illusion can also be seen in the pattern below that is made up of repetitive portraits of Brewster. When we fixate on the same element with both eyes then the pattern appears to lie in the plane of the page. However, if we combine adjacent identical images and maintain them (with the same convergence of the eyes) then the pattern appears to hover above the page or be seen through it. The depth at which the pattern is seen corresponds to the plane at which the eyes converge: the farther apart the combined elements are the greater the apparent depth. When we fixate on the same element with both eyes then the pattern appears to lie in the plane of the page.

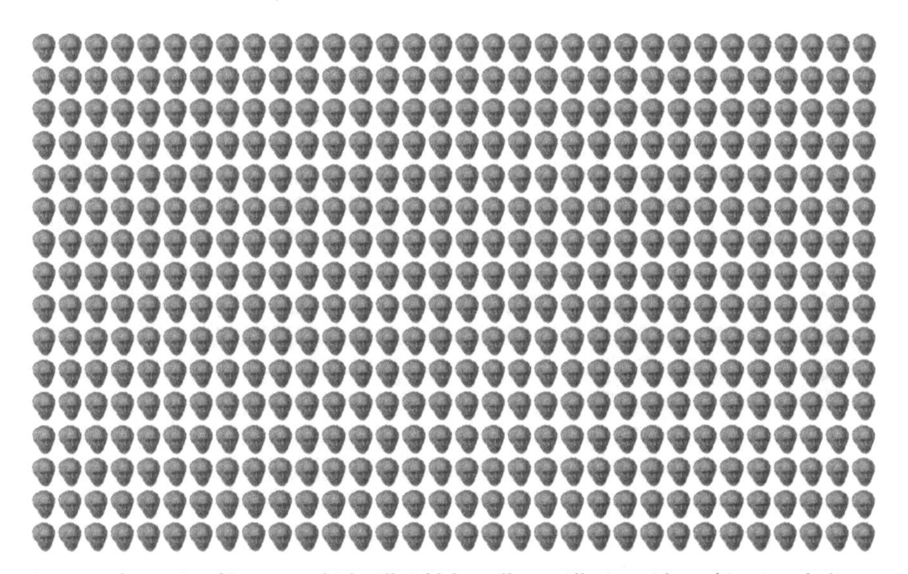

An array of portraits of Brewster which will yield the wallpaper illusion with combination of adjacent images by under- or over-convergence

If the pattern to the left eye is to the right of that in the right eye then the whole pattern will seem to float above the page and vice versa. If slight variations in the locations of the repetitions along rows are introduced then more complex depth planes are visible and aspects of disparity processing become involved.

Autostereograms

Slight variations in the locations of the repetitions along rows of portraits can be introduced (as in the pattern below) then more complex depth planes are visible and aspects of disparity processing become involved. The surface no longer looks flat but stepped in depth from top to bottom.

An array of portraits of Brewster with rows varying systematically in size and separation. Combining adjacent pairs by over- or under-convergence will lead to the appearance of parallel, horizontal humps and hollows

The depth effects can be seen without using red/cyan glasses. When variations in depth are apparent in patterns like this they are called autostereograms.[19] They are typically computer generated patterns that contain many more subtle depth effects and they have also been called 'magic eye' patterns; they were phenomenally popular in the 1990s when many books containing them were published.[20]

[19] See Howard and Rogers (2012), Tyler and Clark (1990).

[20] Baccei and Smith (1993).

Presenting regular and repetitive dot patterns that enable fusion of neighbouring pairs provides the basis of the wallpaper illusion. When equivalent but laterally separated patterns are combined binocularly they seem suspended in the plane of convergence. If slight variations in the locations of the repetitions along rows are introduced then more complex depth planes are visible and aspects of disparity processing become involved. Wallpaper illusions and autostereograms can be seen without the aid of any viewing devise as they only involve converging the eyes to combine neighbouring elements or viewing them with parallel visual axes. More systematic manipulations of repetitions and disparities in autostereograms are complex configurations because they can hide articulated patterns rather than simply depth planes. In the example below, with over- or under-convergence relative to the plane of the paper features of the pattern will gradually articulate to reveal surfaces at different depths; in this case it is a heart shape. Sometimes it requires quite a while for the full depth to be visible.

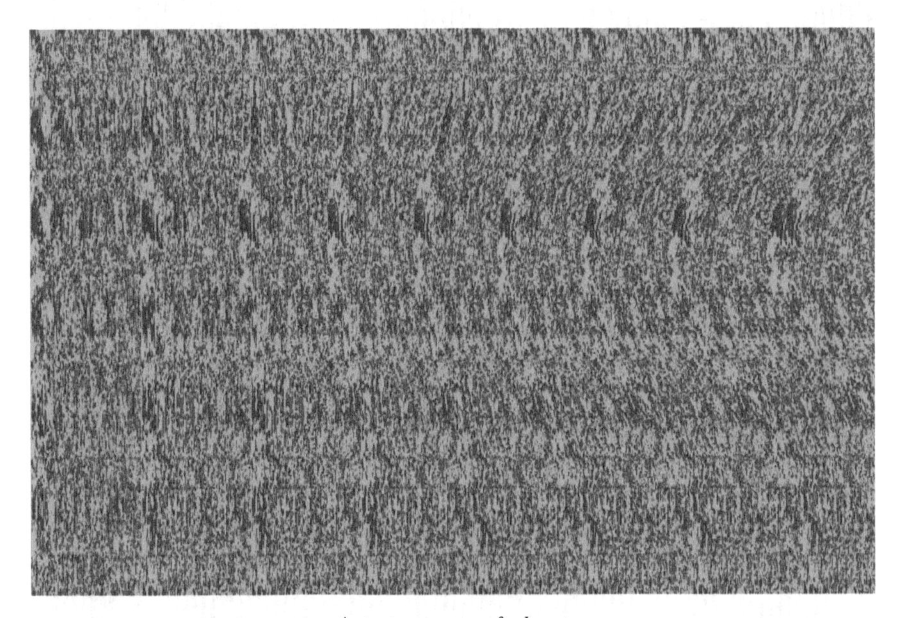

Autostereogram of a heart

As is the case with random dot stereograms, some people find it very difficult to control their over- or under-convergence and so autostereograms can be very frustrating to view, particularly when others can see the effects without any difficulty.

Another stereoscopic technique involving repetitive patterns was introduced in an artistic context by Ludwig Wilding (1927–2010). It bears some resemblance to Cajal's use of transparent surfaces and is based on disparities between moiré fringes generated by the interference of regular repetitive patterns (like gratings) separated slightly in depth. Moreover, the apparent stereoscopic space varies with the viewing distance of the observer (because disparity between the moiré fringes varies with viewing distance). The depth can be produced from curved as well as flat surfaces, and opposite directions of depth are often incorporated in the same work. The relationship between the spatial frequencies of the transparent and printed patterns determines the direction and amount of the depth seen, and they can be given precise mathematical descriptions.[21] Depth can be seen as a consequence of the disparities of the moiré fringes when the head is stationary, and it can be augmented by lateral head movements which yield motion parallax between the moiré patterns.

Ludwig Wilding in combination with a vertical grating

[21] Wilding (2007). See also Wade (2007).

Da Vinci Stereopsis

Da Vinci stereopsis was mentioned in Chap. 2 and a great deal of research has been applied to it in recent years.[22] It refers to the depth experienced with differential occlusion of an object so that one eye sees part of the background not visible to the other. Da Vinci stereopsis was so named in 1990 but there had been experimental research on it long before that, particularly by Adolf von Szily (1848–1920), professor of ophthalmology at Budapest. His research was not restricted to ophthalmology and he made several important contributions to visual science. He studied fluttering hearts, introduced a novel technique for examining stereoscopic depth perception, and conducted experiments on motion aftereffects. He also examined myopia, colour blindness, visual acuity using gratings.

[22] See Tsirlin et al. (2012).

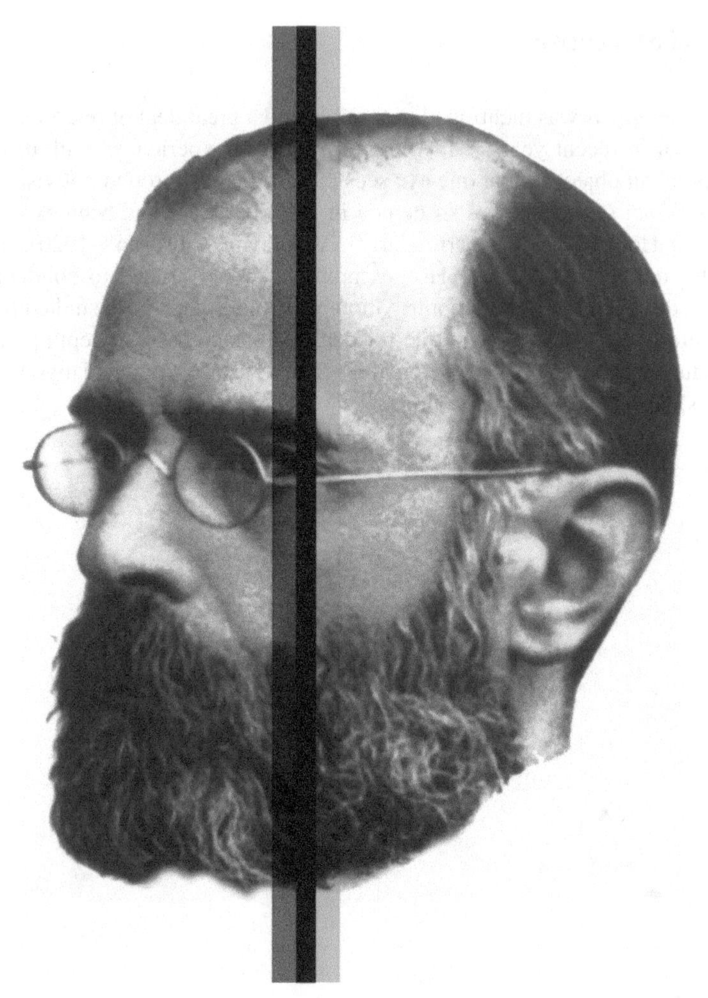

Adolf von Szily behind a vertical band

Von Szily found his inspiration in binocular vision from Panum and Hering, and introduced a novel dimension to the phenomenon—the use of silhouettes. In 1894 von Szily presented a set of figures to a conference in Vienna illustrating how depth can be seen in silhouettes, two of which are shown below. The investigation was reported only as an abstract in the following year. The details of his experiments were not published until after his death and it has been translated into English.[23] Von Szily drew attention not only to the depth seen in the figures but also to the subjective contours that appeared in them.

[23] von Szily (1895), Ehrenstein and Gillam (1998).

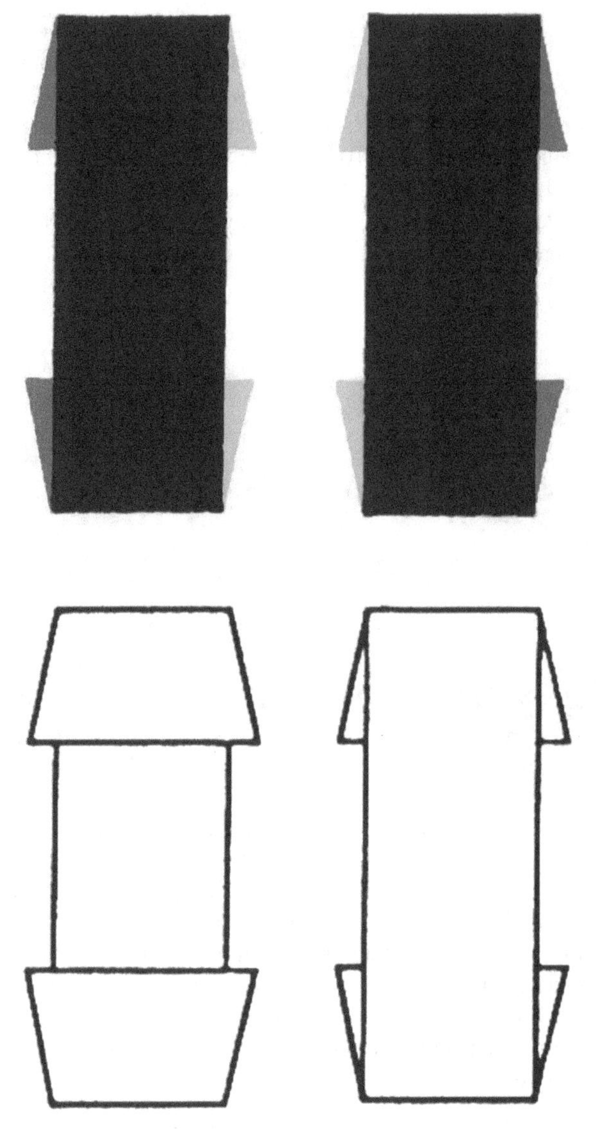

Two of von Szily's stereoscopic silhouettes and outlines of how they appear

Dichoptic Vision

An important aspect of research on binocular vision that was introduced after the invention of the stereoscope involved presenting component of patterns to different eyes. This idea was demonstrated early in the development of stereoscopic investigations. In 1858 Panum presented the images below in which components of a scene were present in separate eyes. His stated desire was to present: "Instead of composite images… I examined the elements from which the image was made".[24] Two of his figures are shown below. In the left one there are no common contours for each eye and so the two elements are unstable. In the right one Panum provided square frames to each eye so that the eyes remained aligned. He produced a range of other stereoscopic images in which elements of more complex patterns were presented to each eye.

Two anaglyphs combined from separate elements illustrated in Panum (1858)

Hering gave a more graphic description of dichoptic viewing relating it to the directions in which objects are seen.[25] One description that is frequently illustrated is that of viewing objects through a window: an object on the right is viewed by the left eye and a mark is made on the window, this mark is then aligned with an object on the left viewed by the right eye alone. When both eyes view the mark the two objects appear straight ahead as if aligned with the cyclopean eye (which he referred to as the imaginary eye). He wrote:

> Let the observer stand about half a metre from a window which affords a view of outdoors, hold his head very steady, close the right eye, and direct the left to an object located somewhat to the right. Let us suppose it is a tree which is well set off from its surroundings. While fixing the tree with the left eye a black mark is made on the window pane at a spot in line with the tree. Now the left eye is closed and the right opened and directed at the spot on the window, and beyond that to some object in line with it, for example, a chimney. Then with both eyes open and directed at the spot, this latter will appear to cover parts of the tree and chimney. Both will be seen simultaneously, now the tree more distinctly, now the chimney,

[24] Panum (1858), p. 1.

[25] Hering (1879).

and sometimes both equally well, according to which eye's image is victor in the conflict. One sees therefore, the spot on the pane, the tree and the chimney in the same direction.[26]

Hering did not illustrate this situation but many have done subsequently; Hering and his illustration is shown below.

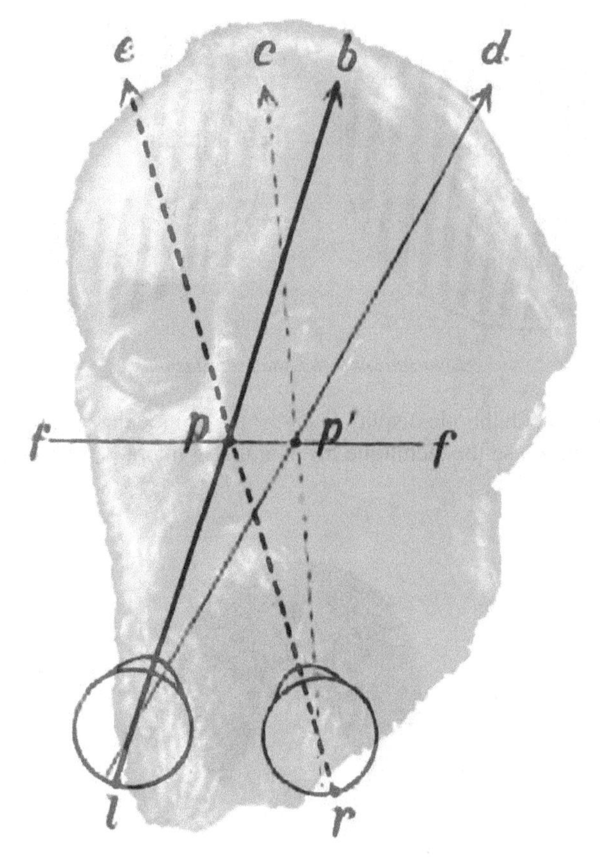

Hering and his illustration of viewing aligned objects through a window; "Let the line ff represent the window pane; p the fixed black spot; on the left, visual line lb is the distant tree; on the right, re the distant chimney"

Certain visual illusions can be presented binocularly or dichoptically. That is, both eyes are presented with the complete figure or different eyes receive the components. The tilt illusion can be seen in the left pattern below. The central lines appear tilted clockwise even though they are physically vertical. The question is, can it be seen in the pattern on the right in which the inducing tilted lines are presented to one eye and the central vertical lines to the other? If the illusion is still visible, is its magnitude the

[26] Hering (1942), p. 38.

same as in the left figure? Experiments have shown that the illusion in the dichoptic condition is about 50% of that measured with both components in the same eye.[27]

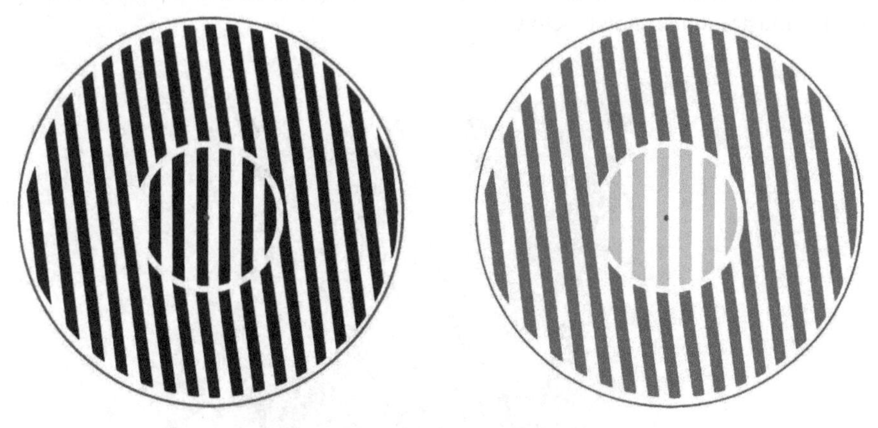

Binocular and dichoptic tilt illusions

If the lines in the dichoptic display intersect then binocular rivalry will result and so it is important to use the technique for examining interactions that do not induce rivalry.

Dichoptic impossible illusory contours

[27] Wade (1980).

Colour Stereoscopy or Chromostereopsis

When long- and short-wave colours (like red and blue) are placed near to one another they can appear to occupy different depths. Usually red seems closer than blue but some people experience the opposite. It is called colour stereoscopy or chromostereopsis and it has a long history in science and art. In his book on the theory of colours Johann Wolfgang Goethe (1749–1832) wrote extensively about colour contrasts. He also made passing reference to differences in the apparent depth of colours: "In looking steadfastly at a perfectly yellow–red surface, the colour seems actually to penetrate the organ… a blue surface seems to retire from us… we love to contemplate blue, not because it advances to us, but because it draws us after it".[28] However, Goethe was referring to viewing large surfaces of a single colour rather than comparing the relative apparent distances of juxtaposed or adjacent colours which are the situations required for chromostereopsis. Goethe is shown below in red and blue on black and white backgrounds. The phenomenon is frequently presented with the colours on a black background and it has been found that the relative depths can reverse when a white background is used.[29] The red and blue portraits of Goethe are the same in the upper and lower illustration below but the recession of the blue one appears greater with the black background. The contrasts between the colours and their backgrounds reverse in these two cases: blue has less contrast relative to black than to white.

[28] Goethe (1810), pp. 294–295/(1840), pp. 310–311.

[29] Vos (2008).

Chromostereoscopic Goethe

A clear description of the association of colour and stereoscopic vision was provided by Brewster who had a better understanding of optics than Goethe.[30] Brewster delivered four papers to the Mathematics and Physics Section of the British Association for the Advancement of Science at the 1848 meeting held at Swansea. One had the title 'On the vision of distance as given by colour' in which he discussed the differences between the apparent distances of adjacent red and blue lines or surfaces:

> When the boundary lines on a map are marked with two lines of different colours, the one rises above or is depressed below the other, and the two lines appear to be placed at different distances from the eye. This remarkable effect is most clearly seen when we look with both eyes through a large reading-glass, spectacles being used along with it by those who require them. The more the two lines differ in refrangibility, the greater, and consequently the more distinctly seen, is the difference of distance at which the lines appear to be placed. The effect is finely seen in the coloured patterns of red and blue paper which Prof. Wheatstone has had executed on paper for exhibiting the mobility or shaking of one part of the pattern. The

[30] Brewster (1849, 1852).

> difference of distance of the colour lines or spaces may be appreciated even with one eye. The explanation of this phaenomenon is very simple. In binocular vision the convergency of the optic axis to different points at different distances corresponding to the different points in the eye, to which the differently coloured rays are refracted, gives us the vision of a different distance from each coloured line, in the same manner as it is given in the stereoscope. In monocular vision the distance is given by an analogous process to that by which the single eye sees distances.[31]

The reference to Wheatstone concerned the 'fluttering heart' phenomenon which will be mentioned later in the chapter. Brewster interpreted the colour-depth difference with the same concepts he adopted for stereoscopic depth perception. Apparent depth was determined by the point at which the two visual axes intersected; in the case of chromostereopsis the convergence of the visual axes for long and short wavelengths of light differed which he argued accounted for the difference in perceived depth. In 1852 Brewster amplified his account a little and referred to the observation of different colours though a large lens as a chromatic stereoscope.

Brewster's colour distances

Chromostereopsis was studied in more detail by Frans Cornelius Donders (1818–1889) and by Ernst Wilhelm Brücke (1819–1892); in 1868 they both related the phenomenon to aspects of chromatic aberration in the eyes.[32] Different wavelengths of light are not focussed equally; typically, the focal plane for red light lies somewhat further from the lens than that for blue light, so that if the eye accommodates to bring one wavelength into focus, others are blurred. This difference in focussing power for different wavelengths is called chromatic aberration, and it is quite marked in the human eye. According to Einthoven, Donders noticed the effect when staying in a hotel in Leipzig; the wallpaper in the room was patterned with a network of

---

[31] Brewster (1849), p. 48.

[32] Donders (1868), Brücke (1868).

fine, bright yellow lines on a dark blue ground.[33] When viewed from about 3 m the yellow grid seemed to lie in front of the wall and it appeared to move with sideways movements of the head. The illusion was so compelling that Donders involuntarily approached the wall upon which the wallpaper appeared flat again. Linking the illusion to accommodation, Donders made large red and blue letters on a black background and noted that the red letters appeared nearer than the blue. He was, however, puzzled that some people saw the opposite of his perception—the blue letters appeared in front of the red ones.

Brücke commenced with the statement that it was common knowledge to artists that some colours seem to approach and others recede. He also remarked that he and Donders had come to the same conclusions about the basis of the phenomenon independently of one another. The stimulus Brücke examined consisted of thin red and blue or green strips of paper on a matte black background with the strips either vertical or horizontal. He found that the red strips appeared closer to him but other observers reported the opposite. He argued that chromostereopsis was a binocular phenomenon because in each eye the location of the fovea is temporal relative to the optical axis of the eye.

[33] Einthoven (1885), p. 212.

Chromostereoptic Brücke and Donders

Willem Einthoven (1860–1927) was a student of Donders and was encouraged to pursue research on chromostereopsis; he published the results of his experiments in 1885. As would be expected, there were many references to the work of Donders but those to Brücke were not to the 1868 paper but to a later textbook. Nonetheless the interpretations of Brücke and Einthoven were very similar: chromostereopsis is a consequence of magnification due to chromatic aberration in the eye and the difference between the visual and optical axes of the eyes; individuals with temporally eccentric pupils see red in front of blue, while with nasally eccentric pupils the relief is reversed. This has been referred to as the traditional theory of chromostereopsis.[34] Einthoven included a figure like that described by Donders and one that has been copied frequently—the word ROTH (red) in red above the word BLAU (blue) in blue both on the same black background. A variant of this is shown below and Einthoven is depicted in the background. Using the coloured words he found that the apparent depth effect (with red closer than blue) disappeared when only one eye was used.

[34] Thompson et al. (1993).

Willem Einthoven and coloured letters

Artists have expressed the phenomenon either explicitly or implicitly for centuries. It can be seen in some stained glass windows. Goethe also drew attention to the visual effects that occur with juxtaposition of red and blue in fabrics: "it may be safely assumed, that a carpet of a perfectly pure deep blue-red would be intolerable".[35] Similar observations were made by Wheatstone and Brewster at a meeting of the British Association for the Advancement of Science held at York in 1844. Wheatstone described a "singular effect of the juxtaposition of certain colours", following which Brewster delivered a paper with the title "On the same subject".[36] It would seem that the phenomenon had been known informally for some time as both Wheatstone and Brewster had taken independent accounts of it to the York meeting. Brewster remarked that it "has been called by some *fluttering hearts,* from one of the colours having the shape of hearts". It was called 'fluttering' because the borders between the two colours are perceptually unstable.[37] Wheatstone observed it in a carpet with a dominant red and green design—the coloured parts appeared to be moving. He found that the effect occurred best with red and green, and that dim lighting was necessary for its observation. According to Helmholtz the effect was most pronounced with juxtaposed, saturated reds and blues; if a card bearing red and

---

[35] Goethe (1840), p. 313.

[36] Wheatstone (1844), Brewster (1844a, b).

[37] Wade (1978).

blue figures was moved to and fro "the figures themselves seem to shift their positions with respect to the paper, and dance about on it". Presumably with stationary figures the movement is provided by the eyes. The effect has been exploited on a large scale by Ellsworth Kelly and more recently in some of Vasarely's coloured paintings. However, it is Marcel Duchamp's delightful visual pun, called *Fluttering Hearts*, which provides the best example: it consists of a blue heart surrounded by a red one, with this sequence repeated. Wheatstone considered that the effect was due to "the eye retaining its sensibility for various colours during varying lengths of time", and Helmholtz adopted essentially the same interpretation.

Fluttering hearts of Duchamp

The increasing use of computers for producing and presenting images has concentrated attention on the characteristics of colour contrasts and combinations. Many references are made to the depth and motion in colour displays but there is a widespread confusion between chromostereopsis, binocular lustre and the fluttering hearts phenomenon. The colour combinations involved in them are usually red and blue but only chromostereopsis and lustre are binocular phenomena. Designs displaying binocular lustre typically employ red and cyan. The differences between the three phenomena are illustrated below with a silhouette profile of Goethe.

Upper, chromostereopsis; the red and blue silhouettes appear at different depth with respect to the black background. Centre, binocular lustre; with the red/cyan glasses the positive and negative portraits and backgrounds have a metallic sheen. Lower, fluttering hearts with the blue/red boundaries appearing unstable particularly if moved

References

Baccei T, Smith C (1993) Magic eye I. A new way of looking at the world. N E Thing Enterprises. Andrews and McMeel, Kansas City

Bergua A, Skrandies W (2000) An early antecedent to modern random dot stereograms—"The secret stereoscopic writing" of Ramón y Cajal. Int J Psychophysiol 36:69–72

Blagden C (1813) An appendix to Mr. Ware's paper on vision. Philos Trans R Soc 103:110–113

Blundell BG (2011) 3D displays and spatial interaction, vol 1. From perception to technology. Walker & Wood, Auckland, NZ

Brewster D (1844a) On the same subject. Report of the British Association for the Advancement of Science. Transactions of the Sections p 10

Brewster D (1844b) On the knowledge of distance given by binocular vision. Trans R Soc Edinb 15:663–674

Brewster D (1849) On the vision of distance as given by colour. Report of the British Association for the Advancement of Science 48

Brewster D (1852) Notice of a chromatic stereoscope. London Edinb Philos Mag J Sci 3:31

Brooks KR (2017) Depth perception and the history of three-dimensional art: who produced the first stereoscopic images? i-Perception 8(1):1–22

Brücke E (1868) Ueber asymmetrische Strahlenbrechung im menschlichen Auge. Sitzungs Berichte der kaiserlichen Akademie der Wissenschaften. Mathematische und Natur Wissenschaftliche Klasse II 58:321–328

Collen H (1854) Earliest stereoscopic portraits. J Photographic Soc 1:200

Donders FC (1868) Invloed der accommodatie op de voorstelling van afstand. Wetenschappelijke bijlage bij het jaarverslag van het Nederlands Gasthuis voor Ooglijders, pp 111–113

Ehrenstein WH, Gillam BJ (1998) Early demonstrations of subjective contours, amodal completion, and depth from half-occlusions: "Stereoscopic experiments with silhouettes" by Adolf von Szily (1921). Perception 27:1407–1416

Einthoven W (1885) Stereoscopie Durch Farbendifferenz. Archiv für Ophthalmologie 19:211–238

Goethe JW (1810) Zur Farbenlehre. Cotta, Tübingen

Goethe JW (1840) Goethe's theory of colours (trans: Eastlake CL). John Murray, London

Hering E (1879) Der Raumsinn und die Bewegungen des Auges. In: Hermann L (ed) Handbuch der Physiologie, vol 3. Vogel, Leipzig, pp 341–601

Hering E (1942) Spatial sense and movements of the eye (trans: Radde A). American Academy of Optometry, Baltimore

Howard IP, Rogers (2012) Perceiving in depth, vol 2, Stereoscopic Vision. Oxford University Press, USA

Julesz B (1960) Binocular depth perception of computer generated patterns. Bell Syst Tech J 39:1125–1162

Julesz B (1971) Foundations of cyclopean perception. University of Chicago Press, Chicago

Klooswijk AIJ (1991) The first stereo photograph. Stereo World May/June, pp 6–11

Ono H, Comerford T (1977) Stereoscopic depth constancy. In: Epstein W (ed) Stability and constancy in visual perception: mechanism and processes. Wiley, New York, pp 91–128

Panum PL (1858) Physiologische Untersuchungen über das Sehen mit zwei Augen. Schwerssche Buchhandlung, Kiel

Papathomas TV, Morikawa K, Wade N (2019) Bela Julesz in depth. Vision 3:8. https://doi.org/10.3390/vision3020018

Rogers B (2017) Perception. A very short introduction. Oxford University Press, Oxford

Rogers B, Graham M (1979) Motion parallax as an independent cue for depth perception. Perception 8:125–134

Thompson P, May K, Stone R (1993) Chromostereopsis: a multicomponent depth effect? Displays 14:227–233

Tsirlin I, Wilcox LM, Allison RS (2012) Da Vinci decoded: does da Vinci stereopsis rely on disparity? J Vision 12(12):2. https://doi.org/10.1167/12.12.2

Tyler CW, Clark MB (1990) The autostereogram. Stereoscopic displays and application. Proc SPIE 1258:182–196

Volpicelli P (1876) Scientific worthies. Nature 13:501–503

von Szily A (1895) Demonstrationen. Verhandlungen der Gesellschaft Deutscher Naturforscher und Ärtze. Vogel, Leipzig, pp 227–228

Vos JJ (2008) Depth in colour, a history of a chapter in physiologie optique amusante. Clin Exp Optom 91:139–147

Wade NJ (1978) Op art and visual perception. Perception 7:21–46

Wade NJ (1980) The influence of colour and contour rivalry on the magnitude of the tilt illusion. Vis Res 20:229–233

Wade NJ (2007) The stereoscopic art of Ludwig Wilding. Perception 36:479–482

Wade NJ (2021) On stereoscopic art. i-Perception 12(3):1–17. https://doi.org/10.1177/20416695211007146

Wheatstone C (1838) Contributions to the physiology of vision—part the first. On some remarkable, and hitherto unobserved, phenomena of binocular vision. Philos Trans R Soc 128:371–394

Wheatstone C (1844) On a singular effect of the juxtaposition of certain colours under particular circumstances. Report of the British Association for the Advancement of Science Transactions of the Sections p 10

Wheatstone C (1852) Contributions to the physiology of vision—part the second. On some remarkable, and hitherto unobserved, phenomena of binocular vision. Philos Trans R Soc 142:1–17

Wilding L (2007) Visuelle Phänomene. Museum für Konkrete Kunst, Ingolstadt

Chapter 6
Binocular Rivalry

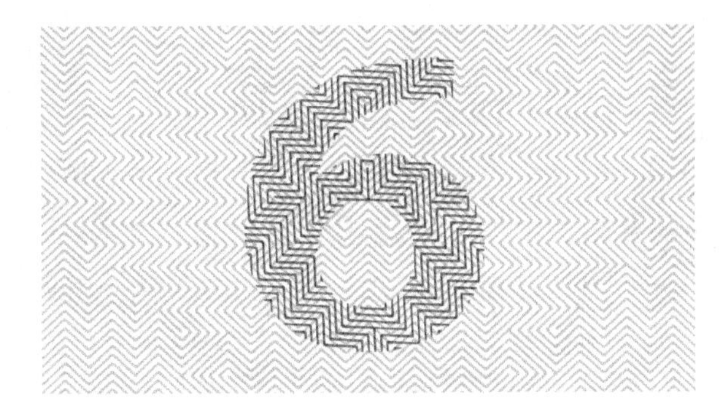

Binocular rivalry occurs when markedly different colours or contours are presented to different eyes. It has a longer descriptive past than does stereoscopic vision but a shorter experimental history. An early concern was whether different colours presented to each eye would combine in the way that they did when presented to a single eye. Most observers found that the colours engaged in competition rather than cooperation, although the outcome often depended on the theoretical position they held. Contour rivalry is much more compelling than colour rivalry; its investigation was greatly facilitated by the stereoscope and in 1838 Wheatstone examined it using different letters. Twenty years later Panum introduced orthogonal gratings as stimuli and these continue to be employed in experiments. He described the dominance of either eye as well as the complex composites made up of local regions of dominance and suppression drawing attention to the importance of the patterns engaging in rivalry as well as the eyes to which they were presented. This contrast between eye and pattern dominance has fueled much contemporary research. Binocular lustre occurs when a positive image is presented to one eye and its negative to the other: surfaces appear to have a metallic sheen. Lustre was described about a decade after the announcement of the stereoscope as well as photography and negative and positive photographic prints were ideal for demonstrating it. Quantitative measures of rivalry durations and alternation rates were recorded by Breese at the end of the nineteenth century and he also examined monocular rivalry. More sophisticated stimuli, measures and analyses were introduced after a resurgence of interest in binocular rivalry after the 1960s. Physiological processes in the visual brain were related to rivalry and it implications in studies of consciousness were explored. The coexistence of rivalry and stereopsis is a matter of theoretical as well as observational importance: can a suppressed stimulus contribute to the experience of depth? The chapter concludes with examples of stereoscopic depth in patterns displaying contour rivalry as well as binocular lustre.

N. Wade, *Vision and Art with Two Eyes*, Vision, Illusion and Perception 3,
https://doi.org/10.1007/978-3-030-77995-5_6

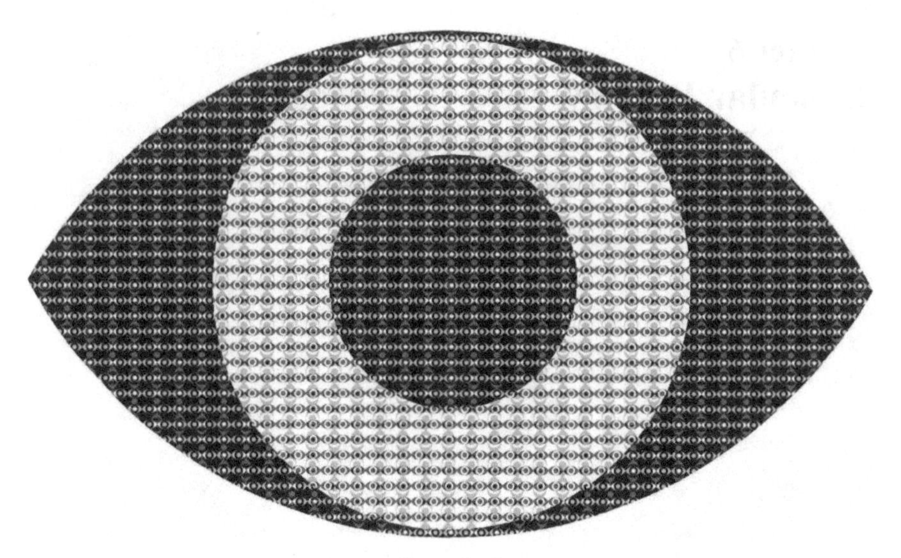

Reyevalry I

Binocular rivalry has a long observational history but its experimental investigation is much more recent. Appreciating the existence of binocular rivalry has its origins in departures from singleness of vision. For example, Aristotle discussed binocular single vision, but it tended to be in the context of its breakdown to double vision (diplopia) either by distorting one eye or in squint or strabismus. A paradox in the history of research on binocularity is that rivalry was described before the involvement of retinal disparity in depth perception was demonstrated. Nor was the stereoscope the first binocular instrument but, as we saw in Chap. 3, earlier ones were enlisted to examine the experience of singleness and rivalry rather than depth. Stereoscopes made the presentation of rivalling stimuli simpler but they did not lead immediately to the quantification of this dynamic and fickle phenomenon. This occurred at the end of the nineteenth century. Thereafter, greater subtlety was involved in the stimuli presented and in the indices of rivalry that could be measured. All was to change again when computers could be enlisted both for stimulus presentation and response measurement.

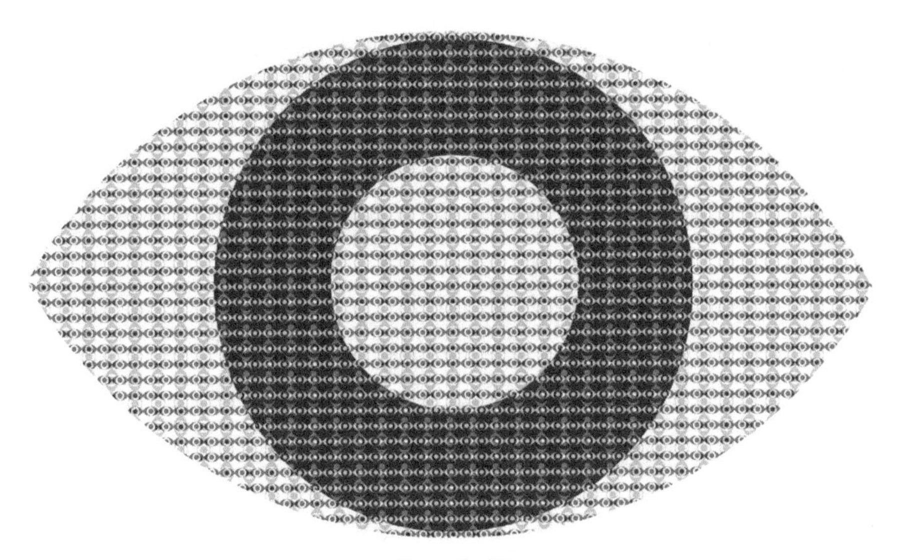

Reyevalry II

Colour Rivalry

The combination of different colours presented to corresponding regions of each retina became an issue of theoretical importance following Isaac Newton's experiments on colour mixing, described in his *Opticks*: are colours combined by either eye as they are when selected from the spectrum? Desaguliers, an advocate of Newtonian optics, was amongst the first to draw attention to the phenomenon of colour rivalry.[1] He applied a method of binocular combination that became widely employed in other studies of binocular vision, namely, placing an aperture in such a position that two adjacent objects were in the optical axes of each eye. In particular, he showed that dichoptically presented coloured lights, or patches of colour, rival with one another rather than combine as in Newton's experiments on colour mixing. The method devised by Desaguliers was applied by Taylor, who added the refinement of placing coloured glasses in front of candle flames; he found that colours combined rather than engaged in rivalry. He wrote:

> ... one takes a piece of blue glass in front the candle E and a piece of red glass in front of the candle D with the intention of distinguishing them one from the other, without changing the quality of the respective images; if then the eyes approach the position that is necessary to look at a distant object, directing the axes of the eyes in the line Gg and Ff one can see a blue candle with the left eye B, and a red candle with the right eye C; and if one looks through the aperture with attention, directing the optic axes in the line dDeE, one sees the blue and

[1] Desaguliers (1719).

red candles together in the aperture E, where they will have the appearance of a candle of the colour purple.[2]

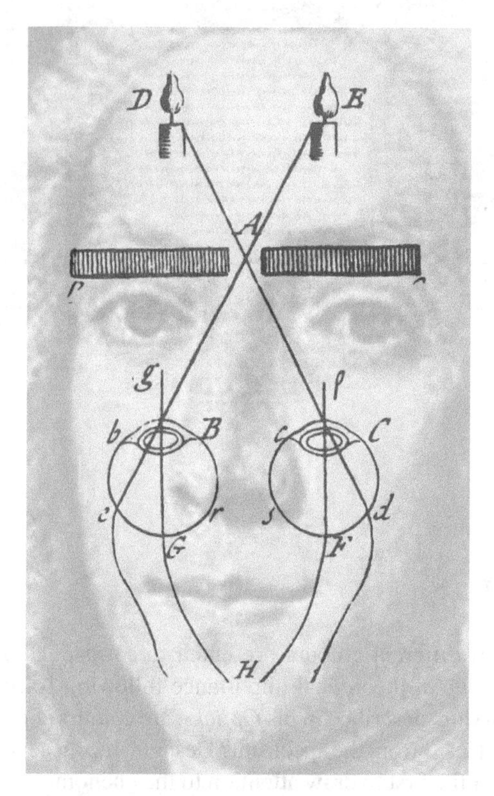

Taylor and his arrangement for observing binocular colour rivalry

Du Tour provided a clear description of binocular colour rivalry.[3] He achieved dichoptic combination by another means: he placed a board between his eyes and attached blue and yellow fabric in equivalent positions on each side, or the fabric was placed in front of the fixation point. When he converged his eyes to look at the patches they did not mix but alternated in colour. That is, his visual awareness fluctuated between the colours presented to the two eyes. Du Tour argued that with slight differences in curvature between the eyes, rather than achieving complete fusion, one of the presented colours (or objects) projected more clearly to one eye at a given time than the other colour to the second eye. During simultaneous presentation, the two different impressions on the retinas would meet in the brain where the optic nerves came together. He proposed that the more distinct impression then affected the mind and became visible, whereas the other less clearly defined impression would

---

[2] Taylor (1738), pp 170–171.

[3] DuTour (1760).

not have such an effect or did not draw the mind's attention. He also argued, in keeping with the suppression theory, that during simultaneous presentation of two different colours, the eyes acted in alternation to see one and then the other image. Such alternate and intermittent action of the two eyes was to avoid fatigue in both simultaneously, through inhibiting one eye while the other eye viewed the presented object.

Rivalling portraits of Du Tour

Du Tour also applied the method of observing the colours through an aperture, as adopted by Desaguliers, and obtained the same outcome. A similar technique was used by Giovanni Battista Venturi (1746–1822), who compared the combination of sounds to two ears with that of colours presented to different eyes. He placed blue and yellow papers next to one another on a table and over-converged his eyes to combine them: "I have repeated this experiment often and with care, and I have never experienced a third colour from the two overlapping colours".[4] This was taken to be evidence that the nerves from the two eyes do not combine in the brain. One of the problems associated with most of the techniques described above is that the observer needed to uncouple accommodation from convergence.

[4] Venturi (1802), p 389.

The complexity of the dynamic appearances during binocular colour rivalry is evident from Johannes Müller's account: sometimes one or the other colour will predominate, whereas at other times "nebulous spots" of one colour are visible on the other.[5] His observation was made without the aid of a stereoscope because it had not been made public at the time Müller was writing this text.

The experiment of looking upon a sheet of white paper through two differently coloured glasses at the same time, may serve as an illustration for the present. The impressions of blue and yellow, for example, are found in such an experiment not to mingle readily; at one moment the blue, at another the yellow is predominant. Sometimes blue nebulous spots are seen upon the yellow field; at other times, yellow spots of varying magnitude upon the blue field: sometimes one colour alone prevails, and has absorbed the other; sometimes the reverse is seen. The appearance of one colour in spots upon a ground of the other colour, shows indeed that the attention can be directed at the same time to one part of one retina, and to the other parts of the other retina.

Müller and translation of text he wrote describing binocular colour rivalry

Müller's conclusion was:

The appearance at one time of one colour in spots upon the ground of the other; the complete disappearance at another time of one colour, simultaneously with its replacement by the other colour; and, lastly, the union of the two colours to form an intermediate tint, though this is observed only occasionally, prove - 1, that at certain moments the two eyes are both in action together; hence the appearance of spots or nebulæ of one colour upon the ground of the other; 2, that at other moments the impression on one retina is entirely, or nearly entirely, obliterated, and that of the other left predominant; and, 3, that at other periods again the impressions on the two eyes combine to produce one sensation. These states, alternating constantly with each other, present the actions of the eyes to us as phenomena of disturbed equilibrium, like the oscillation of a balance. The state of repose or balance of the impressions is attained with difficulty, though it is possible. The equilibrium is disturbed, however, in part by internal influences unknown to us, and in part probably by the mind being directed more especially to the one or the other eye. The phenomena showing this antagonism of the retinæ are, we may mention, quite distinct and vivid in persons whose eyes are perfectly equal in power, as in myself. The appearance of one colour in spots, or as nebulæ, while

[5] Müller (1838/1843).

> all around the other colour prevails, shows moreover the possibility of the different parts of the retina acting unequally; and the phenomenon is, indeed, of the greatest importance, as affording information with respect to the internal conditions of the retina.[6]

Wheatstone dispensed with difficulties of aligning the eyes by viewing different colours with the stereoscope. The outcome was binocular rivalry, but this was either of the whole monocular stimulus or local parts of the two coloured discs:

> If a blue disc be presented to the right eye and a yellow disc to the corresponding part of the left eye, instead of a green disc which would appear if these two colours had mingled before their arrival at a single eye, the mind will perceive the two colours distinctly one or the other alternately predominating either partially or wholly over the disc.[7]

Like Müller, Wheatstone was describing fluctuations in the visibility of the different colours while no changes took place in the colours themselves. The changes in visual awareness corresponded to his inferential theory of vision in which binocularity was considered to be cognitive rather than physiological. Helmholtz followed Wheatstone theoretically and he also embraced colour rivalry as evidence in its favour.[8] On the other hand, Hering argued for a physiological interpretation of rivalry and much of the dispute surrounded the visibility of yellow from dichoptic combinations of red and green.[9] Hering did establish a number of stimulus parameters that influenced rivalry: small colour patches yielded more clear-cut rivalry and brief stimulus presentation favoured combination rather than competition.

---

[6] Müller (1843), p 1209.

[7] Wheatstone (1838), pp 386–387.

[8] Helmholtz (1867).

[9] Hering (1861).

Helmholtz and Hering in rivalry

In the early 1860s, binocular rivalry with colour images was also investigated by Joseph Towne, a medical sculptor at Guy's Hospital, London who had a lasting interest in binocular vision. In particular, he first sought to understand the relationship between corresponding retinal areas during rivalry using complementary stimulus pairs. Colour pairings were introduced to the complementary stimuli. With vertically split discs, the outer semicircle viewed by each eye was the same colour while the inner semicircles were another colour. When red and black were used, the colours combined and alternations were not observed. With blue and yellow however, he observed interocular grouping during binocular rivalry, or stimulus rivalry:

> In the next experiment, the two temporal halves of the retinæ are submitted to yellow, while the two nasal halves are submitted to blue; these colours being disharmonic, it follows that disharmonic colours fall on corresponding parts of the two retinæ, the result being that the colours antagonise, first one, then the other being seen, with now or then an iridescent appearance over the whole field, occasioned by the gleaming of one colour over the surface of the other colour. The disturbance occasioned by this means is sometimes very striking; the four halves of the retinæ appear to be thrown into separate action, so that their respective images are brought together, but not united; and for a second they appear as if struggling, so to speak, for their respective places. Under these circumstances, the changes which occur are remarkable; at one moment there will be a complete disc of yellow, the blue being altogether lost; at the next the yellow will entirely disappear over exactly one half of the disc, so that

> the resultant image is one half yellow, the other half blue, then the reverse, and so on, with constant alternations.[10]

A double portrait of Towne and the coloured stimuli he used to examine binocular rivalry

With the observations gained from such stimulus presentation, Towne acknowledged the essential utility provided by the stereoscope in their demonstration. In further experimentation, one eye viewed an image of a star composed of yellow dots in one vertically symmetrical half and blue dots in the remaining half, while the other eye viewed the complementary image. What ensued was concurrent binocular combination and rivalry between local zones, rather than regular alternations between global coherent percepts as might be expected from his previous observations:

> Over the greater portion of the resultant image, the spots will appear double, the blue and yellow dots lying side by side, while in other parts of the figure there will be constant alternation; *but in no part will the discordant colours blend, or the two images be permanently superposed;* in short, the result may be thus broadly stated—that we have in this instance a double star, or rather a star composed of double rows of spots, excepting where the images occasionally alternate.[11]

Towne appreciated that rivalry could not be considered to involve competition between one eye and the other alone, local factors needed to be considered, too. Parts of the pattern or colour in one eye compete with corresponding parts in the other rather than the whole stimulus in one eye alternating in visibility with that in the other. Rarely is it the case that colour rivalry operates in isolation. For this to happen the whole fields in both eyes would need to be exposed to different colours. Thus, the experiments by Towne confounded colour with contour rivalry as did most other researchers who followed.

---

[10] Towne (1863), p 121.

[11] Towne (1864), pp 129–130.

While many early twentieth century psychologists discussed consciousness and its investigation in their writings, very few did so specifically in relation to rivalry. An exception to this was William McDougall (1871–1938). He had a special interest in biology which was expressed in experimental work on vision. In studies using ambiguous figures and binocular colour rivalry, he found similarities in their (successive) interval durations, and in the effect of voluntary control. From this background, McDougall hypothesized that such phenomena represented:

> ...a principal condition of the alternating appearance in consciousness of two objects, while the impression made on the sense-organ remains unchanged, is fatigue of the cortical tract concerned in the perception of either object, a fatigue which is induced during the period of perception, and which rapidly passes away during the period of rest in which the other object is present to consciousness.... [T]he paths which suffer fatigue and become alternately active and passive or, in other words, alternately transmit the stream of nervous energy coming in from the sense-organ and cease to transmit it while it is diverted to the alternative path, these paths are, in the case of the binocular rivalry of colours, paths of the sensory area of the cortex, paths of the second level in the scheme...; while in the case of ambiguous figures...the nervous energy coming in from the sense-organ is continuously transmitted by the paths of the sensory level while it penetrates alternately to one or other of two higher-level paths.[12]

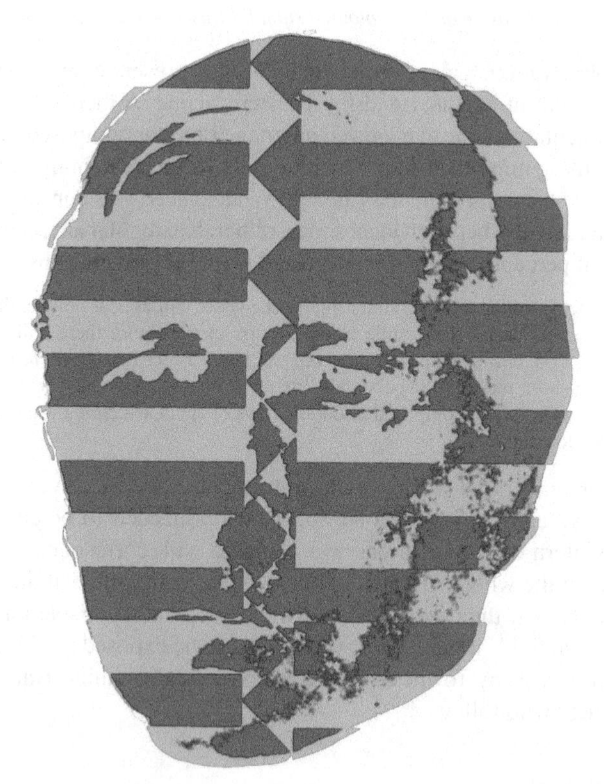

McDougall related binocular rivalry to the perceptual alternations seen in ambiguous figures like the arrows defining his facial features

[12] McDougall (1906), pp 346–347.

From this passage, it is clear that McDougall provided an early multi-level neurophysiological account of perceptual rivalry (i.e., binocular rivalry, ambiguous figures and bi-stable motion illusions). This account is also the key forerunner to the principle of neural dissociation that is now common in modern mechanistic studies of rivalry (i.e., distinguishing neural activity mediating the alternate percepts from that corresponding to the constant sensory input). McDougall's work was the earliest specifically to compare binocular rivalry with other types of alternating figures. His account regarding visual rivalry was also notable for articulating elements of Gestalt satiation theory which would be formulated decades later. In addition, it has been argued that his conceptualization of both neural fatigue and reciprocal inhibition between neurons provided a clearer explanation for Sherrington's finding of rhythmic alternations between antagonistic muscle groups.[13] McDougall also applied his approach to binocular rivalry, thereby enunciating key aspects of modern reciprocal inhibition models of the phenomenon. Moreover, he first reported observations of another phenomenon closely related to binocular rivalry, now known as flash suppression:

> ...if, when a red field is presented to one eye and a blue field to the corresponding area of the other eye, one eye be closed or covered for a brief period – one second will suffice – the colour presented to that eye always predominates over and inhibits the colour presented to the other eye as soon as the eye is uncovered, *i.e.,* the rested tract predominates over the relatively fatigued tract, even if the period of rest be not more than one second.[14]

From the outset, the stimulus for studying binocular colour combination has been the characteristics of monocular colour mixing. Desaguliers sought to determine whether Newton's investigations of mixing lights had a parallel in binocular vision. With the proposals of trichromatic and opponent process theories of colour vision concern has been directed to binocular colour combination. Much hinged on whether a combination of red to one eye and green to the other would yield the perception of yellow. It has been shown that binocular colour combination can be achieved when achromatic fusion aids are present in both colour fields.[15]

Colour rivalry is most easily examined with optical stereoscopes and it is with these that experiments examining it have been used. Anaglyphs are not suitable for such demonstrations unlike the case for contour rivalry.

Contour Rivalry

Many of the previous illustrations in this book have involved rivalry between different contours presented to the eyes. Binocular contour rivalry is more evident and compelling than colour rivalry and it can readily be observed with different stimuli in a stereoscope. Wheatstone examined rivalry between the letters A and S each surrounded by similar circles and reported that:

---

[13] Sherrington (1906).

[14] McDougall (1901), p 598.

[15] De Weert and Wade (1988).

> The common border will remain constant, while the letter within it will change alternately from that which would be perceived by the right eye alone to that which would be perceived by the left eye alone. At the moment of change the letter which has just been seen breaks into fragments, while the fragments of the letter which is about to appear mingle with them, and are immediately after replaced by the entire letter. It does not appear to be in the power of the will to determine the appearance of either of the letters, but the duration of the appearance seems to depend on causes which are under our control: thus if the two pictures be equally illuminated, the alternations appear in general of equal duration; but if one picture be more illuminated than the other, that which is less so will be perceived during a shorter time.[16]

The letters used by Wheatstone to examine binocular contour rivalry

Brewster also examined binocular rivalry, initially with the same letters as used by Wheatstone. The name Brewster gave to the binocular competition—ocular equivocation—did not appeal to his scientific colleagues and was not adopted. He wrote: "The *ocular equivocation*, as it may be called, which is produced by the capricious disappearance and reappearance of images formed on nearly corresponding points of each eye, is placed beyond a doubt by Mr Wheatstone's own experiments".[17]

Letters were recognised as complex patterns and simpler stimuli were soon enlisted. One of the first systematic studies of rivalry was published by Panum. The stimuli he introduced have dominated the study of rivalry ever since—orthogonal gratings. He also described what is seen when viewing the stimuli with two eyes as 'rivalry'. With regard to orthogonal gratings (and those below are derived from his illustrations) he wrote:

---

[16] Wheatstone (1838), p 386.

[17] Brewster (1844), p 359.

> The rivalry of contours is at its strongest if the different lines in the two images are as equal as possible with regard to thickness and light intensity... The resulting composite image in the joint visual field cannot easily be drawn due to its constant restless variation; at one moment the diagonal lines of one side appear alone, at another those of the other side, but mostly some lines of both stimuli are present, so that in one place the inclined lines from one predominate and in others those of the other places, and both are visible in some locations though weaker and similarly washed out or blurred.[18]

Peter Ludvig Panum introduced orthogonal gratings in the study of binocular rivalry

Panum can be seen in the illustration he drew of the mixtures or composites that he saw with orthogonal gratings. He also refers to these mixtures as 'mosaic-like'.

[18] Panum (1858), p 38.

Panum and a pattern of mixtures or composites seen during rivalry between orthogonal gratings

By drawing attention to the dynamic variations and to the mixtures or composites that are seen Panum sought to interpret the phenomenon in physiological terms rather than the psychological factors proposed by Wheatstone and later by Helmholtz. Panum wrote:

> Following the lawful rules that have been stated, the mosaic-like filling of the general visual field, combined with partial fusion of the impressions taking place, arises from neither psychological causes, attention, imagination or the like, nor from any dread of double images, nor from a total alternative paralysis of the two retinas, but from very characteristic means of perception or sensory energies emanating from the simultaneous action of the excitation of corresponding retinal points on the central organ of vision (in the brain).[19]

Panum examined a range of rivalry figures which were used by others later, like broad vertical and horizontal lines intersecting to form a cross and pairs of thin vertical and horizontal lines.

Wilhelm Wundt (1832–1920) was quick to appreciate the import of Panum's work and incorporated it into his book on sensory perception published in 1862. In his survey of binocular vision Wundt included a section on rivalry of perception (Wettstreit der Wahrnehmungen) and presented figures similar to those employed by Panum, like pairs of vertical and horizontal lines as well as vertical and horizontal gratings. Wundt drew attention to the effects of eye movements on the visibility of such patterns. He also emphasised the mixtures that are visible under normal

[19] Panum (1858), pp 93–94.

conditions of viewing: "With ordinary steady viewing... you always see in places the horizontal, and in places the vertical lines merge into the collective image".[20] He interpreted rivalry in terms of alternations of attention.

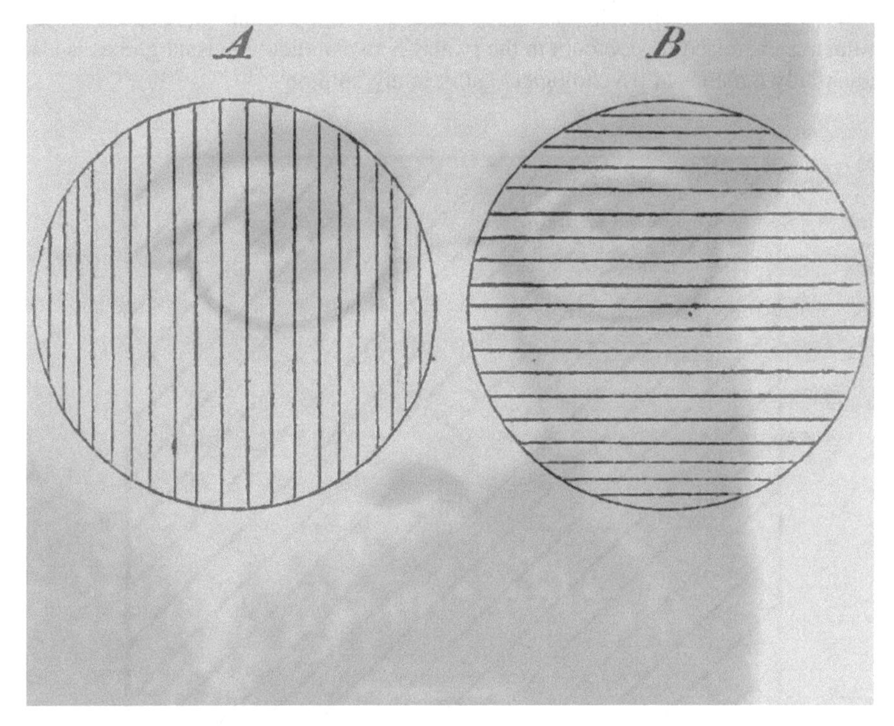

Wilhelm Wundt and stimuli (vertical and horizontal gratings) that he used to examine binocular rivalry

It is significant that Wundt modified Panum's 'rivalry between the visual fields' to 'rivalry of perception' because it reflected a distinction that remains to this day: is rivalry a psychological or a physiological process? Panum and Hering considered that it was physiological whereas Wheatstone, Helmholtz and Wundt maintained that it was psychological. In his *Handbuch* of 1867 Helmholtz discussed rivalry in some detail and also emphasised that changing, complex mixtures of the two stimuli tend to be visible most of the time with only occasional periods in which the stimulus in one eye alone dominates:

> ...in the various parts of the field, one image will prevail more than the other, whereas in other parts the other image will predominate. Sometimes there will be alternations, so that, where for a while only parts of one image were visible, presently parts of the other image will emerge and suppress portions of the first image. This fluctuation, in which parts of the two images mutually supplant each other, either side by side, or one after the other, is what is usually meant by *the rivalry between the visual globes*.[21]

[20] Wundt (1862), p 366.

[21] Helmholtz (1925), p 494.

Like Wundt, Helmholtz placed great importance on eye movements in rivalry and made a modification to Panum's orthogonal gratings configurations: he placed two small squares at the centre of both gratings to facilitate common fixation by each eye because otherwise "it is difficult to concentrate the attention on one of the systems of lines".[22] Thus, Helmholtz also assigned rivalry to attention: "The extraordinary influence exercised by contours in the rivalry between the two visual globes is also essentially a matter of psychological habit, in my opinion".[23]

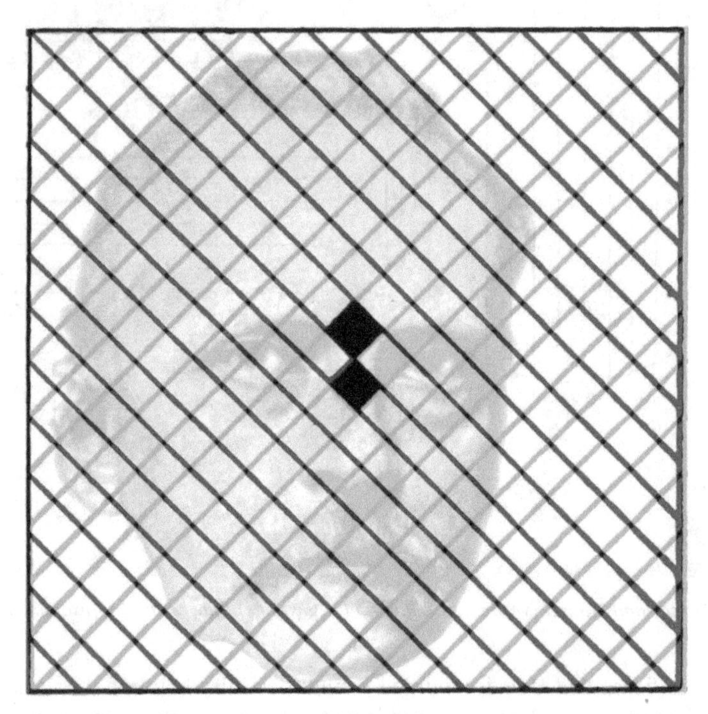

Helmholtz in combination with the crossed gratings he illustrated in his Handbuch

The crossed diagonal figure was used by Helmholtz to support his theory that rivalry is a psychological rather than a physiological process because he could control which stimulus was visible:

> These experiments show that man possesses the faculty of perceiving the images in each eye separately, without being disturbed by those in the other eye, provided it is possible for him... to concentrate his whole attention on the objects in this one field. This is an important fact, because it signifies, that *the content of each separate field comes to consciousness without being fused with that of the other field by means of organic mechanisms;* and that, therefore, *the fusion of the two fields in one common image, when it does occur, is a psychic act.*[24]

---

[22] Helmholtz (1925), p 498.

[23] Helmholtz (1925), p 501.

[24] Helmholtz (1925) p 499.

This view led to a bitter dispute with Hering, as will be described in Chap. 7.

Rivalry between photographs of faces was examined by Francis Galton (1822–1911). The impetus for his experiments on composite faces[25] stemmed from presenting different photographic portraits in a stereoscope: "At first, for obtaining pictorial averages I combined pairs of portraits with a stereoscope, with more or less success".[26] When Galton devised a method for making multiple exposures of different faces on a single photographic plate, he dispensed with the use of the stereoscope. Similar stereoscopic combinations of different portraits were conducted independently at around the same time (1877) and communicated to Galton's half-cousin, Charles Darwin, who passed on a letter he received from A. L. Austin of New Zealand. Galton commented that the letter afforded: "another of the many curious instances of two persons being independently engaged in the same novel inquiry at nearly the same time, and coming to similar results".[27] Austin's description concentrated on the blending of the two portraits, whereas Galton drew attention to the rivalry between them:

> Convenient as the stereoscope is, owing to its accessibility, for determining whether any two portraits are suitable in size and attitude to form a good composite, it is nevertheless a makeshift and imperfect way of attaining the required result. It cannot of itself combine two images; it can only place them so that the office of attempting to combine them may be undertaken by the brain. Now the two separate impressions received by the brain through the stereoscope do not seem to me to be relatively constant in their vividness, but sometimes the image seen by the left eye prevails over that seen by the right and vice versa. All the other instruments I am about to describe accomplish that which the stereoscope fails to do; they create true optical combinations. As regards other points in Mr. Austin's letter, I cannot think that the use of a binocular camera for taking the two portraits intended to be combined into one by the stereoscope would be of importance. All that is wanted is that the portraits should be nearly of the same size. In every other respect I cordially agree with Mr. Austin.[28]

The purpose of combining different portraits was not to investigate binocular rivalry but to derive some composite average of human types.[29] Nonetheless, it is clear from Galton's description that rivalry did occur and it was for this reason that composite photographs were preferred. Paired photographic stimuli did not have the impact on studies of binocular rivalry than was the case for stereoscopic vision. Galton is shown in a composite image where the two views are in rivalry with one another; he is also shown in full face and profile view, as was used in later 'mug shots' of criminals.

---

[25] Galton (1878a, 1878b, 1879).

[26] Galton (1908), p 259.

[27] Galton (1878b), p 98.

[28] Galton (1878b), pp 98–99.

[29] See Wade (2016).

Face recognizer—Francis Galton

From the end of the nineteenth century, neural theories of consciousness based on rivalry experiments were advanced by Burtis Burr Breese (1867–1939). In a long article under the title 'On inhibition' published in 1899 he introduced measurements of the periods of visibility and the frequencies of dominance of rivalling stimuli.[30] These were two squares with 45° and 135° black lines on them; they were on backgrounds that were either green or red, or they were both the same colour. He found no differences in either the predominance durations or the periods of visibility under these conditions. It should, however, be noted that the comparisons were taken from different experiments and were based upon his own observations alone. There is a further problem in interpreting Breese's findings: he used gratings of unspecified visual subtense, and recorded the phenomenal alternation between the two gratings. In view of the history of experiments on contour rivalry it is unlikely that there was simply alternation between the two gratings. Almost all previous investigators noted

[30] Breese (1899).

that fragments of the two monocular stimuli appear simultaneously in different parts of the field. It has since been found that these fragments or composites are visible for around 30% of the observation period for rivalry between gratings.[31] Perhaps Breese observed such composites and categorised them in terms of the dominance of one of the monocular fields, although he made no mention of this.

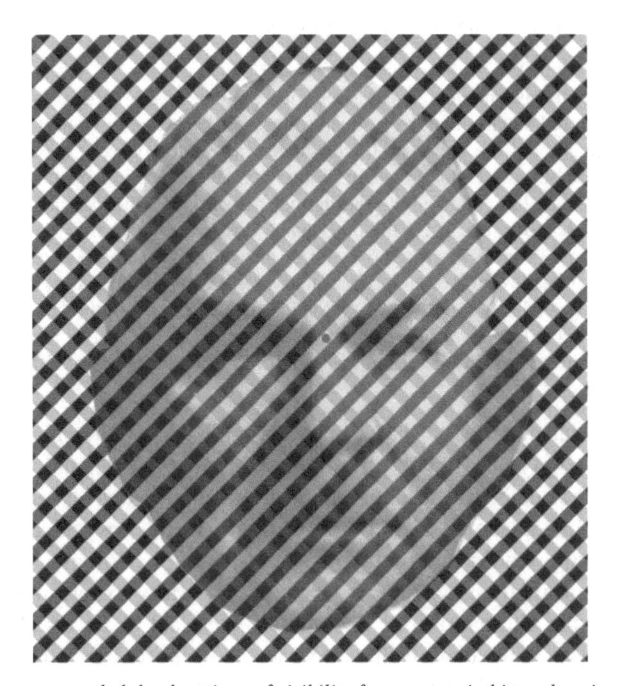

Breese recorded the durations of visibility for patterns in binocular rivalry

Along with examining the effects of motor inhibition on memory, Breese used binocular rivalry as a paradigm to examine the inhibition of sensations and argued that consciousness had a sensorimotor basis. More importantly, he made the first quantitative measures of binocular rivalry. He examined: (i) the effect of stimulus strength changes (e.g., motion, size, luminance) on perceptual predominance and alternation rate; (ii) individual variation in alternation rate; (iii) the effect of unilateral motor activity on predominance; (iv) rivalry between after-images and their slower alternation rate compared to real stimuli; and (iv) the phenomenon of monocular rivalry. He also investigated the influence of willpower on binocular rivalry. Subjects could voluntarily hold attention on one image with the (inadvertent) use of eye movements, but without eye movements such voluntary control was limited. Breese argued that binocular rivalry could not be explained by purely mental conditions because complete control over the alternations could not be demonstrated.

[31] Wade (1974).

Correspondingly, physical conditions such as retinal adaptation could not solely explain the effect of different brightness levels on rivalry rate, which may have also been due to greater attention directed towards the brighter image. Instead, he concluded that the phenomenon "would be at once 'psychical' and 'physiological' in that it is dependent upon central processes, and is affected by the nature of motor adaptations".[32] Breese subsequently elaborated on the distinction between consciousness and attention during rivalry, along with postulating their associated activity in the brain:

> ...in the case of the alternating red and green squares in the stereoscope, if, when I give my attention to the red square, the green square displaces it in sensory consciousness, I may still be thinking about the red square, attending to its quality, size, et cetera, so that the change in consciousness is not a change of attention. What really happens in this case is that part of the time I am attending to the sensory perception of the red square, and part of the time to its memory image. On the other hand, the green square may not occupy clear or attention-consciousness at all... So we infer that the brain activities corresponding to attention-consciousness involve larger areas than those corresponding to mere sensation or perception. We conclude, therefore, that attention may change with the changes of sense-stimuli or may act independently of them. Shifting of the attention involves more factors than those involved in the fluctuation of minimal sensations or in the rivalry of objects presented to the sense-organs.[33]

The theoretical conclusion that Breese reached was: "Binocular rivalry, then, would be at once 'psychical' and 'physiological' in that it is dependent upon central processes, and is affected by the nature of the motor adaptations".[34] He was reflecting on the debate that separated Helmholtz, who was in the "psychical" camp, from Panum and Hering whose interpretations were more "physiological". In 1909, ten years after his original study, Breese repeated the rivalry experiments on himself and noted that his alternation rate was almost identical.[35] This within-individual retest reliability of binocular rivalry rate was also reported by others, as was his earlier finding of individual variation in alternation rate. The quantitative rivalry experiments conducted by Breese and his interpretation of their findings reflected the broader development of psychology into a scientific discipline as well as interest in the neural basis of attention and consciousness. Both notions of attention and consciousness had been central to early views of the mind, thus not surprisingly the advent of experimental psychology also marked the beginnings of a science of consciousness.

Experimental studies of binocular rivalry were relatively sparse in the first half of the twentieth century but there were some significant studies. One by Emilio Diaz-Caneja (1892–1967) concerned the influence of organization on rivalry[36]; it was largely forgotten until it was translated in 2000. The portrait of Diaz-Caneja is contained within rivalry patterns like those he used—patterns in which each half was

[32] Breese (1899), p 48.

[33] Breese (1917), pp 74–75.

[34] Breese (1899), p 48.

[35] Breese (1909).

[36] Dias-Caneja (1928).

formed from red horizontal lines and green semicircles. He found that rivalry was not solely between the eyes but also between the patterns. That is, horizontal lines or concentric circles were also seen so that the components from different eyes were integrated: "At a particular moment, the lines and circles are mixed; an instant after, we can see lines everywhere or circles everywhere; more rarely, we see either the right half or the left half of the card".[37] Diaz-Caneja was demonstrating that rivalry can be between figures, parts of which are in opposite eyes as well as by figures in each eye. This is an issue that has continued to reverberate throughout rivalry research.

Emilio Diaz-Caneja and his rivalry patterns

The pace of research on rivalry quickened in the second half of the century. It was boosted in the 1960s by three researchers—Robert Fox in America, William (Pim) Levelt in The Netherlands, and Paul Whittle in Britain.[38] All brought an added precision to the recording and analysis of rivalry sequences. In 1965 Levelt presented propositions regarding rivalry which have stood the test of time.[39] A key concept was that the strength of the stimuli in each eye determined the characteristics of the rivalry that ensued. This could be supported by examining the individual rivalry dominance and suppression durations and describing the distributions mathematically. Levelt is shown in text taken from the monograph.

---

[37] Alais et al (2000), p 1444.

[38] See Blake (2005).

[39] Levelt (1965), see also Brascamp et al. (2015).

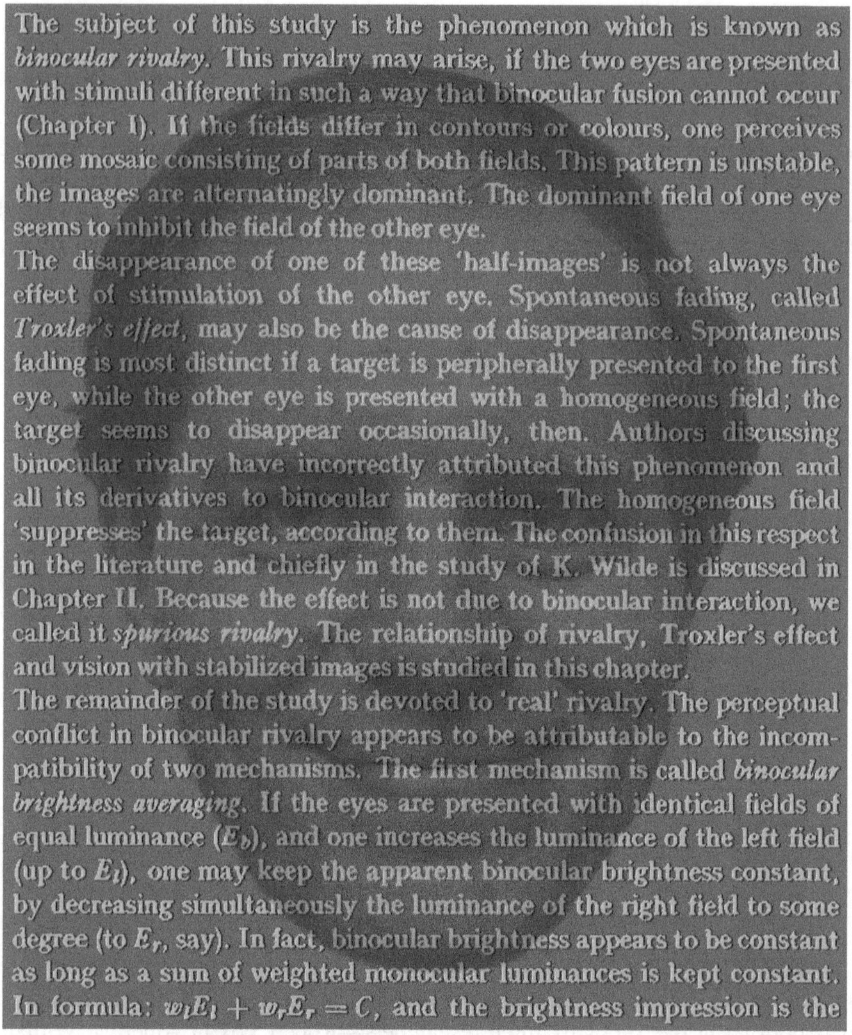

The subject of this study is the phenomenon which is known as *binocular rivalry*. This rivalry may arise, if the two eyes are presented with stimuli different in such a way that binocular fusion cannot occur (Chapter I). If the fields differ in contours or colours, one perceives some mosaic consisting of parts of both fields. This pattern is unstable, the images are alternatingly dominant. The dominant field of one eye seems to inhibit the field of the other eye.

The disappearance of one of these 'half-images' is not always the effect of stimulation of the other eye. Spontaneous fading, called *Troxler's effect*, may also be the cause of disappearance. Spontaneous fading is most distinct if a target is peripherally presented to the first eye, while the other eye is presented with a homogeneous field; the target seems to disappear occasionally, then. Authors discussing binocular rivalry have incorrectly attributed this phenomenon and all its derivatives to binocular interaction. The homogeneous field 'suppresses' the target, according to them. The confusion in this respect in the literature and chiefly in the study of K. Wilde is discussed in Chapter II. Because the effect is not due to binocular interaction, we called it *spurious rivalry*. The relationship of rivalry, Troxler's effect and vision with stabilized images is studied in this chapter.

The remainder of the study is devoted to 'real' rivalry. The perceptual conflict in binocular rivalry appears to be attributable to the incompatibility of two mechanisms. The first mechanism is called *binocular brightness averaging*. If the eyes are presented with identical fields of equal luminance (E_b), and one increases the luminance of the left field (up to E_l), one may keep the apparent binocular brightness constant, by decreasing simultaneously the luminance of the right field to some degree (to E_r, say). In fact, binocular brightness appears to be constant as long as a sum of weighted monocular luminances is kept constant. In formula: $w_lE_l + w_rE_r = C$, and the brightness impression is the

Pim Levelt and text from his book 'On binocular rivalry'

In common with Levelt, Robert Fox (1922–2018) made detailed analyses of the durations of dominance and suppression in sequences of rivalry. He also drew attention to the piecemeal aspects of binocular rivaly, particularly when large stimuli are presented to an observer, as well as the influence of stimulus motion.[40] He developed techniques that presented brief probes to an eye when the stimulus to that eye was suppressed and found that visual sensitivity was reduced, unlike the same presentation when the pattern was dominant. Moreover, quite marked changes in the stimulus presented to the suppressed eye can go undetected suggesting that rivalry is

[40] Fox (2005), Fox and Hermann (1967), Fox and Check (1968), Blake and Fox (1974), Lehmkuhle and Fox (1975).

between the eyes rather than the particular stimuli. Fox introduced a technique that has had considerable experimental traction: can a suppressed stimulus influence the characteristics of a pattern presented subsequently? More specifically, can a visual aftereffect be induced by a pattern in the suppressed eye? In general, the answer from many studies has been in the positive.[41]

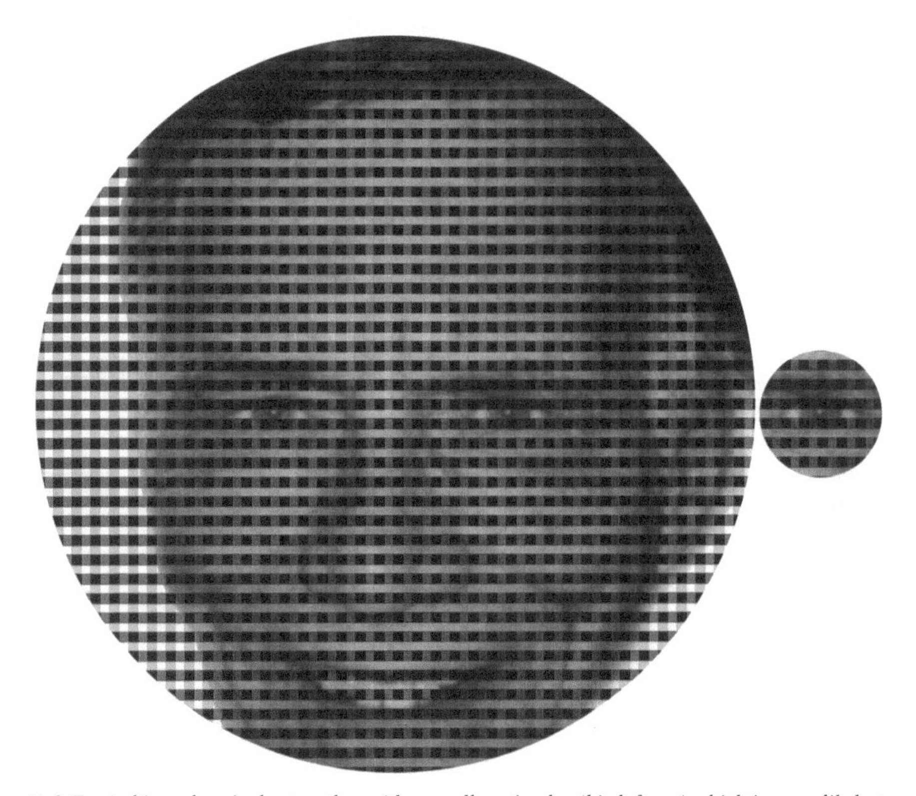

Bob Fox in binocular rivalry together with a smaller stimulus (his left eye) which is more likely to alternate between the two gratings

A modification of the view that rivalry is a low level aspect of binocular interaction was provided by Paul Whittle (1938–2009); he examined aspects of figural organization and contrast in binocular rivalry.[42] That is, he sought to distinguish between local and global interactions in rivalry. For example, when segments of a figure viewed by different eyes engage in rivalry they do so in a similar way to the segments presented to the same eye. This applies whether the stimuli are viewed in

---

[41] O'Shea and Crassini (1981), Wade (1980), Wade and Wenderoth (1978).

[42] Whittle (1965), Whittle et al (1968).

a stereoscope or as afterimages.[43] The issue of local and global interactions features in a wide range of visual phenomena and their extension to binocular vision was a significant advance.

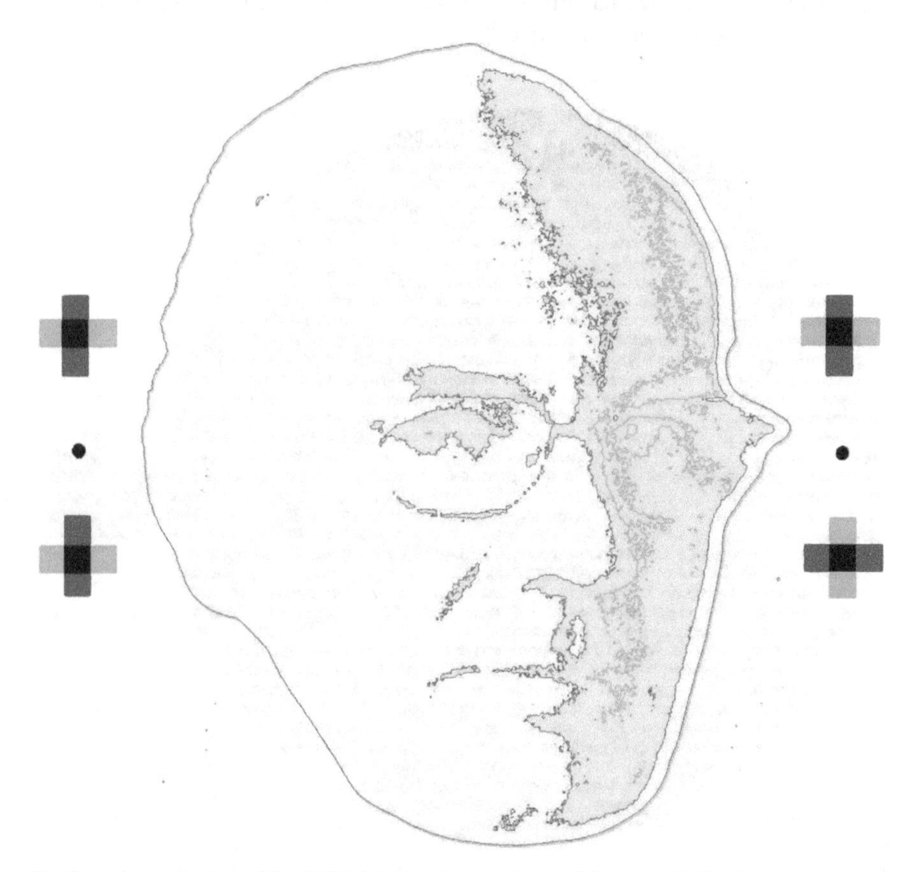

The figural organization of Paul Whittle's portrait together with line stimuli like those he examined

Much more has been learned about responses to patterns in the mammalian visual cortex since the 1960s as well as novel techniques for recording from the human brain. Accordingly, concerted efforts have been made to link binocular rivalry using a wider range of stimulus patterns with measures of cortical activity. Recordings from single cells in monkey visual cortex can be related to stimuli they see.[44] A combination introduced by Tong and colleagues has been used extensively to illustrate such links[45]; it consists of a picture of a face presented to one eye and of a house to the other, like that shown below. They found that the periods when the face was dominant corresponded to increased activity in a brain region known to be related

[43] Wade (1973).

[44] Leopold and Logothetis (1996), Logothetis et al (1996).

[45] Tong et al (1998), Tong (2005).

to face processing (fusiform face area) whereas those when the house was dominant led to similar effects in an area concerned with coding location (parahippocampal place area).

Face/place rivalry combination. The house is located in Charleston, South Carolina, and it belonged to the Wells family in the eighteenth century; William Charles Wells was a pioneer of research on binocular vision

Much of the subsequent research has examined the distinction between theories proposing a low level basis for binocular rivalry or whether higher level processes are involved in it.[46] Investigations of binocular rivalry have broadened enormously in the last decades and much attention is directed at determining the neurophysiological underpinnings of dominance and suppression as well as their interplay. Increasingly,

[46] Alais and Blake (2005), Blake (2017), Blake and Logothetis (2002), Crick and Koch (1995).

computational models of the perceptual oscillations have been developed. Moreover, advances in computer controlled stimulus presentation and analysis have extended the subtleties of the experimental manipulations that are available. Individual differences in the temporal dynamics of rivalry has also come to the forefront, either by relating the pattern of rivalry sequences to known dimensions of personality (like introversion/extraversion) or to underlying physical conditions (like vestibular disorders).[47] Rivalry has been used as a way of investigating consciousness and its neural correlates.[48] Indeed it has been said that: "Binocular rivalry is a popular tool in the scientific study of consciousness because it dissociates stable, unchanged, visual stimulation from fluctuations in visual awareness".[49] Reviews of this research can be found in numerous articles and books[50] and Robert O'Shea has assembled a bibliography of rivalry research up to 2001.[51]

Monocular Rivalry

Rivalry between patterns can occur when using one eye alone. Fixating steadily the center of either image below for at least 30 s without the red/cyan glasses will produce monocular rivalry. Initially the two gratings appear superimposed, but then the clarity (the perceived contrast) of one orientation diminishes while the clarity of the other enhances for a second or so, and then they reverse. Occasionally one grating alone will be visible. More frequently, combinations of the gratings are seen, in which, for example, the red lines are clearer than the cyan. Transitions among the various states can also be piecemeal and dynamic.

[47] Ngo et al (2013).

[48] Miller (2013), Blake et al (2014).

[49] Klink et al (2013), p 323.

[50] Alais and Blake (2005), Alais and Blake (2015), Miller (2013).

[51] https://sites.google.com/site/oshearobertp/publications/binocular-rivalry-bibliography.

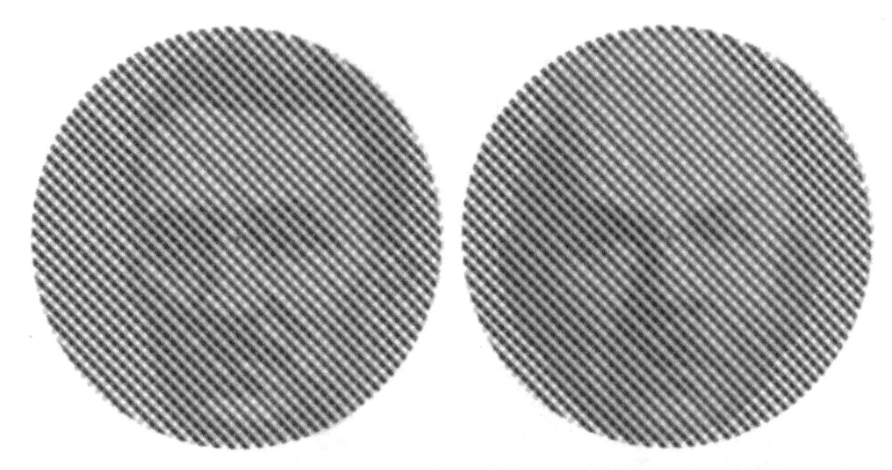

Monocular rivalries. Superimposed orthogonal red and green gratings combined with portraits of Tscherning on the left and Breese on the right, both of whom described monocular rivalry at the end of the nineteenth century

Breese called this 'monocular rivalry' in order to contrast it with binocular rivalry in which each component image is presented to a different eye, and on which he conducted a long series of studies. You can see binocular rivalry if the patterns are observed with the red/cyan glasses. Monocular rivalry occurs between any superimposed images. They can be simple, such as gratings, lines, or bars. Images can also be complex, such as of faces and houses. The images can be achromatic, but the phenomenon is more pronounced if the component images differ in colour. Breese's section on monocular rivalry came at the end of his experimental studies on binocular rivalry and he used the same paired stimuli—orthogonal, oblique green and red gratings on a black background. These were superimposed using either a prism or by transmission and reflection from a glass plate. He described monocular rivalry thus:

> Now the interesting part of the experiment is that if the center of the field was fixated, a rivalry of colours was perceptible. Neither disappeared entirely: but at times the red would appear very distinctly while the green would fade; then the red would fade and the green appear distinctly. The two sets of lines showed the same fluctuation, keeping pace with the changing intensities of the colors. Sometimes one of them would disappear altogether. The rivalry of the colors and of the lines was much slower than the rivalry in binocular vision. A very slight movement of the eye to the right or left would cause changes in the intensities of the fields [beause of techical limitations from Breese's method of superimposing the images]; so that extreme care was necessary in order to avoid the movements, or, at least, not to confuse changes caused by them with the rivalry of the fields, which is independent of eye movements.[52]

Breese did not apply the systematic measurements of predominance durations and alternation frequencies to monocular rivalry as he had to binocular rivalry, possibly because of the evanescent nature of the former.

[52] Breese (1899), pp 43–44.

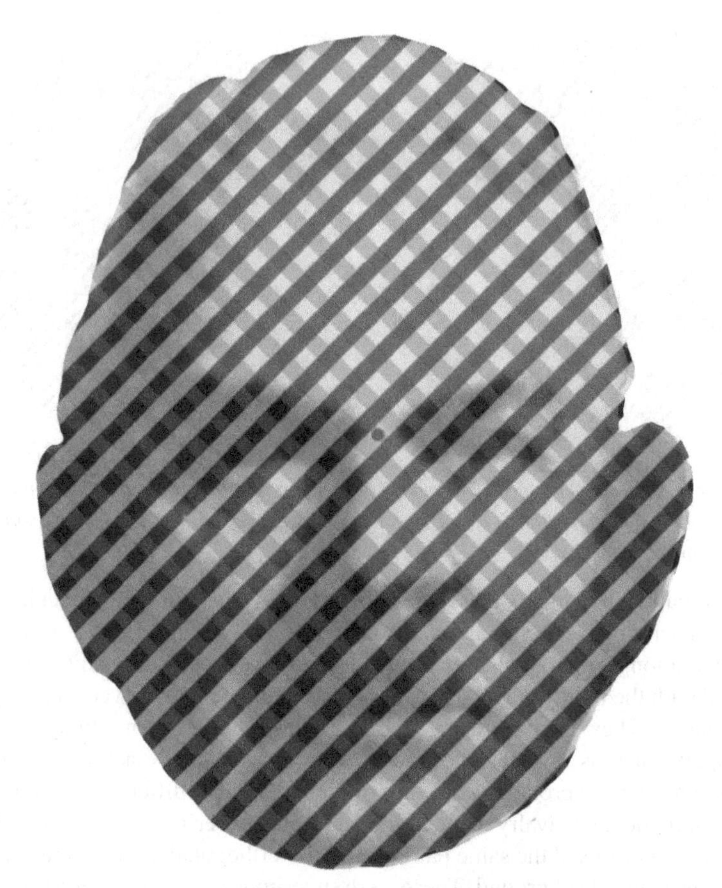

Breese examined the variations in visibility of crossed gratings that occurs without the use of red/cyan glasses; he called it monocular rivalry

One year earlier, Marius Hans Erik Tscherning (1854–1939) described similar observations when viewing a grid of vertical and horizontal lines[53]; he noted the pattern alternations in the course of studying Troxler fading—the disappearances of elements in low contrast patterns. Tscherning was a student of Panum in Copenhagen before going on to train in ophthalmology. Tscherning's interests in myopia overlapped significantly with those of Louis Émile Javal (1839–1907) who invited Tscherning to visit him at the Sorbonne. Tscherning took up the position of Javal's assistant and remained at the Sorbonne for the next 26 years, where he became director of the ophthalmological laboratory when Javal retired because of the onset of his blindness from glaucoma. In his *Optique physiologique* Tscherning covered topics such as the optics of light and lenses, dioptrics, optical aberrations, entoptical phenomena, accommodation, ophthalmoscopy, and the psychophysics of light and of color. In his chapter on "The Form Sense" the final section was concerned with

[53] Tscherning (1898, 1900).

Troxler's phenomenon—the disappearance and reappearance of stimuli away from a fixation point. Tscherning wrote:

> For my eye, the phenomenon begins after fixing the image for eight or ten seconds, that is to say at the moment when fixation begins to be less sure. From this moment, the figure shows perpetual changes: sometimes part of the figure disappears, sometimes another. What is curious is that most often the scotomata are not absolute: sometimes the horizontal lines disappear in one place, while the vertical lines persist, sometimes the opposite takes place. These phenomena are very reminiscent of what has been described as the antagonism of visual fields [binocular rivalry], and is observed, for example, by presenting, in a stereoscope, horizontal lines to one eye, and vertical lines to another.[54]

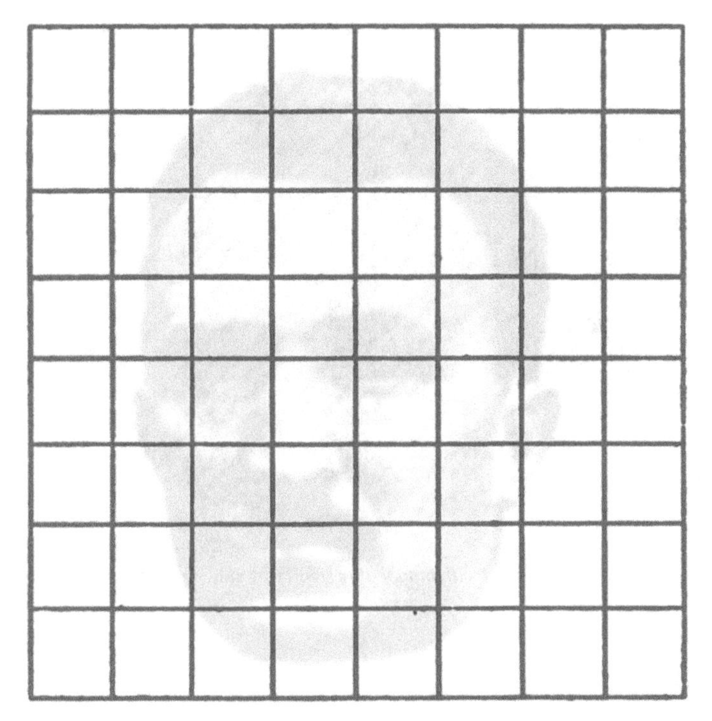

Tscherning and the pattern in which he observed monocular rivalry

Breese's main interest was in binocular rivalry, but he was able to modify his stereoscope so both rival images were delivered to the same eye. This modification required careful fixation—something that appears to be necessary for monocular rivalry to occur.

The phenomenon has been rediscovered several times since Tscherning and Breese, often by accident.[55] Experiments comparing monocular and binocular rivalry

---

[54] Tscherning (1898), pp 264–265.

[55] O'Shea et al. (2017).

indicate that they follow different time courses.[56] Dominance in binocular rivalry results in the visibility of one pattern alone; this occurs rarely in monocular rivalry, where dominance is associated with the increased distinctiveness of one pattern and the reduced clarity of the other. The differences can easily be compared by viewing the figure below with and without the red/cyan glasses.

Binocular rivalry can be compared with monocular rivalry by viewing the pattern with and without red/cyan glasses

Binocular Lustre

With the marriage between the stereoscope and photography it was possible to present a positive image to one eye and a negative of the same image to the other. However, the initial studies of binocular lustre used drawings of geometrical figures either as black on white or white on black. Such combinations were investigated in the 1850s, first by Dove followed by Brewster and others.[57] Dove's article was made available to an English reading audience with its translation in 1852; Dove wrote:

[56] Wade (1975).

[57] Dove (1851, 1852), Brewster (1861).

The projection for one eye was drawn in white lines upon a black ground, and for the other eye with black lines upon a white ground. A most remarkable result was obtained by the stereoscopic combination of both. The relief started into existence with surfaces which shone like graphite, having their edges formed of dazzling white and deep black lines which run parallel and in contact with each other throughout.[58]

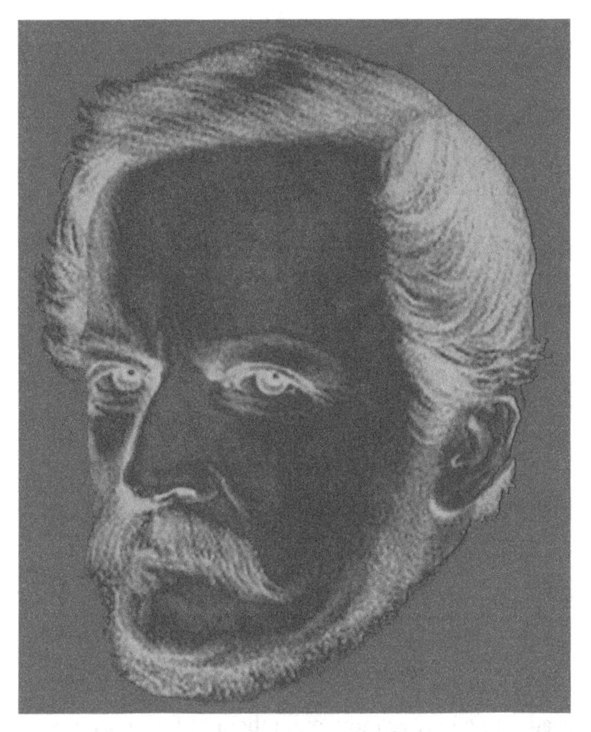

Dove described binocular lustre from combining positive and negative images in the two eyes

Dove noted that the lustre disappears when there is strong rivalry between the eyes so that one monocular stimulus alone is visible. Accordingly, stimuli that have relatively few contours display the phenomenon more effectively. In 1861 Brewster maintained that he made his first positive and negative drawings of geometrical figures in 1843 but he did not publish an account of them. He suggested that:

Binocular lustre arises from a physiological and not from a physical cause, we must look for this cause in the operations which take place in the eyes of the observer when binocular lustre is distinctly seen. These operations are of two kinds. First, in combining geometrical or other figures to represent solids whose parts are at different distances from the eye, the optic axes are in constant play, not only in varying the distance of their focus of convergence, to unite similar points at different distances in the two diagrams, but in maintaining the unity of the picture by rapidly viewing every point of its surface. Secondly, when the two surfaces have different shades or colours, the retina of one eye is constantly losing and recovering the

[58] Dove (1852), p 242.

vision of one of them. Each optic nerve is conveying to the brain the sensations of a different tint or colour. The brain is therefore agitated sometimes with one of these sensations and sometimes with the other, and sometimes with both of them combined, and it is therefore not an unreasonable conclusion that, in the dazzle produced by this struggle of flickering sensations, something like lustre may be produced.[59]

Brewster in binocular lustre

Brewster was adopting his eye movement theory of stereoscopic vision to account for lustre. This was discounted by Helmholtz who found that lustre is still visible when the stimuli are illuminated with an electric spark, thereby preventing any eye movements over the pattern. Helmholtz considered that stereoscopic lustre (as he called it) provided important clues to binocular interaction. The combination of colours demonstrated a fundamental difference between binocular and monocular vision:

The phenomenon of stereoscopic lustre is particularly important in connection with the theory of the activity of the retinas of the two eyes. The statements of various observers as to the result of the binocular fusion of unlike images are so different, that, were it not for this phenomenon, doubtless, we never should have known positively that the visual impression produced by the action of two different kinds of light on corresponding places of the two retinas was absolutely different from that produced by the action of two homogeneous kinds of light on the same retinal places.[60]

---

[59] Brewster (1861), pp 30–31.

[60] Helmholtz (1925), p 514.

The stimuli Helmholtz used to describe lustre even warranted a figure in his *Handbuch*. Binocular lustre continues to pose problems for models of binocular interaction.[61]

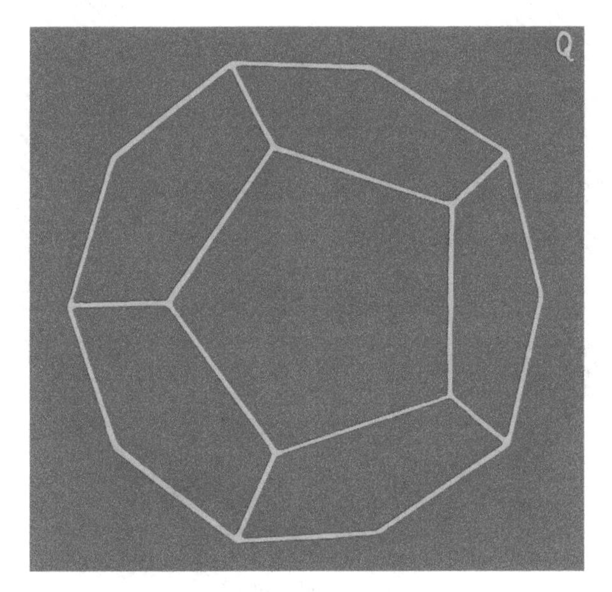

Helmholtz's figure for demonstrating stereoscopic lustre

Rivalry and Stereopsis

Both stereoscopic vision and binocular rivalry are robust phenomena but can they co-exist at the same time in the same place? Scientists have examined this in some detail because of its theoretical significance.[62] While it is evident that rivalry and depth can be experienced in the same stimulus pair at different locations the evidence is less clear that they can be experienced simultaneously at the same location. The following images show that they can. For example, the rivalry in the outer annulus of the image below is restricted to corresponding regions of each eye whereas that in the central annulus is seen both in rivalry and depth. The rivalry remains when the red/cyan glasses are reversed as does the depth, but its sign changes. That is, with red/LE and cyan/RE the central annulus appears closer than the surround whereas it looks more distant with the cyan/LE and red/RE combination.

[61] Muryy et al (2016), Wendt and Faul (2019).

[62] Wolfe (1986), Howard and Rogers (2012).

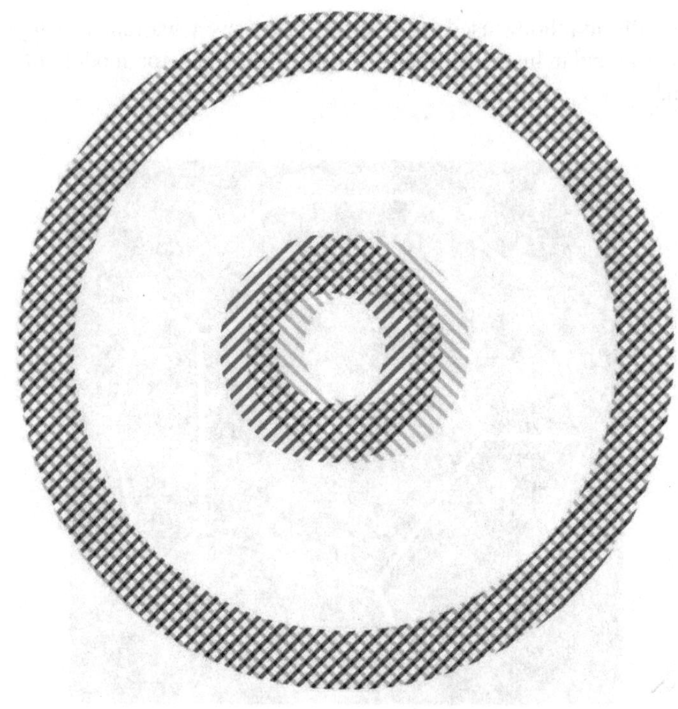

Depth and rivalry I

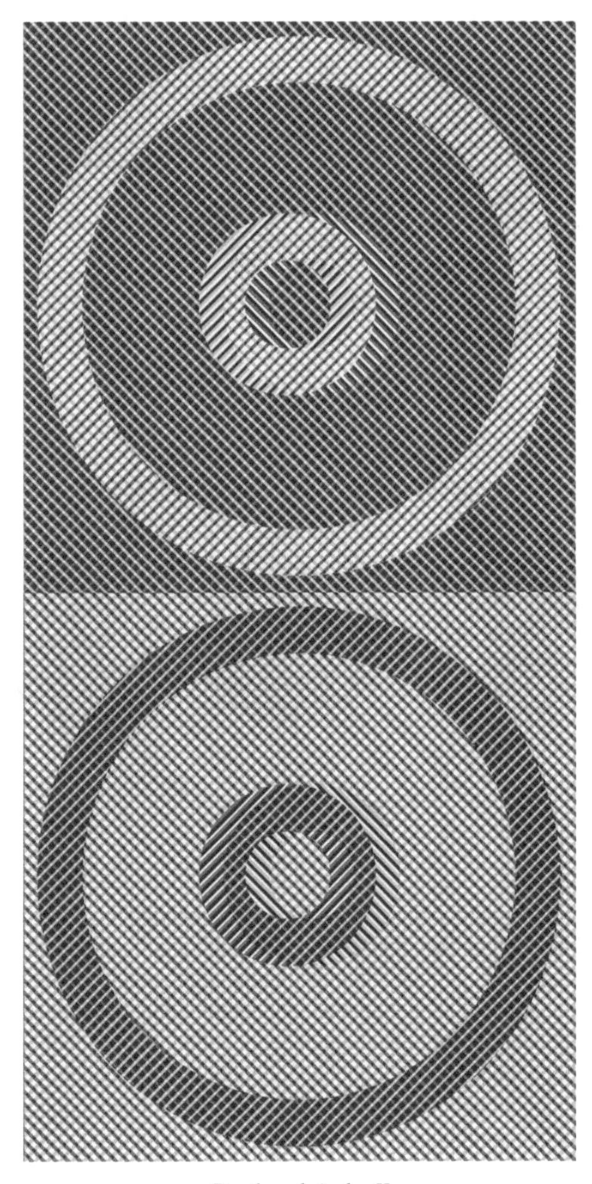

Depth and rivalry II

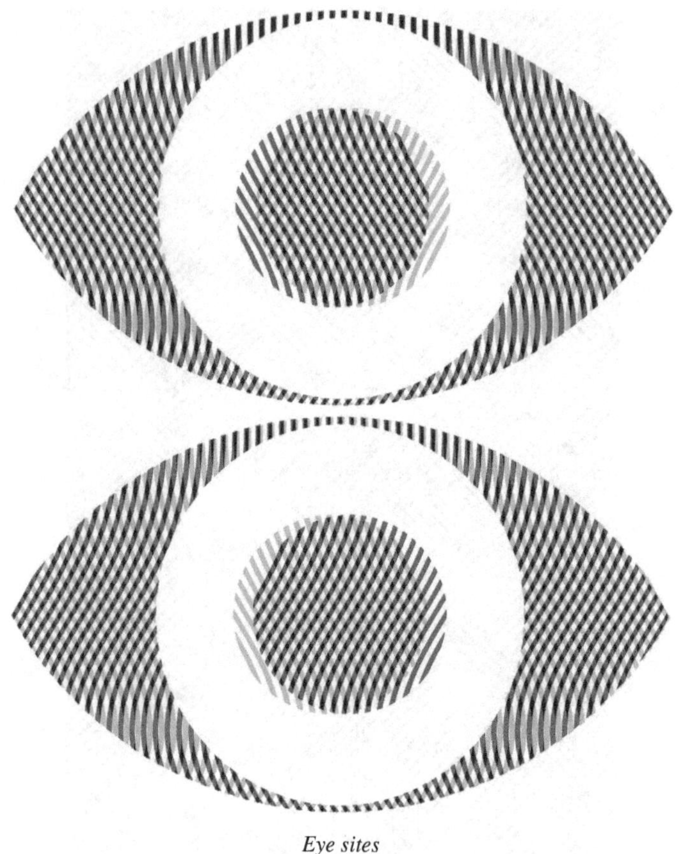

Eye sites

Viewed with one eye the pattern below appears distorted. Even though the squares are vertical and horizontal they appear tilted clockwise and counterclockwise relative to their neighbours. They appear tilted in the opposite directions with the other eye. With two eyes the squares do not appear as distorted. The illusion in the monocular patterns is due to the diagonal lines intercepting the vertical and horizontal ones. However, the diagonals are in opposite orientations in each eye and engage in rivalry when both eyes are used. The perceived depth is a consequence of small lateral displacements in the component patterns.

Concentric squares illusions

The remaining images present designs, manipulated photographs or conventional photographs that exhibit both depth and rivalry when viewed through the red/cyan glasses.

Wavering contours

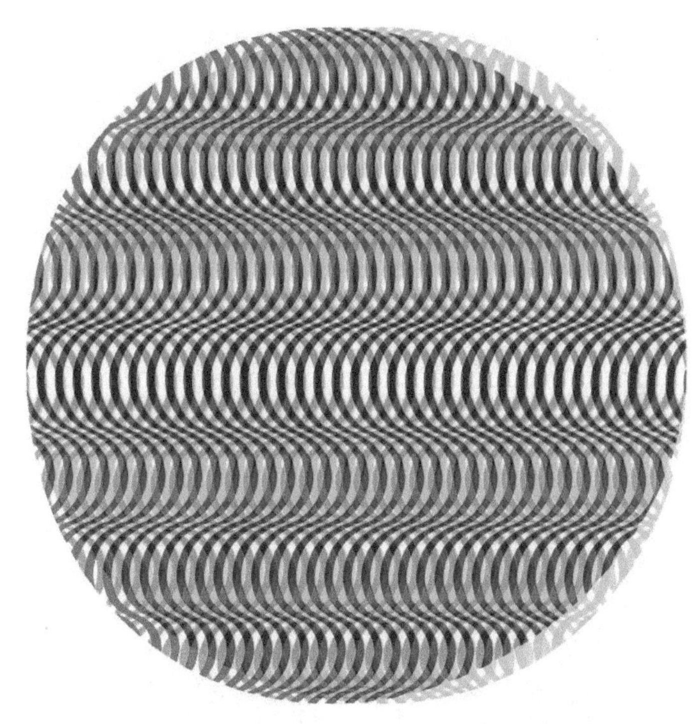

Ambiguities in depth and rivalry

The effects seen in *Depth and rivalry I* at the beginning of this section could be related to disparities between the subjective contours formed from the extremities of the lines in the central annuli which are visible in each eye. A more stringent test would be to present monocular images that did not display the disparate parts, as in *Depth and lustre I* below. The upper half is a stereoscopic pair with a central circle visible in depth; the lower half is the same pairing but one is a negative image to produce binocular lustre. Rivalry is produced from lustre rather than from differences in contour orientation. No disparate areas can be seen in the monocular views of either the upper or lower half. However, depth is clearly visible in the upper half when the red/cyan glasses combine the two monocular images, unlike what is seen in the lower half. An impression of depth is evident in the centre which reverses with reversing the filters but it does not have the clearly circumscribed boundary that is present in the upper half. The distinction is more marked when the stereoscopic depth is defined by changes over the whole surface rather than enclosed parts of it, as is evident in the upper and lower halves of *Depth and lustre II.*

Depth and lustre I

Depth and lustre II

References

Alais D, Blake R (2005)(eds) Binocular rivalry. MIT Press, Cambridge, MA
Alais D, Blake R (2015) Binocular rivalry and perceptual ambiguity. In: Wagemans J (ed) The Oxford handbook of perceptual organization. Oxford University Press, Oxford, pp 775–798
Alais D, O'Shea RP, Mesana-Alais C, Wilson IG (2000) On binocular alternation. Perception 29:1437–1445

Blake R (2005) Landmarks in the history of binocular rivalry. In: Alais D, Blake R (eds) Binocular rivalry. MIT Press, Cambridge, MA, pp 1–27
Blake R, Fox R (1974) Adaptation to invisible gratings and the site of binocular rivalry suppression. Nature 249:488–490
Blake R (2017) Binocular rivalry. The illusion of disappearance. In: Shapiro AG, Todorovic D (eds) The Oxford compendium of visual illusions. Oxford University Press, Oxford, pp 721–725
Blake R, Logothetis N (2002) Visual competition. Nat Rev Neurosci 3:13–21
Blake R, Brascamp J, Heeger DJ (2014) Can binocular rivalry reveal neural correlates of consciousness? Philos Trans Royal Soc London Ser B Biol Sci 369:20130211
Brascamp JW, Kink PC, Levelt WJM (2015) The 'laws' of binocular rivalry: 50 years of Levelt's propositions. Vis Res 109:20–37
Breese BB (1899) On inhibition. Psychol Monogr 3:1–65
Breese BB (1909) Binocular rivalry. Psychol Rev 16:410–415
Breese BB (1917) Psychology. Scribner's, New York
Brewster D (1844) On the knowledge of distance given by binocular vision. Trans Roy Soc Edin 15:663–674
Brewster D (1861) On binocular lustre. Report of the British Association for the Advancement of Science, pp 29–31
Crick F, Koch C (1995) Are we aware of neural activity in primary visual cortex? Nature 375:121–123
Desaguliers JT (1716) A plain and easy experiment to confirm Sir Isaac Newton's doctrine of the different refrangibility of the rays of light. Philos Trans Royal Soc 34:448–452
Desaguliers JT (1719) Lectures of experimental philosophy. Mears. Creake, and Sackfield, London
de Weert CMM, Wade NJ (1988) Compound binocular rivalry. Vis Res 28:1031–1040
Diaz-Caneja E (1928) Sur l'alternance binoculaire. Annales d'Oculistique 165:721–731
Dove HW (1851) Ueber die Ursache des Glanzes und der Irradiation, abgeleitet aus chromatischen Versuchen mit der Stereoskop. Annalen der Physik und Chemie 83:169–183
Dove HW (1852) On the stereoscopic combination of colours, and on the influence of brightness on the relative intensity of different colours. London Edinburgh Dublin Philos Mag J Sci 4:241–249
Du Tour E-F (1760) Discussion d'une question d'optique. Académie des Sciences. Mémoires de Mathématique et de Physique Présentés pars Divers Savants 3:514–530
Fox R (2005) Forward. In: Alais D, Blake R (eds) Binocular rivalry. MIT Press, Cambridge, MA, pp vii–xix
Fox R, Check R (1968) Detection of motion during binocular rivalry suppression. J Exp Psychol 78:388–395
Fox R, Hermann J (1967) Stochastic properties of binocular rivalry alternations. Percep Psychophys 2:432–436
Galton F (1878a) Address to the Department of Anthropology. Report of the Forty-Seventh Meeting of the British Association for the Advancement of Science. 1877. Transactions of the Sections, pp 94–100
Galton F (1878b) Composite portraits made by combining those of many different persons into a single figure. Nature 18:97–100
Galton F (1879) Composite portraits, made by combining those of many different persons into a single resultant figure. J Anthropol Inst G B Irel 8:132–144
Galton F (1908) Memories of my life. Methuen, London
Helmholtz H (1867) Handbuch der physiologischen Optik. In Karsten G (ed) Allgemeine Encyklopädie der Physik, vol 9. Voss, Leipzig
Helmholtz H (1925) Helmholtz's Treatise on Physiological Optics, vol 3. (trans: Southall JPC). Optical Society of America, New York
Hering E (1861) Beiträge zur Physiologie. I. Vom Ortsinne der Netzhaut. Engelmann, Leipzig
Howard IP, Rogers (2012) Perceiving in depth. Volume 2. Stereoscopic vision. Oxford University Press, USA

Klink PC, van Wezel RJA, van Ee R (2013) Reaching through a window on consciousness. In: Miller SM (ed) The constitution of visual consciousness. Lessons from binocular rivalry. Benjamins, Amsterdam, pp 305–332
Lehmkuhle SW, Fox R (1975) Effect of binocular rivalry suppression on the motion aftereffect. Vis Res 15:855–859
Leopold DA, Logothetis NK (1996) Activity changes in early visual cortex reflect monkeys' percepts during binocular rivalry. Nature 379:549–553
Levelt WJM (1965) On binocular rivalry. Institute for Perception, Soesterberg, The Netherlands
Logothetis NK, Leopold DA, Scheinberg DL (1996) What is rivalling during binocular rivalry? Nature 380:621–624
McDougall W (1901) On the seat of the psycho-physical processes. Brain 24:579–630
McDougall W (1903) The nature of the inhibitory processes within the nervous system. Brain 26:153–191
McDougall W (1906) Physiological factors of the attention-process (IV) Mind 15:329–359
Miller SM (2013) (ed) The constitution of visual consciousness. Lessons from binocular rivalry. Benjamins, Amsterdam
Müller J (1838) Handbuch der Physiologie des Menschen, vol 2. Hölscher, Coblenz
Müller J (1843) Elements of physiology, vol 2 (trans: Baly W). Taylor and Walton, London
Muryy AA, Fleming RW, Welchman AE (2016) 'Proto-rivalry': how the binocular brain identifies gloss. Proc R Soc B 283:20160383
Ngo TT, Barsdell WN, Law PCF, Miller SM (2013) Binocular rivalry, brain stimulation and bipolar disorder. In: Miller SM (ed) The constitution of visual consciousness. Lessons from binocular rivalry. Benjamins, Amsterdam, pp 211–252
O'Shea RP Binocular rivalry bibliography. https://sites.google.com/site/oshearobertp/publications/binocular-rivalry-bibliography
O'Shea RP, Crassini B (1981) Interocular transfer of the motion after-effect is not reduced by binocular rivalry. Vis Res 21:801–804
O'Shea RP, Roeber U, Wade NJ (2017) On the discovery of monocular rivalry. i-Perception, Nov–Dec, 1–12. https://doi.org/10.1177/2041669517743523
Panum PL (1858) Physiologische Untersuchungen über das Sehen mit zwei Augen. Schwerssche Buchhandlung, Kiel
Sherrington CS (1906) Integrative action of the nervous system. Yale University Press, New Haven, CT
Taylor J (1738) Le mechanisme ou le nouveau traité de l'anatomie du globe de l'oeil, avec l'usage de ses différentes parties, & de celles qui lui sont contigues. David, Paris
Tong F (2005) Investigations of the neural basis of binocular rivalry. In: Alais D, Blake R (eds) Binocular rivalry. MIT Press, Cambridge, MA, pp 63–80
Tong F, Nakayama K, Vaughan JT, Kanwisher N (1998) Binocular rivalry and visual awareness in human extrastriate cortex. Neuron 21:753–759
Towne J (1863) The stereoscope, and stereoscopic results—Section III. Guy's Hosp Rep 9:103–126
Towne J (1864) The stereoscope, and stereoscopic results—Section V. Guy's Hosp Rep 10:125–141
Tscherning MHE (1898) Optique physiologique. Masson, Paris
Tscherning M (1900) Physiologic optics: dioptrics of the eye, functions of the retina, ocular movements, and binocular vision (trans: Weiland C). Keystone Publishing, Philadelphia, PA
Venturi JB (1802) Betrachtungen über die Erkenntniss der Entfernung, die wir durch das Werkzeug des Gehörs erhalten, Archiv für die Physiologie 5:383–392
Wade NJ (1973) Contour synchrony in binocular rivalry. Percept Psychophys 13:423–425
Wade NJ (1974) Some perceptual effects generated by rotating gratings. Perception 3:169–184
Wade NJ (1975) Monocular and binocular rivalry between contours. Perception 4:85–95
Wade NJ (1980) The influence of colour and contour rivalry on the magnitude of the tilt illusion. Vis Res 20:229–233
Wade NJ (2016) Faces and photography in 19th-century visual science. Perception 45:1008–1035

Wade NJ, Wenderoth P (1978) The influence of colour and contour rivalry on the magnitude of the tilt after-effect. Vis Res 18:827–835
Wendt G, Faul F (2019) Differences in stereoscopic luster evoked by static and dynamic stimuli. i-Perception 10(3):1–26
Wheatstone C (1838) Contributions to the physiology of vision – Part the first. On some remarkable, and hitherto unobserved, phenomena of binocular vision. Philos Trans Roy Soc 128:371–394
Whittle P (1965) Binocular rivalry and the contrast at contours. Q J Exp Psychol 17:217–226
Whittle P, Bloor DC, Pocock S (1968) Some experiments on figural effects in binocular rivaly. Percept Psychophys 4:183–188
Wolfe JW (1986) Stereopsis and binocular rivalry. Psychol Rev 93:269–282
Wundt W (1862) Beitrage zur Theorie der Sinneswahrnehmung. Winter, Leipzig

Chapter 7
Binocular Controversies

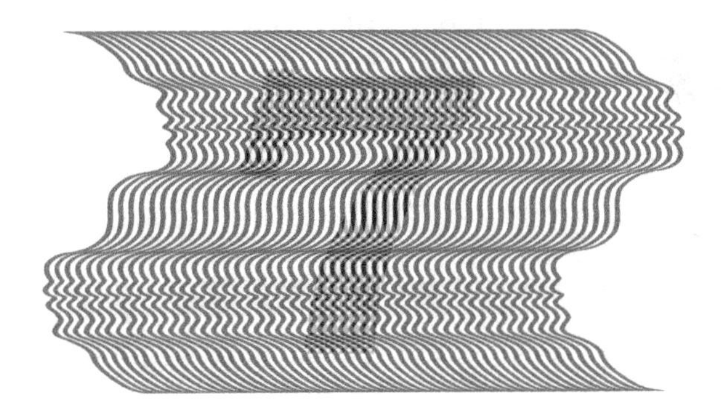

Rivalry has not been restricted to eyes or patterns but it has enveloped personalities, too. The pioneers of stereoscopy, Wheatstone and Brewster, clashed over the history of research on binocular vision, priorities of the invention of the stereoscope, interpretations of stereoscopic depth perception and lens separations for binocular photographs. Both were what we would now call physicists and both were essentially experimentalists rather than theorists. These rivalries have been evident from earlier chapters but they are examined here with particular reference to Brewster's attempts to wrest the priority of inventing the stereoscope from Wheatstone and of the notorious controversy over whether two drawings from the sixteenth century by Jacopo Chimenti were intended for a stereoscope. Brewster bridled under the impression that, as an authority on optics, he rather than Wheatstone should have invented the first stereoscope. That he did not was in part because of his theoretical outlook—the phenomena of binocular vision should comply with the laws established for vision with one eye. Wheatstone was not so constrained and explored the world of binocular vision with wonder and insight. It could be said that Brewster represented the theoretical past whereas Wheatstone pointed the way to the future of research on vision with two eyes. Unlike Brewster, Wheatstone argued that the phenomena of stereoscopic depth perception were in the domain of psychology rather than physics, of observation rather than optics. This contrast was mirrored in the conflicts between Helmholtz and Hering. Helmholtz expressed his theoretical debts to Wheatstone and adopted a similar cognitive/inferential analysis of vision with two eyes. Hering can be seen in the line of Brewster but Hering replaced reliance on physics to one on physiology. Both sets of rivalries drew stark attention to the psychology and sociology of science. For vision with two eyes personalities as well as patterns could engage in rivalry. Wheatstone was reluctant to entangle with "so disputatious an antagonist" as Brewster and wisely considered that his time was better employed in investigating new facts. He did, however, state that Brewster had viewed history through a pseudoscope!

N. Wade, *Vision and Art with Two Eyes*, Vision, Illusion and Perception 3,
https://doi.org/10.1007/978-3-030-77995-5_7

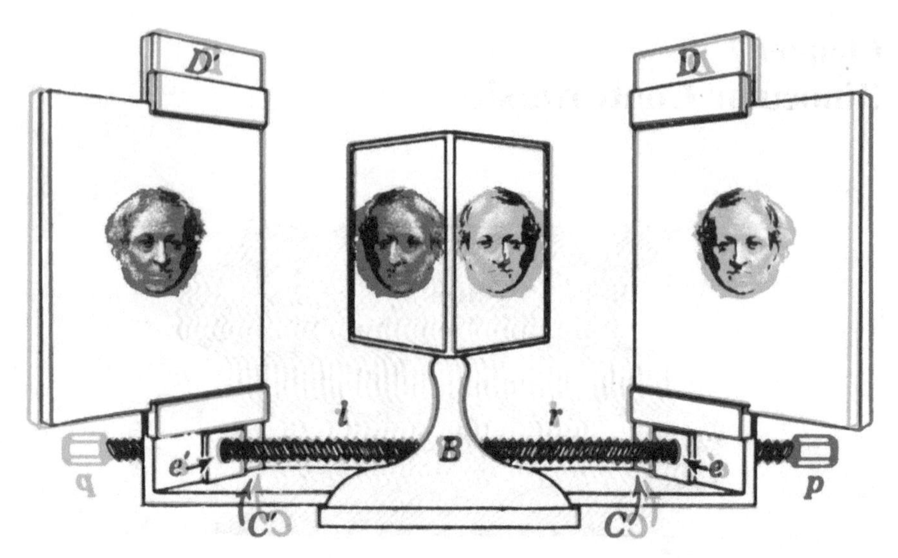

Binocular rivalries

From the very beginnings of stereoscopy rivalries have surfaced over priorities, perceptions and processes. Wheatstone and Brewster were pioneers of research on binocular vision but they did not see eye-to-eye on interpretations of stereoscopic vision nor on the history of its study. It is fitting, therefore, that they both examined binocular rivalry with the aid of Wheatstone's mirror stereoscope. There were no disputes about the optics of projections to the two eyes. Brewster made several such diagrams from 1830 onwards, and particularly in *The Stereoscope*; Wheatstone made few illustrations of this kind.

THE

STEREOSCOPE

ITS HISTORY, THEORY, AND CONSTRUCTION

WITH ITS APPLICATION TO THE FINE AND USEFUL ARTS
AND TO EDUCATION.

BY

SIR DAVID BREWSTER, K.H., D.C.L., F.R.S., M.R.I.A.,

VICE-PRESIDENT OF THE ROYAL SOCIETY OF EDINBURGH, ONE OF THE EIGHT
ASSOCIATES OF THE IMPERIAL INSTITUTE OF FRANCE, OFFICER OF THE
LEGION OF HONOUR, CHEVALIER OF THE PRUSSIAN ORDER OF MERIT,
HONORARY OR CORRESPONDING MEMBER OF THE ACADEMIES
OF PETERSBURGH, VIENNA, BERLIN, COPENHAGEN,
STOCKHOLM, BRUSSELS, GÖTTINGEN, MODENA,
AND OF THE NATIONAL INSTITUTE OF
WASHINGTON, ETC.

WITH FIFTY WOOD ENGRAVINGS.

LONDON:
JOHN MURRAY, ALBEMARLE STREET.
1856.

[The Right of Translation is reserved.]

Brewster and the title page of his book on the stereoscope

Their theoretical disputes about binocular rivalry were concerned with the sites at which depth and rivalry took place. Wheatstone adopted the methods of the physical sciences to support empiricist theories of perception which were later pursued by Helmholtz. Brewster also employed the methods of physics but his theories were physically based, too. He had little sympathy for inferential approaches to perception. With Brewster's background in optics, he would have been the one expected to devise a stereoscope and this was probably a factor in his hostility towards Wheatstone. Brewster's views did not change with knowledge of stereoscopic phenomena because for him the laws of binocular vision had to comply with those of monocular vision. Whereas Brewster was confounded by the phenomenal complexities of binocular rivalry they were seen as no impediment to Wheatstone's inferential theory. Similar theoretical issues can be found in the rivalries between Helmholtz and Hering.

Wheatstone and Brewster

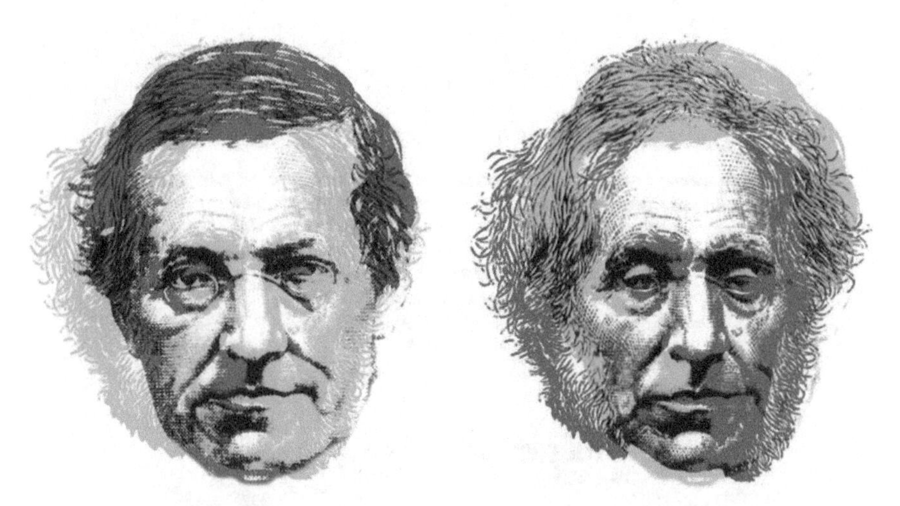

The binocular rivalries between Wheatstone and Brewster

The rivalries between Wheatstone and Brewster were personal as well as perceptual. In the 1830s both Wheatstone and Brewster came to stereoscopic vision armed with their individual histories of research on the senses. Brewster was an authority on physical optics and had devised the kaleidoscope; Wheatstone extended his research on audition to render acoustic patterns visible with his kaleidophone or phonic kaleidoscope. Both had written on subjective visual phenomena, a topic upon which they first clashed at the inaugural meeting of the British Association for the Advancement of Science in 1832 (the year Wheatstone made the first stereoscopes). When Wheatstone published his account of the mirror stereoscope Brewster's initial reception of it was glowing. However, he later disputed Wheatstone's theoretical interpretation as well as his priority regarding the invention of the stereoscope. They both described investigations of binocular contour rivalry but their interpretations diverged. As was the case for stereoscopic vision, Wheatstone argued for central processing whereas Brewster's analysis was peripheral and based on visible direction. They later clashed over Brewster's claim that drawings made by Jacopo Chimenti were made for a sixteenth century stereoscope.

Brewster was the older man and was generally considered as one of the foremost authorities in Britain on physical optics and on polarisation in particular. His analyses of binocular optics were based on the concept of visible direction—alignment with the point stimulated on the retina. He wrote a long article on optics[1] for the *Edinburgh Encyclopædia* and this formed the basis for his *Treatise of Optics* which ran to many editions.[2] He invented the kaleidoscope about which he wrote a book[3] and he published descriptions of a variety of visual phenomena, like afterimages, the colours and pattern distortions seen in finely ruled black-and-white gratings and the reversed depth seen in hollow objects such as masks.[4]

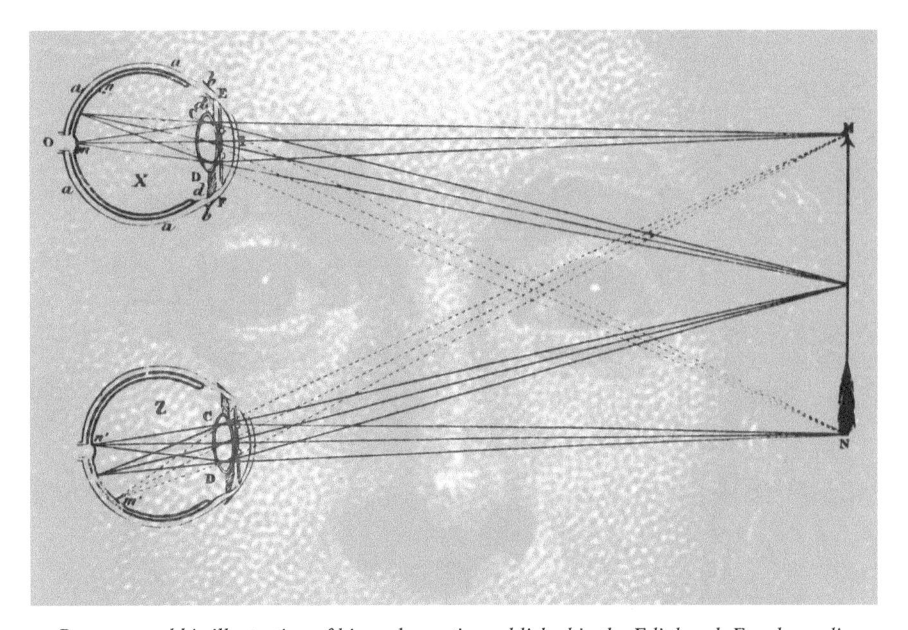

Brewster and his illustration of binocular optics published in the Edinburgh Encylopaedia

The Wheatstone family business, initially located in Gloucester and later in London, was concerned with the manufacture of musical instruments.[5] Wheatstone

---

[1] Brewster (1830).

[2] Brewster (1831).

[3] Brewster (1819).

[4] See Wade (1983) in which Brewster's publications on vision are reprinted.

[5] See Bowers (2001) for an excellent biography of Wheatstone.

was led to the study of vision through the visual expression of acoustic phenomena. Indeed, his first scientific paper was on acoustical figures and he later expressed these with a philosophical toy of his invention, which he called the kaleidophone or phonic kaleidoscope.[6] It enabled an observer to see the paths of rapidly vibrating rods. Wheatstone wrote: "In the property of 'creating beautiful forms,' the Kaleidophone resembles the celebrated invention of Dr. Brewster".[7] Around 1816 Brewster invented a devise for reflecting and rotation shapes with mirrors in a tube; he called it a kaleidoscope because it enabled beautiful patterns to be seen. The kaleidophone was an extension of a method described by Thomas Young,[8] in which silvered wire was attached to a piano string so that its vibration could be observed with the aid of a magnifying glass. Wheatstone constructed the kaleidophone to amplify the vibrations so that they could be seen by the naked eye. Silvered glass beads were attached to the ends of rods having different cross-sections and shapes; when the rods were bowed or struck complex figures could be seen in the light paths traced by reflections from the beads.

[6] Wheatstone (1823, 1827).

[7] Wheatstone (1827), p 344.

[8] Young (1800).

Kaleidoscopic Brewster and kaleidophonic Wheatstone

Wheatstone came to vision from his work on acoustics for the family business of musical instrument manufacture. This appealed to Wheatstone's mechanical ingenuity, and among his inventions was the concertina. His first scientific paper was on acoustical figures in which he also investigated binaural hearing.[9] He extended his research on audition to render acoustic patterns visible with his kaleidophone or phonic kaleidoscope. He provided a translated summary of Purkinje's book on subjective visual phenomena and described a better method of rendering visible shadows of the retinal blood vessels.[10] However, the rivalry between Wheatstone and Brewster remained theoretical until after the invention of the stereoscope when Brewster appreciated the implications of stereoscopic phenomena to his theory of vision. Thereafter Brewster sought every opportunity to diminish the importance of Wheatstone's invention. Their portraits are presented embedded in rivalling gratings.

[9] See Wade (2005), Wade and Ono (2005), Wade and Deutsch (2008).

[10] Wheatstone (1830).

Grating personalities

Brewster's positive reception of Wheatstone's mirror stereoscope turned to bitter acrimony with publication of Brewster's book on *The stereoscope*. He tried to wrest the invention of the stereoscope from Wheatstone, claiming that an 'ocular stereoscope' had been invented in 1834 by James Elliot, a teacher of mathematics in Edinburgh.[11] It consisted essentially of a box with a septum and two apertures to view two disparate drawings of a landscape. Elliot had been encouraged to make his claim by Brewster who went further and maintained that "Previous to or during the year 1834, he [Elliot] had resolved to construct an instrument for uniting two dissimilar pictures, or of constructing a stereoscope – but he delayed doing so till the year 1839".[12] Elliot claimed to have "constructed a stereoscope, in everything

[11] See Gill (1969), Klooswijk (1989, 2010), Wade (1983).

[12] Brewster (1856), p 19.

but name, more than thirteen years ago".[13] Thus, Elliot did not construct a stereoscope in 1834, but only "resolved to". Brewster reproduced the stereoscopic drawings attributed to Elliot to bolster his claim and an anaglyph of it is shown below. The claim was repeated in an anonymous letter to *The Times* later in 1856. The correspondence between Wheatstone and Brewster did have the virtue of establishing that both mirror and prism stereoscopes were made for Wheatstone in 1832, While Elliot retracted the claim made on his behalf, Brewster was not so repentant.[14]

Elliot's stereoscopic landscape

The ghost of Elliot continues to haunt the history of stereoscopes but in a slightly different guise—that of Helioth! Despite claims that Helioth invented a stereoscope in the 1830s, the first indications of his existence are found in the mid-twentieth century. In his book *Stereoptics* Dudley described early stereoscopes: "Although the invention of the stereoscope is usually attributed to Wheatstone, it is to be noted that the instrument which bears his name was preceded four years earlier by a stereoscope devised by Helioth. The instrument, however, incorporated no optical elements to assist convergence".[15] Since no bibliographical information is provided

[13] Elliot (1852), p 397.

[14] The *Times* correspondence between Wheatststone and Brewster is reprinted in Wade (1983).

[15] Dudley (1951), p 22.

by Dudleyabout Helioth's article, the source from which the statement derived cannot be checked. Helioth does not appear in accounts of the stereoscope and its history before Dudley.[16]

The term 'Helioth-Wheatstone stereoscope' was coined by Kaufman[17] citing Dudley as his source of information concerning Helioth. Kaufman's diagram was of a Wheatstone mirror stereoscope in spite of the fact that Dudleystated that Helioth's instrument contained no optical elements (lenses, prisms or mirrors). Subsequently, in a personal communication, Kaufman regretted that he had released this spectre. He planned to correct it in the next edition of *Sight and Mind*, but no second edition was published. Despite Kaufman's contrition, Schiff[18] repeated the phrase 'Helioth-Wheatstone' on Kaufman's authority, and the phrase has resurfaced in a flurry of recent reports that cite either Kaufman or Dudley or both.[19]

The non-existence of an individual is always difficult to determine, but in this case one can speculate on the reason for Dudley's shoddy scholarship. While it seems unlikely that Helioth ever existed, someone with a similar name did, and this person claimed to have considered making a stereoscope, without any optical elements, four years before Wheatstone published his memoir. That is the same year that Helioth is said to have devised his stereoscope. It seems likely that Dudley had learned about Elliot's claim but had either misheard the name or misconstrued the details to fabricate the spectre of Helioth. As with many of the distortions in the early history of stereoscopy, Brewster was at its heart, but it has been twisted further by Dudley and amply reflected by Kaufman. The claim that Helioth was the first person to construct a stereoscope in 1834 is without foundation for a further reason: Wheatstone constructed both reflecting and refracting stereoscopes in 1832, six years before he published his memoir (see Chap. 4).

Wheatstone and Brewster also clashed over the separations between lenses required for stereoscopic photographs. Brewster maintained that the separation (as in his binocular camera) should correspond to the separation between the eyes. Wheatstone was more pragmatic and provided a table of convergence angles for objects at different distances.[20] An example of this difference in approach can be seen with the stereophotographs of Brewster's statue which stands at King's Buildings at Edinburgh University. Like all my other stereoscopic photographs in this book they were taken with a single camera. The distance from the statue was about 6 m and the camera separation for the left pair was 6.5 cm and for the right pair is was 100 cm. The statue by William Brodie was erected in 1871 and unseen from the camera angles adopted a small stereoscope protrudes beneath Brewster's robes.

[16] Brewster (1856), Clay (1928), Helmholtz (1896), see also Gill (1969), Howard and Rogers (1995).

[17] Kaufman (1974).

[18] Schiff (1980).

[19] See Wade (2012).

[20] Wheatstone (1852).

Stereophotographs of Brewster's statue taken with camera separations of 6.5 and 100 cm

Chimenti Drawings

Not content with raising Elliot's spurious claim to cast doubt on Wheatstone's invention of the stereoscope, Brewster returned to the fray when he was informed of two sketches of a young man holding a compass and a plumb line by Jacopo Chimenti (c 1551–1640), an artist from Empoli in Tuscany.[21] Chimenti spent most of his life in Florence and painted altarpieces in Tuscany; he was noted for his skills as a draughtsman, particularly in still-life drawings. In addition to his altarpieces, his paintings hang in the major galleries of Europe. He produced numerous drawings throughout his life, but was forced to sell them in old age. Art historical accounts of Chimenti's work seldom mention specific drawings, but two of the sketches, when rediscovered in the nineteenth century, stirred the world of visual science. They were exhibited in the Musée Wicar, at Lille, mounted separately and side by side. Their dimensions are approximately 30 cm × 22 cm, and copies of photographs of the drawings (taken by and kindly provided by Arthur Gill) have been made into an anaglyph which is shown on the left below. Gill stated that there was no evidence that they had ever been mounted as a pair.[22] Woodcuts from photographs were published by Joseph Bancroft Reade (1801–1870) in 1862 and are reproduced as an anaglyph

[21] See Wade (2003, 2019).

[22] Gill (1969).

below.[23] Note that Reade's woodcuts did not include the rectangular frames present in the originals and also that copying errors would have been involved in making the woodcuts.

The Chimenti drawings derived from photographs taken by Arthur Gill (left) and from the woodcuts published by Reade in The Photographic Journal (right)

The drawings were seen by Alexander Crum Brown (1838–1922) on a visit to the museum in 1859. Crum Brown combined the drawings binocularly by converging his eyes to a point in front of them, and described the depth he saw. He considered that this union could only have been achieved if the drawings had been intended for binocular viewing. Crum Brown conveyed his observations on the Chimenti drawings in a letter to James David Forbes (1809–1868) who had succeeded Brewster as Principal of St Andrews University.[24] A small diagram was included in the letter, indicating how the eyes should converge in front of the pictures in order to combine them. Forbes immediately passed on Crum Brown's letter to Brewster who returned it two days later. In his reply Brewster indicated that he had requested a photograph of the Chimenti drawings. Brewster was President of the Photographic Society of Scotland and he read the letter at its monthly meeting, held in George Street Hall, Edinburgh, on 10 March, 1860. Crum Brown's letter was printed, with slight modifications, in the May issue of *The Photographic Journal* by Brewster. Brewster added his own commentary on its remarkable contents, concluding: "This account of the two drawings is so distinct and evinces such knowledge of the subject, that we cannot for

---

[23] Reade (1862). See also Brooks (2017).

[24] Brown (1860).

a moment doubt that they are binocular drawings intended by the artist to be united into relief either by the eye or by an instrument".[25]

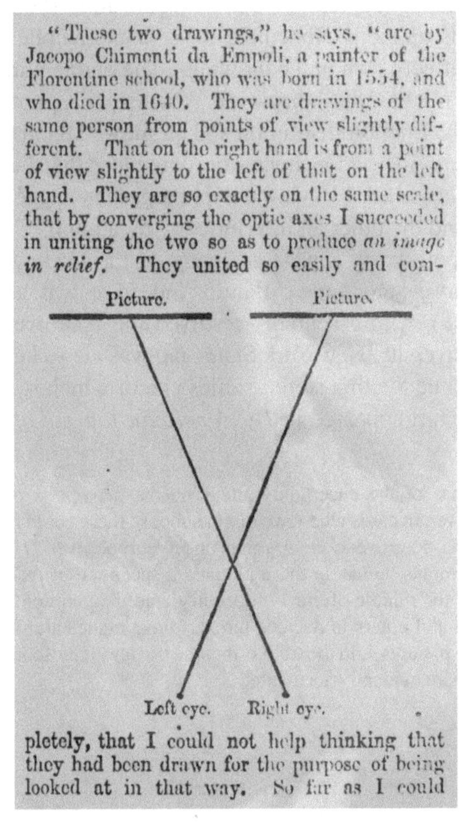

"These two drawings," he says, "are by Jacopo Chimenti da Empoli, a painter of the Florentine school, who was born in 1554, and who died in 1640. They are drawings of the same person from points of view slightly different. That on the right hand is from a point of view slightly to the left of that on the left hand. They are so exactly on the same scale, that by converging the optic axes I succeeded in uniting the two so as to produce *an image in relief*. They united so easily and com-

pletely, that I could not help thinking that they had been drawn for the purpose of being looked at in that way. So far as I could

Alexander Crum Brown and part of Brewster's second publication (from The Photographic Journal, 1862) of his letter describing the Chimenti drawings

Brewster's pursuit of photographic copies of the drawings did not prove immediately possible because the director of the museum informed him that "in consequence of certain abuses having taken place, [he] had resolved not to allow any copies to be taken of the pictures and drawings".[26] Thus, Brewster's claim that the drawings were stereoscopic was made on the basis of Crum Brown's description alone; he had not seen the Chimenti pictures himself, but this did not inhibit him including reference to the paired drawings in his entry on the 'Stereoscope' for the eighth edition of the *Encyclopædia Britannica*.[27] Brewster even went on to refer to the technique of over-convergence for uniting binocular pictures as 'the method of Chimenti'! One cannot

[25] Brewster (1860a), p 233.

[26] Brewster (1860a), p 233.

[27] Brewster (1860b).

but be impressed by the scale of Brewster's imagination based on a description of two drawings he had not even seen.

Brewster's chagrin must have been heightened on discovering that Wheatstone had obtained photographic copies of the drawings in June 1860 although he did not publish his comments on them. Nonetheless, his London colleagues did cast doubt on the stereoscopic effects seen in the drawings. The eminent microscopist William Benjamin Carpenter (1813–1885) displayed Wheatstone's photographic copies in a lecture delivered at the London Institution and he showed them to "some of the most eminent photographers" in London; none were convinced that they were stereoscopic. Artists who Carpenter enlisted to view the photographs suggested that one "is the work of the master and the other an inferior copy by a pupil".[28]

The 'Chimenti controversy' ensnared many of the leading visual scientists in Europe and even those across the Atlantic. Indeed it was an American, Edwin Emerson (1823–1908), who essentially resolved the controversy. Emerson was a professor at Troy University, New York State, and was also a keen photographer. He took what might well be the first stereo-selfies one of which is shown below.[29]

Emerson's first report appeared in *The American Journal of Science and Arts* in 1862. He noted:

> A remarkable instance of the uncertainty attending the perception or non-perception of stereoscopic relief, even in cases where we might suppose there could be no want of knowledge is shown by the controversy now going on in Europe over *The Chimenti pictures.* Sir David Brewster thinks he has in these pictures a specimen of real stereoscopic drawings produced about the middle of the 17th century; and this opinion is endorsed by Prof. Tait, Prof M'Donald and others in decided terms. I have made careful examination of the photographs of these pictures, and the truth is that the trifling stereoscopic and pseudoscopic qualities about them are evidently accidental.[30]

Self-stereoscopic portrait of Edwin Emerson

[28] Carpenter (1862), p 11.

[29] See Wade (2014).

[30] Emerson (1862), p 315.

Emerson went on to describe an experiment of making a sketch and then trying to copy it exactly noting that it will generate differences that are both stereoscopic and pseudoscopic: "you have done that very thing that Sir David Brewster has repeatedly declared was quite beyond human skill! If Prof. Wheatstone gets no heavier blow than this, his fame as discoverer is secure".[31] Brewster's ire was raised when Emerson's article was reprinted in the *Philosophical Magazine* the next year, and he responded in a typically robust manner, casting aspersions on Emerson's sentience and on his science.[32]

Brewster's riposte elicited four replies from Emerson. In the first he stated that: "I felt myself unable to permit even so distinguished a man as Sir David Brewster to make, unchallenged, the insinuations of unfair dealing, poor logic, &c, by which he gives a character to his letter",[33] and proceeded to refute Brewster's criticisms of his earlier article. The second reply highlights the improbabilities of the drawings being made to produce a stereoscopic effect. These included the dimensions of the drawings, the inappropriateness of the subject matter, and the absence of any written account about them. Brewster had discounted Wheatstone's observations on the Chimenti drawings because he was an interested party, presumably assuming that Brewster himself was not! Emerson enlisted four impartial observers (of whom one was Ogden Rood, an authority on vision and art) to combine the pictures; none of them described any stereoscopic effects. Emerson concluded: "The foregoing is probably sufficient to satisfy unprejudiced readers that, as far as competent, independent testimony can settle the character of the Chimenti drawings, they have no claim to the importance with which it would suit Sir David Brewster to invest them".[34] Ogden Rood's expertise was enlisted again for the third retort[35]; he was asked to copy one of the drawings as accurately as he could, and Emerson did the same. There was a slight stereoscopic effect when they were combined. Emerson's final reply contained the most convincing evidence. He pursued the approach initially proposed by Carpenter—he measured the dimensions of the two drawings, and a colleague did so independently. These were of internal features of each drawing, like "From the highest point of the head to the point of the left foot". The outcome was clear:

> As Sir David Brewster requires particulars, we enumerate:- A stereoscopic left knee, a pseudoscopic dress hanging over it, a stereoscopic left arm, but a pseudoscopic back, a pseudoscopic stool, and a pseudoscopic left foot, and a right foot still more pseudoscopic relatively to the left foot and the stool; and so we might go over the whole picture and show a *mélange* of pseudoscopic and stereoscopic lines, producing precisely the commingled and uncertain effect which a drawing and an ordinary copy of it would produce if adjusted for the stereoscope. And yet these are the pictures that Dr. Crum Brown 'succeeded in uniting so as to produce *an image in relief*' and they united for him '*easily and completely!*' These are the pictures which Sir David Brewster claims, gravely and persistently, a high 'degree of

[31] Emerson (1862), p 315.

[32] Emerson (1863), Brewster (1864).

[33] Emerson (1864a), p 111.

[34] Emerson (1864b), p 133.

[35] Emerson (1864c).

> stereoscopic effect,' and on that account boldly proceeds to give the seventeenth century the high honour rightly belonging to one of the most beautiful discoveries of our own age![36]

Emerson's conclusion has been confirmed by Brooks[37] who presented copies of the Chimenti drawings to naïve observers and found no consistent ratings of depth. The most likely interpretations of the paired pictures are that Chimenti either made two drawings of the boy in sequence or made a copy of the sketch. In both cases disparities would have been introduced by the inaccuracies of drawing but they would not have been consistent.

Whether Brewster saw the error of his ways we do not know, but he did not re-enter the fray. The controversy was at its height when Helmholtz was writing the third volume of his *Handbuch der physiologischen Optik*, which treated stereoscopic vision. By that stage, photographic copies of the Chimenti sketches were available, as well as the woodcuts from *The Photographic Journal*, although Helmholtz believed that the two drawings were on the same sheet of paper. He gave a sober description of the arguments, and provided another possible interpretation:

> D. Brewster conjectured that the pictures may have been made by Chimenti to test Porta's theory, which was published in 1593. Since then photographic reproductions of these pictures have been made for sale. The two pictures of the man were certainly made from different positions, but I must admit that it seems to me very unlikely that Chimenti intended them for a stereoscopic experiment, because the stool, the dividers, and the plumb line, which could easily have been drawn correctly, are treated as unessentials and all drawn so irregularly and so differently that they cannot be combined. Had the artist desired to test a theory, it is more likely that he would have drawn the easy things correctly and the difficult parts, such as the man, more inaccurately. It seems more probable to me that the artist was not quite satisfied with the first figure and did it over again from another point of view, using the same sheet of paper quite by accident.[38]

A few years after writing this, in 1871, Helmholtz visited Scotland and played golf with Crum Brown, Peter Guthrie Tait and Thomas Huxley at St Andrews; he also spent some time sailing with Crum Brown and William Thompson off the West coast of Scotland.[39] It is interesting to speculate whether they discussed these controversial drawings during their round voyage!

In *The Times* correspondence over the invention of the stereoscope, Wheatstone wrote: "I have hitherto avoided entangling myself in the meshes of controversy with so disputatious an antagonist as Sir David Brewster. I have always thought myself more usefully employed in investigating new facts, than in contending in respecting errors which time will inevitably correct".[40] In the cases of both Elliot's stereoscope and the Chimenti drawings time has indeed found in Wheatstone's favour. Perhaps the most fitting epitaph to this sorry saga was delivered by Reade in 1862: "The eye is a treacherous guide when fortified by a little previous theory".[41]

---

[36] Emerson (1864d), p 204.

[37] Brooks (2017).

[38] Helmholtz (1925), p 363.

[39] See Wade and Swanston (2001).

[40] Reprinted in Wade (1983), p 181.

[41] Reade (1862), p 29.

This cautionary tale of the unbridled search for support of an initially untestable and ultimately untenable theory should not be considered as an historical oddity. It is a consequence of associating theories with their protagonists, and continues to be a feature of the scientific enterprise. In this regard, a quotation from the editor of the *British Journal of Photography*, George Shadbolt, in August 1860 (when the controversy was in its infancy) is most apposite:

> It is very unfortunate that when an announcement of any supposed fact is once made, and subsequently proved to be erroneous, it is almost impossible to correct the false impression as thoroughly as is desirable, because there must always exist many persons who read the assertion but not the contradiction, while those who see the contradiction without the previous erroneous statement can play but a very unimportant part in its rectification... We think it is fair therefore to presume that, whatever may have been the object proposed by the artist in executing the two similar pictures, it was certainly not from any knowledge of the stereoscopic phenomenon, and that Sir David Brewster was in this instance wrong in his conjecture.[42]

Helmholtz and Hering

The binocular rivalries between Helmholtz and Hering

There are similarities between the rivalry involving Wheatstone and Brewster with the later one that enmeshed Helmholtz and Hering. Helmholtz was the older and established scientist and the conflicts involved binocular vision (as well as

[42] Shadbolt (1860), p 232.

colour perception). Moreover, Helmholtz adopted a theoretical stance similar to that advanced by Wheatstone whereas Hering, like Brewster, employed concepts of visual direction and eye movements. Hering, however, placed greater emphasis on physiological mechanism than Brewster. Steven Turner has written an excellent book on the rivalry between Helmholtz and Hering in the context of vision.[43]

The contributions made by Helmholtz to visual science are legion,[44] but his most lasting impact was his theory of perception. For Helmholtz the brain only had indirect access to the external world, via the senses, and it could only process messages in the language of nerve impulses. This realization made any equation of the retinal image with perception unnecessary. Helmholtz argued that this only created a problem if there was a picture in the retina that required further perception. This applied to stereoscopic vision in the same way as it did to other aspects of space perception. His empiricist theory and the experiments he used to bolster it was expressed in his major work on physiological optics, the three volumes of which were published together in 1867. Hering published his theory of binocular vision in the following year. It contrasted his nativist theory of vision with the empiricism of Helmholtz.

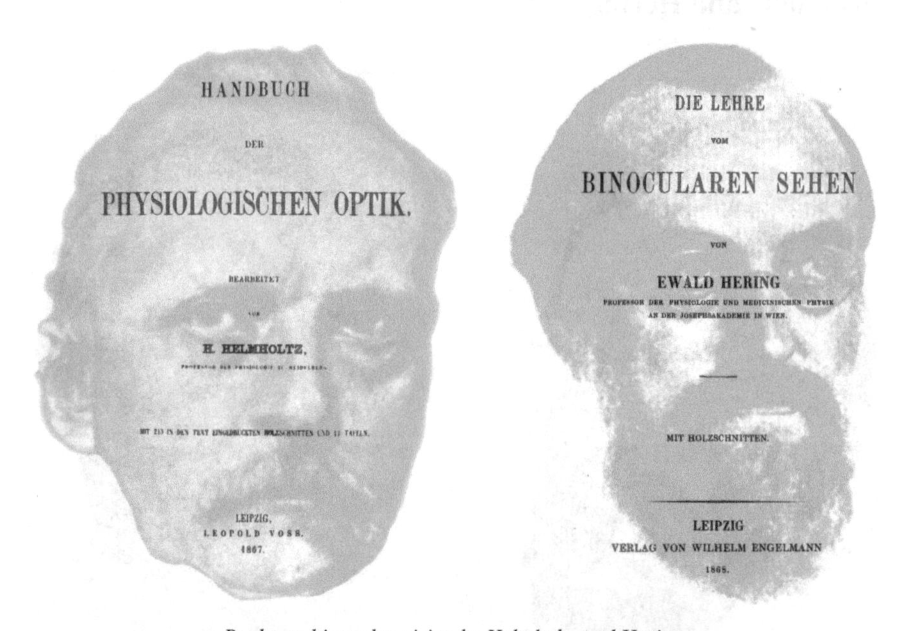

Books on binocular vision by Helmholtz and Hering

Helmholtz acknowledged that little he wrote on theories of vision was novel, but he marshalled the arguments over a wider range of phenomena than others had done before. By adopting a starkly empiricist interpretation of perception, and by

---

[43] Turner (1994). See also Howard (1999), Wade (2021).

[44] See Cahan (1993, 2018).

contrasting it so sharply with nativism, he reopened a debate that has reverberated throughout perception ever since. The debate was personified in the conflict between Helmholtz and Hering, and the main battle-grounds were colour vision and stereoscopic depth perception. Helmholtz summarized his position succinctly:

> *The sensations of the senses are tokens for our consciousness, it being left to our intelligence to learn how to comprehend their meaning....* Evidently, any other sensations, not only of sight but of other senses also, produced by a visible object when we move our eyes or our body so as to look at the object from different sides or to touch it, etc., may be learned by experience. The content of all these possible sensations united in a total idea forms our *idea* (*Vorstellung*) of the body; and we call it *perception* when it is reinforced by actual sensations.... The only psychic activity required for this purpose is the regularly recurrent association between two ideas which have often been connected before. The oftener this association recurs, the more firm and obligatory it becomes.[45]

Hering was a physiologist whose psychology was in the tradition of Goethe. He represented the phenomenological and nativist position in studying perception. His work in vision concerned space perception, colour vision and contrast phenomena.[46] Like Goethe, Hering stressed the subjective dimension of colour, and he based his opponent-process theory on colour appearances rather than on mixing lights of different wavelengths after the manner of Young, Maxwell, and Helmholtz. He adopted the procedure of presenting coloured papers to observers and asking them to name the colours from which they were mixed. Red, green, blue, and yellow were not said to be mixtures of any other colours, nor were black and white. He also examined simultaneous and successive colour contrast phenomena. Together these led him to propose a theory of colour vision based on three oppositional pairs: red-green, blue-yellow, and white-black. He speculated that there are three retinal pigments that are either built up or broken down by light to yield the six elements.

In the area of space perception Hering utilized the concept of local sign; each retinal point was considered to have a local sign for height, width, and depth. This conflicted with Helmholtz's emphasis on learning to interpret the retinal signals, largely via information from eye movements. Hering did investigate binocular eye movements and argued that the two eyes move as a single unit: his law of equal innervation states that when one eye moves the other moves with equal amplitude and velocity, either in the same or the opposite direction. Eye movements were also implicated in visual direction: "For any given two corresponding lines of direction, or visual lines, there is in visual space a single visual direction line upon which appears everything which actually lies in the pair of visual lines".[47] The centre of visual direction was called the imaginary central eye by Hering and the cyclopean eye by Helmholtz. Hering carried out experimental investigations of spatial illusions, at least two of which bear his name—the Hering illusion and the Hering grid[48] both of which are shown below. Hering also attracted a large body of able and loyal students who would carry the nativist banner into the twentieth century.

---

[45] Helmholtz (1925), pp 533–534.

[46] See Baumann (2002).

[47] Hering (1868), p 17.

[48] Hering (1861, 1907). See also Wade (2014, 2016).

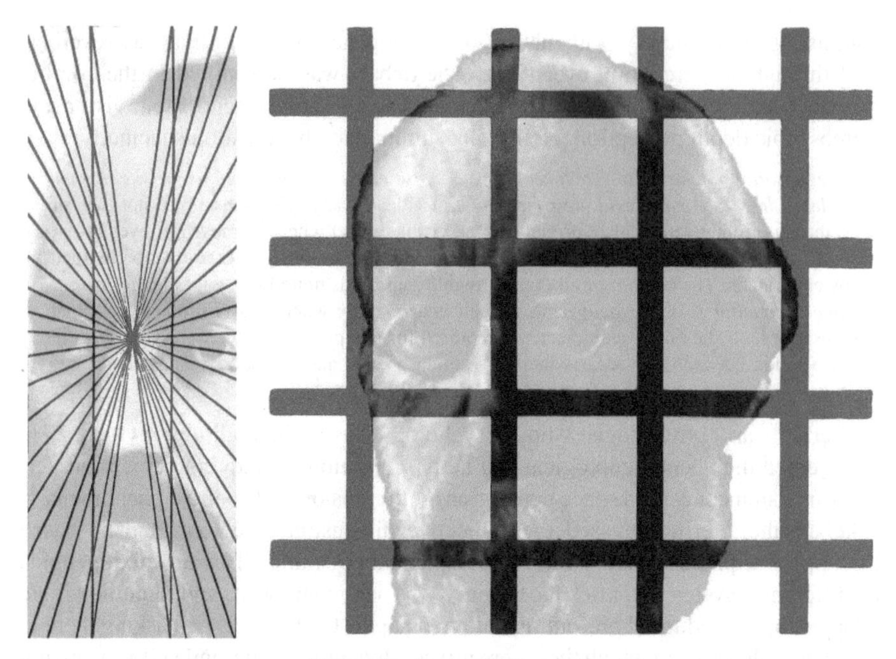

Hering illusion and Hering grid

The critical issue in the conflict between Helmholtz and Hering was whether local signs were adaptable or fixed and how they served stereoscopic vision. "The Empirical Theory regards the local signs (whatever they really may be) as signs the signification of which must be learnt, and is actually learnt, in order to arrive at a knowledge of the external world… The Innate Theory, on the other hand, supposes that the local signs are nothing else than direct conceptions of differences in space as such, both in their nature and their magnitude".[49] Hering had argued that each retinal point has signs for height, breadth, and depth, and that binocular fusion was physiological. Helmholtz considered that each monocular view was available to consciousness and that fusion was a psychical act. The stereoscope proved to be the critical instrument for examining the predictions from the two theories, and one of Wheatstone's figures—a thick vertical line in one eye and a thin vertical and thick inclined line in the other—was of cardinal importance; Hering referred to it as the 'Wheatstone experiment'.[50] Wheatstone said that the vertical line in one eye combined with the inclined line in the other to be seen in depth. That is, fusion of the thick lines involved non-corresponding points whereas the corresponding vertical lines were seen double. Helmholtz improved on the stimulus by presenting two lines to each eye (Plate III of the *Handbuch*) and reached a similar conclusion, even when

[49] Helmholtz (1873), p 275.

[50] Hering (1862).

the stereopairs were illuminated by an electric spark to exclude the effects of eye movements. It is not surprising that Helmholtz concluded "The invention of the stereoscope by Wheatstone made the difficulties and imperfections of the Innate Theory of sight much more obvious than before".[51]

The psychology of the senses led Helmholtz into the metaphysical domain he had assiduously avoided in his physical and physiological endeavours. The philosophical problems remain a matter of constant revision and reanalysis, but many of the issues concerned with the senses that were debated in Helmholtz's day became the topics of experimental enquiry in the then new discipline of psychology.

Optics and Observation

The history of research on binocular vision reflects a rivalry between two approaches—one based on optics and the other on observation. The former is concerned with optical projections and the latter with visual directions as they are observed. The origins of both approaches can be traced to antiquity.[52] The optical tradition was most clearly enunciated by Euclid who reduced space perception to the geometry of projections from the eye; the observational tradition was espoused by Aristotle, but its ablest early protagonist in binocular vision was Ptolemy.[53] Euclid examined binocular vision in the context of optical projections to spheres differing in diameter with respect to the interocular separation. While Euclid's analysis of binocular vision was geometrical, it was also cursory; he examined three dimensions of spheres that could be observed by two eyes, and simply related them to the amount of the spheres in the optical projections. Ptolemy carried out controlled observations of the perceived locations of vertical cylinders; from these he specified the conditions for singleness of vision, the distinction between crossed and uncrossed disparities, and the direction in which objects are seen with two eyes. Ptolemy's analysis was extended by Alhazen.

The modern optical tradition was based upon Kepler's analysis of retinal image formation, and the specification of direction in terms of lines of projection to the retina. Brewster's adherence to the concept of visible direction placed him squarely in this optical line. By means of the stereoscope, Wheatstone was able to present defined horizontal retinal disparities to yield predictable relative depth perception. That is, differing visual directions were associated with depth rather than diplopia, providing that the disparities were not too large. Wheatstone initiated the union between the optical and observational traditions. He drew a distinction between physical philosophy (optical projection) and mental philosophy (observation), and placed stereoscopic depth perception in the domain of the latter.

---

[51] Helmholtz (1873), p 274.

[52] See Wade and Swanston (1996).

[53] See Howard and Wade (1996).

None the less, both Wheatstone and Brewster relied on their own observations in deciding on the possible stereoscopic status of the Chimenti drawings. They did not apply the methods that were emerging in Germany to measure perception with the precision associated with physics. In the year that the Chimenti drawings were made a *cause célèbre*, Gustav Theodor Fechner (1801–1887) published his account of psychophysics.[54] However, the resolution of the controversy was a consequence of Emerson's recourse to physics rather than psychology, to measurement of the stimulus rather than the response. The history of science may look upon this controversy of observation as a triumph for optics!

[54] Fechner (1860).

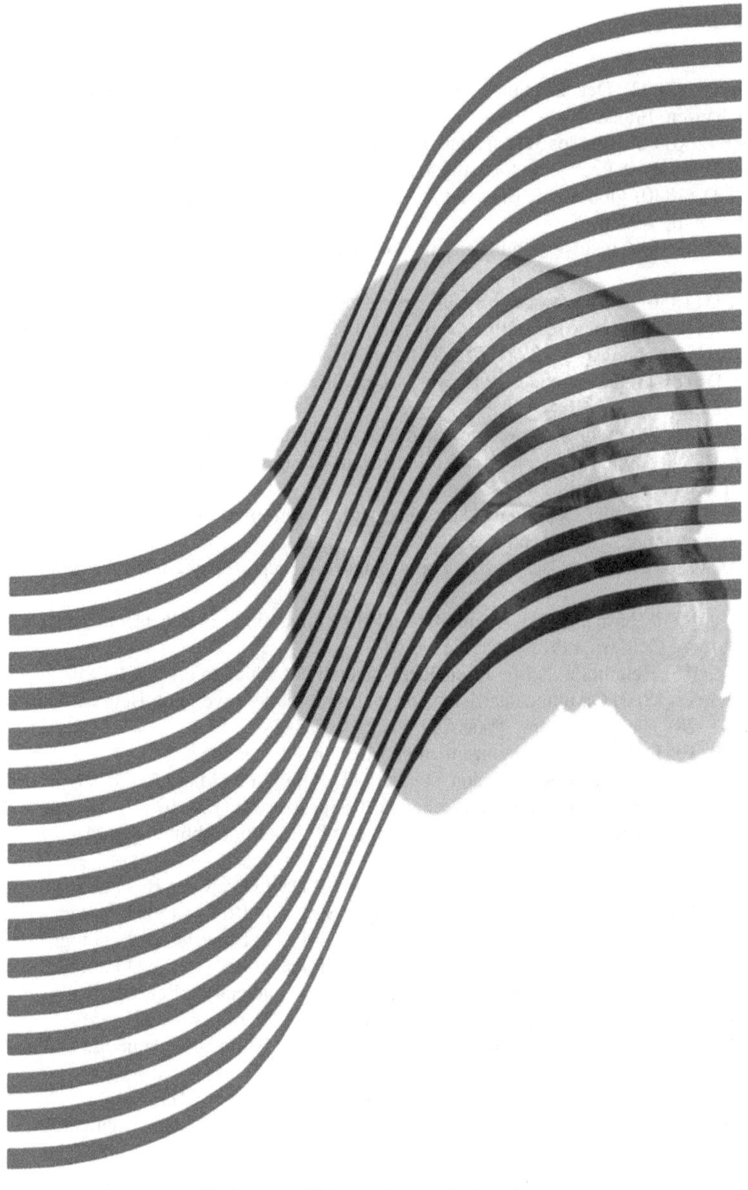

Fechner and his psychometric functions

References

Baumann C (2002) Der Physiologe Ewald Hering (1834–1918). Curriculum Vitae. Hänsel-Hohenhausen, Frankfurt
Bowers B (2001) Sir Charles Wheatstone. Her Majesty's Stationery Office, London
Brewster D (1819) A treatise on the kaleidoscope. Constable, Edinburgh
Brewster D (1830) Optics. In: Brewster D (ed) Edinburgh Encyclopædia, vol 15. Blackwoods, Edinburgh, pp 460–662
Brewster D (1831) A treatise on optics. Longmans, Rees, Orme, Brown and Green and John Taylor, London
Brewster D (1856) The stereoscope. Its history, theory, and construction. John Murray, London
Brewster D (1860a) Notice respecting the invention of the stereoscope in the sixteenth century, and of binocular drawings, by Jacopo Chimenti da Empoli, a Florentine artist. Photogr J 6:232–233
Brewster D (1860b) Stereoscope. In: Encyclopaedia Britannica, vol 20, 8th edn. Adam and Charles Black, Edinburgh, pp 684–691
Brewster D (1862) On the stereoscopic pictures executed in the 16th century. Photogr J 8:9–12
Brewster D (1864) On the stereoscopic relief in the Chimenti pictures. London, Edinburgh and Dublin Philos Mag J Sci 27:1–3
Brooks K (2017) Depth perception and the history of three-dimensional art: who produced the first stereoscopic images? i-Perception 8(1):1–22
Brown AC (1860) Letter to J D Forbes. Forbes papers, msdep 7/25 a & b (St Andrews University Library)
Cahan D (1993) Hermann von Helmholtz and the foundations of nineteenth century science. University of California Press, Berkeley, CA
Cahan D (2018) Helmholtz: A life in science. University of Chicago Press, Chicago
Carpenter WB (1862) On binocular vision and the stereoscope. Liverpool, Henry Greenwood
Clay RS (1928) The stereoscope. Trans Opt Soc 29:149–166
Dudley LP (1951) Stereoptics: an introduction. MacDonald, London
Elliot J (1852) The stereoscope. London Edinburgh Dublin Philos Mag J Sci 3:397
Emerson E (1862) On the perception of relief. Am J Sci Arts 34:312–316
Emerson E (1863) On the perception of relief. London Edinburgh Dublin Philos Mag J Sci 26:125–130
Emerson E (1864a) The Chimenti pictures: a reply to Sir David Brewster. Br J Photogr 11:111–112
Emerson E (1864b) The Chimenti pictures: a reply to Sir David Brewster. Br J Photogr 11:132–133
Emerson E (1864c) The Chimenti pictures: a reply to Sir David Brewster. Br J Photogr 11:167–169
Emerson E (1864d) The Chimenti pictures: a reply to Sir David Brewster. Br J Photogr 11:202–204
Fechner GT (1860) Elemente der Psychophysik. Breitkopf and Härtel, Leipzig
Gill AT (1969) Early stereoscopes. Photograph J 109:546–559, 606–614, 641–651
Helmholtz H (1873) Popular lectures on scientific subjects. Atkinson E Trans. Longmans Green, London
Helmholtz H (1896) Handbuch der physiologischen Optik ed 2. Voss, Hamburg
Helmholtz H (1925) Helmholtz's treatise on physiological optics, vol 3 (trans: Southall JPC). Optical Society of America, New York
Hering E (1861) Beiträge zur Physiologie. I. Vom Ortsinne der Netzhaut. Engelmann, Leipzig
Hering E (1862) Beiträge zur Physiologie. II. Von den identischen Netzhautstellen. Engelmann, Leipzig
Hering E (1868) Die Lehre vom Binokularen Sehen. Engelmann, Leipzig
Hering E (1907) Vom simultanen Grenzkontrast. In: Saemisch T (ed) Graefe-Saemisch Handbuch der gesamten Augenheilkunde. Engelmann, Leipzig, pp 135–141
Howard IP (1999) The Helmholtz-Hering debate in retrospect. Perception 28:543–549
Howard IP, Rogers BA (1995) Binocular vision and stereopsis. Oxford University Press, Oxford
Howard IP, Wade NJ (1996) Ptolemy's contributions to the geometry of binocular vision. Perception 25:1189–1201

Kaufman L (1974) Sight and mind. Oxford University Press, Oxford
Klooswijk A (1989) Elliot's inventions. Bull Stereoscopic Soc 24:11
Klooswijk A (2010) Was wist James Elliot over de stereoscopie. 3-D Bulletin (NL) 187:32–36, 188:27–33
Reade JB (1862) The Chimenti pictures. Photogr J 8:29–30
Turner RS (1994) In the eye's mind. Vision and the Helmholtz-Hering controversy. Princeton University Press, Princeton, NJ
Shadbolt G (1860) Editorial. Br J Photogr 7:232
Schiff W (1980) Perception: an applied approach. Houghton Mifflin, Boston
Wade NJ (1983) Brewster and Wheatstone on vision. Academic Press, London
Wade NJ (2003) The Chimenti controversy. Perception 32:185–200
Wade NJ (2005) Sound and sight: acoustic figures and visual phenomena. Perception 34:1275–1290
Wade NJ (2012) The ghost of Helioth and his stereoscope: the return of a phantom. Perception 41:1001–1002
Wade NJ (2014) Geometrical optical illusionists. Perception 43:846–868
Wade NJ (2016) Art and illusionists. Springer, Heidelberg
Wade NJ (2019) Ocular equivocation: the rivalry between Wheatstone and Brewster. Vision. https://doi.org/10.3390/vision3020026
Wade NJ (2021) Helmholtz at 200. i-Perception 12(4):1–19. https://doi.org/10.1177/20416695211022374
Wade NJ, Deutsch D (2008) Binaural hearing—before and after the stethophone. Acoust Today 4:16–27
Wade NJ, Ono H (2005) From dichoptic to dichotic: historical contrasts between binocular vision and binaural hearing. Perception 34:645–668
Wade NJ, Swanston MT (1996) A general model for the perception of space and motion. Perception 25:187–194
Wade NJ, Swanston MT (2001) Helmholtz on golf. Perception 30:1407–1410
Wheatstone C (1823) New experiments on sound. Ann Philos 6:81–90
Wheatstone C (1827) Description of the kaleidophone, or phonic kaleidoscope: a new philosophical toy, for the illustration of several interesting and amusing acoustical and optical phenomena. Q J Sci Lit Arts 23:344–351
Wheatstone C (1830) Contributions to the physiology of vision. No. 1. J Royal Institute 1:101–117
Wheatstone C (1852) Contributions to the physiology of vision—part the second. On some remarkable, and hitherto unobserved, phenomena of binocular vision. Philos Trans R Soc 142:1–17
Young T (1800) Outline of experiments and enquiries respecting sound and light. Philos Trans R Soc 90:106–150

Chapter 8
Binocular Art

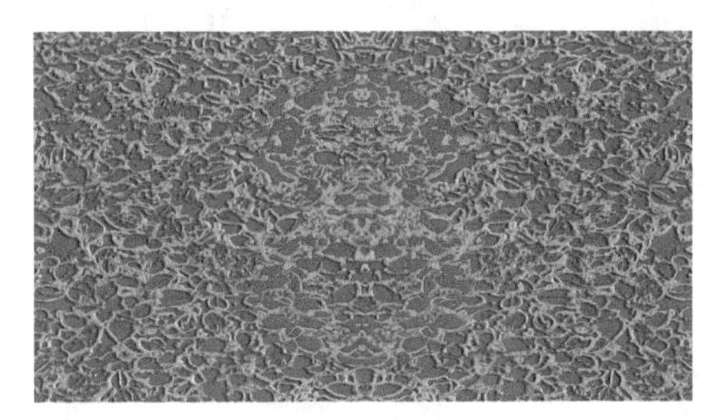

Stereoscopes provided not only a precise instrument for investigating the science of vision with two eyes but they also opened up a new world of art. Stereoscopic pictures could supply a dimension missing from those that preceded them—depth. The new art presented graphical difficulties for artists but these were overcome by the novel processes of photography. As Wheatstone astutely observed "What the hand of the artist was unable to accomplish, the chemical action of light, directed by the camera, has enabled us to effect". Many cameras have achieved this effect since then and the majority of binocular art is photographic. Photographs are shown in this chapter but they are not confined to stereoscopic effects nor are they concerned solely with capturing scenes. Many binocular phenomena are celebrated in the anaglyphic images that follow—stereoscopic and rivalling designs as well as photographs. Indeed, manipulating photographs and designs offers freedoms not available to conventional binocular photographs: a specified viewpoint is not required nor are there limits to the disparities between the two components. In fact they are intentionally extreme for rivalling images. Moreover, stereoscopic photographs of scenes yield apparent depth with a fixed combination of eye and filter so that each eye views the scene from the appropriate direction; reversing the filters does not reverse the apparent depth. Such constraints do not apply to stereoscopic or rivalling designs and readers are encouraged to vary the eye/filter combinations for viewing them. The science of stereoscopy was advanced when random dot displays were devised so that depth perception could be investigated independently of object recognition. Here a wider array of carrier patterns for disparities has been devised so that the patterns have an appeal of their own. Photographs of natural shapes like leaves, flowers, branches and stones can be manipulated graphically to create complex patterns of symmetry; disparities can be introduced in these that cannot be detected until combined with similar patterns to yield apparent depth. Rivalry art has been neglected relative to stereoscopic art. This chapter tries to restore some balance by presenting patterns, photographs and portraits that display complex instabilities when viewed through red/cyan glasses. Thus, binocular art is broader than stereopsis because it can address the competition between the eyes in rivalry and lustre as well as their cooperation in yielding stereoscopic depth.

N. Wade, *Vision and Art with Two Eyes*, Vision, Illusion and Perception 3,
https://doi.org/10.1007/978-3-030-77995-5_8

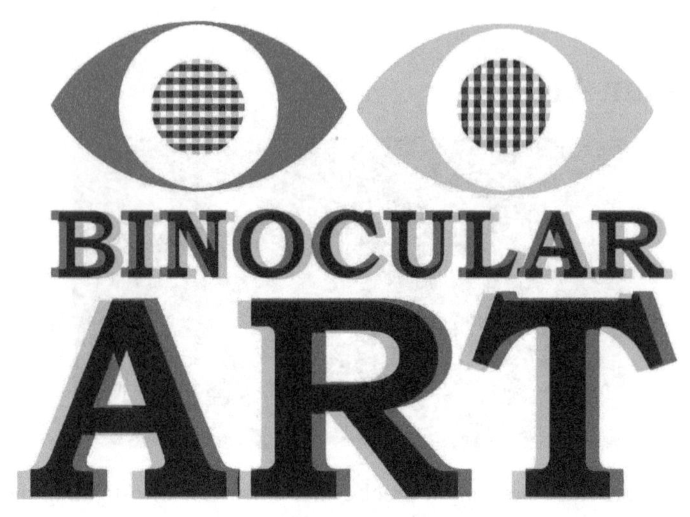

Deyeoptic Art

The term 'binocular art' is assigned to those pictorial works which are dependent on the operation of two eyes either in cooperation (stereoscopic or dichoptic) or in competition (rivalry or lustre). There has been much sterile debate about whether photography is pictorial art and I do not wish to add to it. However, the debate is essentially pointless in the context of binocular pictorial art as it did not exist prior to the almost synchronous inventions of stereoscopy and photography. Most pictures made for two eyes are stereoscopic and there are many books devoted to them. The emphasis in this chapter will be on stereoscopic images and the neglected aspect of vision with two eyes, namely rivalry.

Stereoscopic Art

Paintings in two-dimensions allude to the depth that they do not contain: a variety of stratagems is enlisted to convey the impression that surfaces on the picture plane are at different distances from the viewer. The lengths to which artists have gone to fool the eye attest to the difficulty of achieving this. Another strategy is to produce two paintings with defined horizontal disparities between them and viewed with the aid of a stereoscope. If the paintings are large then they are typically placed facing one another and combined by means of a mirror stereoscope. Such paired pictures are individually flat and the depth is determined by disparity. Few artists have had the skill and patience to pursue this strategy. Much of the stereoscopic art is now photographic or computer based. Despite the ease with which stereoscopic photographs can be taken it is remarkable that there are excellent examples of stereoscopic paintings, particularly those by Salvador Dali.[1]

[1] Martinez-Conde et al. (2015), Wade (2021a).

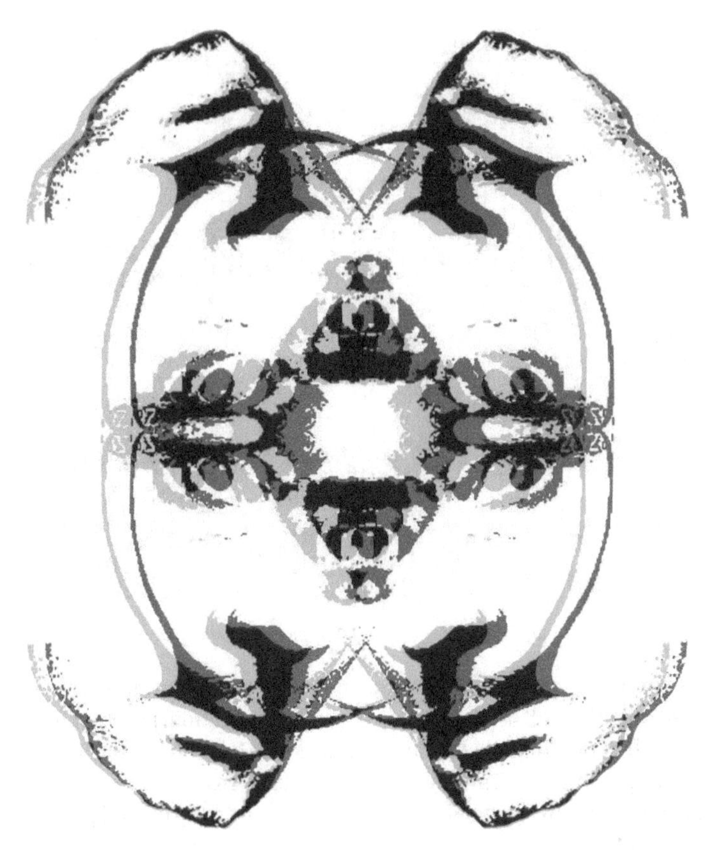

Stereoscopic Salvador Dali

Ludwig Wilding described himself as a concrete artist.[2] His work is geometrical and abstract with a close affinity to the processes of perception. Indeed, some of his early works were anaglyphic. As was mentioned in Chap. 5, he devised a new way of creating disparities with superimposed and separated gratings. Disparities were between the relative locations of moiré fringes in each eye and the stereoscopic depth changed with the movements of the observer towards or away from the works.

[2] See Wade (2007a, b, c).

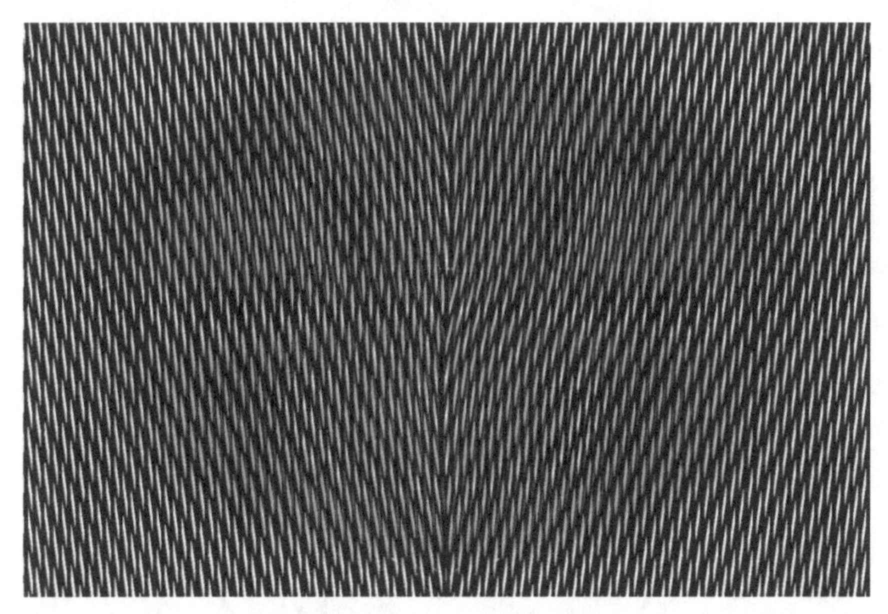

Ludwig Wilding—stereo moiré master

Calum Colvin commenced his artistic endeavours by painting scenes over three-dimensional objects with a view from a particular station point. Alignment was maintained so that the final photographic image was taken from the only point from which this was possible. Essentially, it is like the view from a *camera obscura*. When viewing his photographs, the solid scene is initially overlooked and pictorial flatness dominates perception. With more protracted viewing, the objects in the scene emerge and a strange tension is introduced between the solid and the flat, between the scene and the seen. It could be considered as turning *trompe l'oeil* on its head—objects in three dimensions are intentionally merged in the picture plane—and their identity is recognised after the flatness is transcended. His more recent works represent explorations of new dimensions. By adopting two viewpoints, neither of which will yield perfect alignment between the contours painted on the solid objects, retinal disparity is introduced. The clues to the objects are given visually rather than conceptually. Even so, disparity takes time to develop and our familiarity with pictorial images tends determine the initial visual victory.

In fact, the subjects of some of his stereoscopic works have been Wheatstone and Brewster.[3] Depth derived from disparity vies with pictorial depth, so that the art is not stereoscopic in the narrow sense but the works display a dynamic duel between the pictorial and binocular cues to depth. A dual duel is implied with rivalry not only between the slight contour misalignments but also between selected elements within each stereoscopic image. In one sense the binocular works of Calum Colvin revert to the approach adopted by Wheatstone and Brewster, but in another they add a delightful twist to this stereoscopic tale. By incorporating rivalry with stereoscopy he

[3] Colvin (2009). See also Wade (2009).

is extending the art of the third dimension. Accordingly, it is appropriate to represent this natural magician in a manner that draws on the marvels of binocular combination.

Calum Colvin in Caledonia

Terry Pope is a constructivist artist who has designed new forms of hyper- and pseudo-scopes, as was described in Chap. 4. He has also made three-dimensional stimuli for viewing with these devices. Accordingly, he is shown below together with a *Stereocube* of his devising. Both the *Stereocube* and the portrait are in stereo but the former was photographed with an excessive separation between the two views.

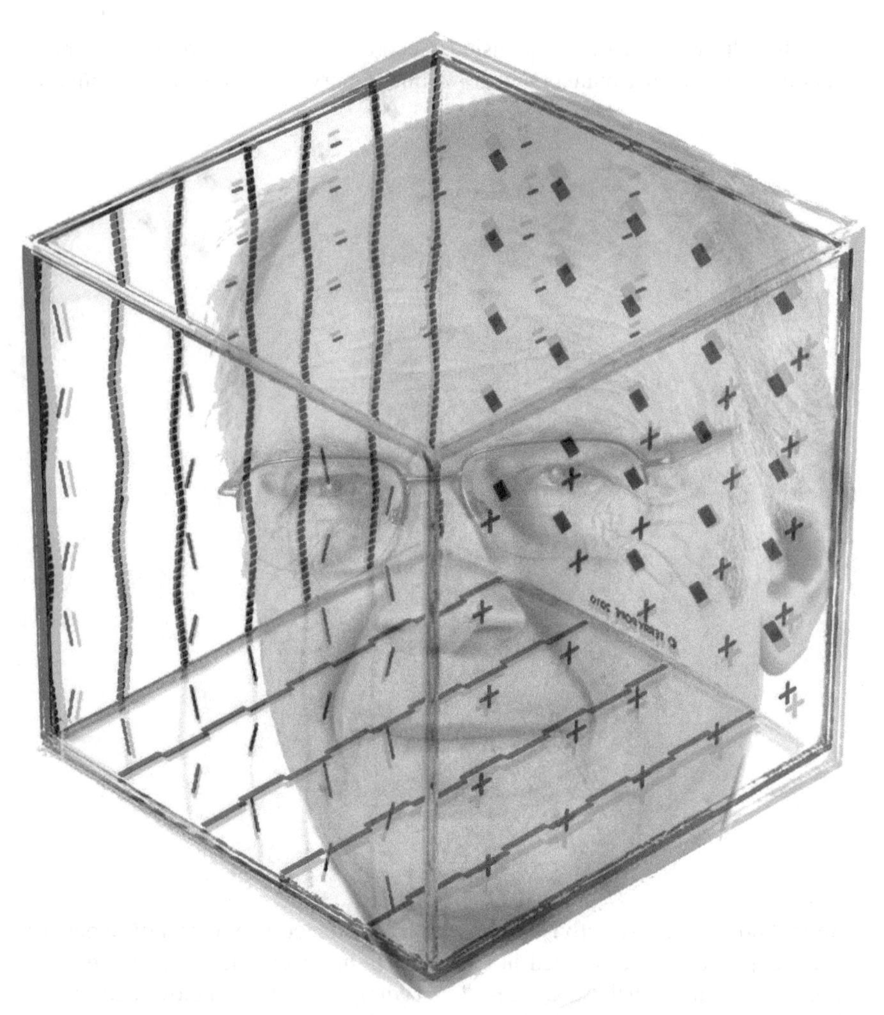

Terry Pope and his Stereocube

Stereoscopic art is undergoing a resurgence of interest in artistic circles but not in terms of the conventional representation of depth from disparity. An example of this is the work of Sara Le Roy who combines narrative painting with stereoscopic imagery.[4] Much of her work interweaves dark and surreal themes with anaglyphic depth so that the stereoscopic effects are but a part of a partially painted scene. The paintings are made up of layers, starting from the most distant and photographs are taken of each layer. The photographs are assembled to produce anaglyphs and then printed on canvas which is painted yet again. Sara Le Roy is shown below in anaglyphic combination with conventional portraits embedded in a surreal background. The

[4] https://saraleroy.com/.

background is derived from a flowing abstract painting that was photographed and digitally manipulated.

Surreal Sara Le Roy

There is relatively little abstract stereoscopic painting but it is possible to introduce stereoscopic depth to abstract geometrical paintings, as in the example below which contains several depth planes. Not only does the surface appear in curved depth but the discs in each quadrant are also at different stereoscopic depths.

Twisted

The remaining parts of this chapter contain examples of designs and photographs exhibiting the many ways in which different patterns presented to the two eyes can complement or compete with one another.

Photographs

The traditional method for introducing apparent depth or distance into a design is linear perspective. Visual angles are captured to signify relative distances—for objects of the same physical size the smaller their representations the greater their apparent distance. Another powerful source of spatial suggestion is occlusion—figures that partially obscure parts of others are seen as nearer. However, these are monocular cues to distance. Here we are concerned with those that require the use of two eyes. All the stereoscopic photographs shown below were taken with a single-lensed camera with a time difference between the two components. The spatial separations between the two exposures varied according to the distances of objects in the scene. Therefore some disparities are a consequence of objects moving between the two exposures as well as their spatial separation.

The centre of each member of a stereopair of photographs is typically a specific point in the scene. When combining the two stereoscopic images this point can be common to both and superimposed or other features can be arranged in correspondence. The examples below are combined from the same two photographs of an overgrown, ruined stone cottage. In the upper one the stonework of the arch is in correspondence whereas the lower one has been arranged to superimpose the trunk of the central tree.

Woodland cottage

The examples above have used conventional photographs to produce the stereoscopic effects but the photographic components themselves can be manipulated prior to stereoscopic combination. Most of the detail in a photograph can be removed by extracting the boundaries between the lighter and darker regions in it. That is, just the contours in the image remain rather than the full range of colours and lightnesses as is shown in the tree scene below.

Trees, Scotscraig

Another simple modification of a photograph is to render it like a silhouette, with only the light and dark areas present. This can be seen in the photograph of Trafalgar Square in which Nelson's Column and the equestrian state of King George IV are the prominent features.

Trafalgar Square

Designs

Most graphic designs are two dimensional; they are produced to display the interactions of contours and colours. Stereoscopic designs do not need to comply with the requirements of stereoscopic photography where objects in space do have defined distances from the camera capturing them. Variations in structure between two designs can introduce depth in ways that are difficult, if not impossible, to achieve with photographs of objects. In all the images shown in this section the component patterns can readily be seen by closing each eye in turn so that the red/cyan separations are visible. In every case, the starting point was a two-dimensional pattern which was either a graphic design or a photograph which was then manipulated in the dark room or with computer graphics. Often the stereoscopic effect is not visible initially and so some patience might be required for the depth to emerge. Unlike conventional stereoscopic photographs, reversing the filters will reverse the depth seen.

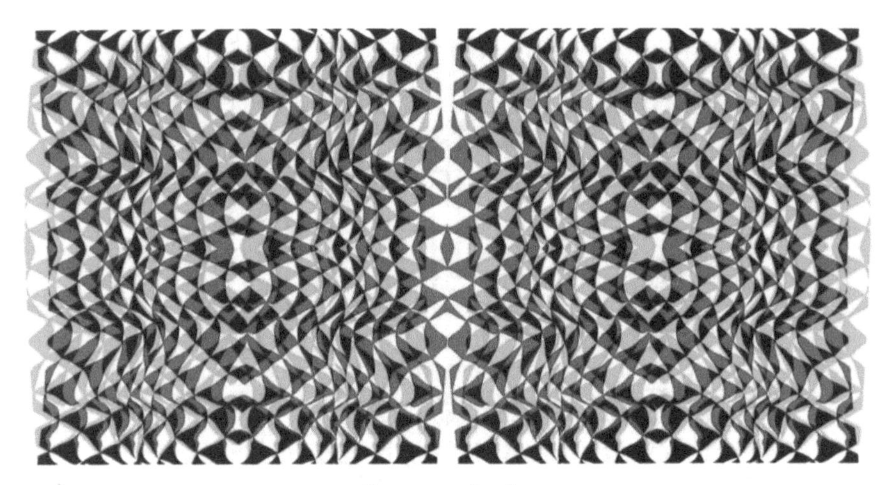

Curvaceous borders

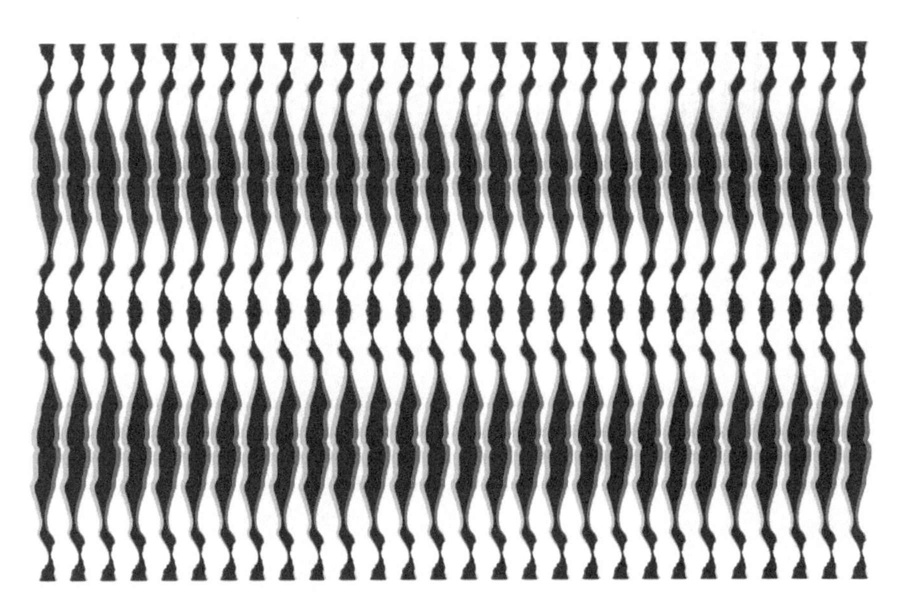

Chorus lines

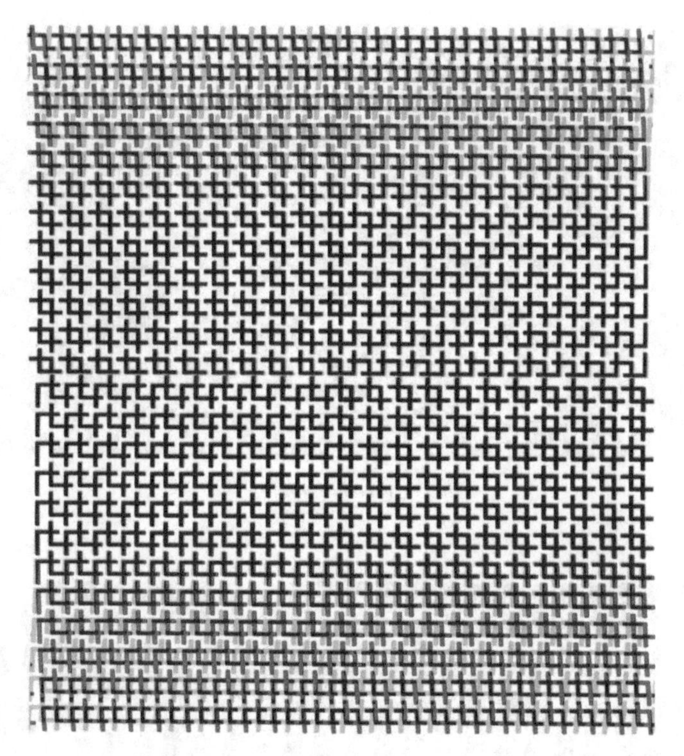

Curved space

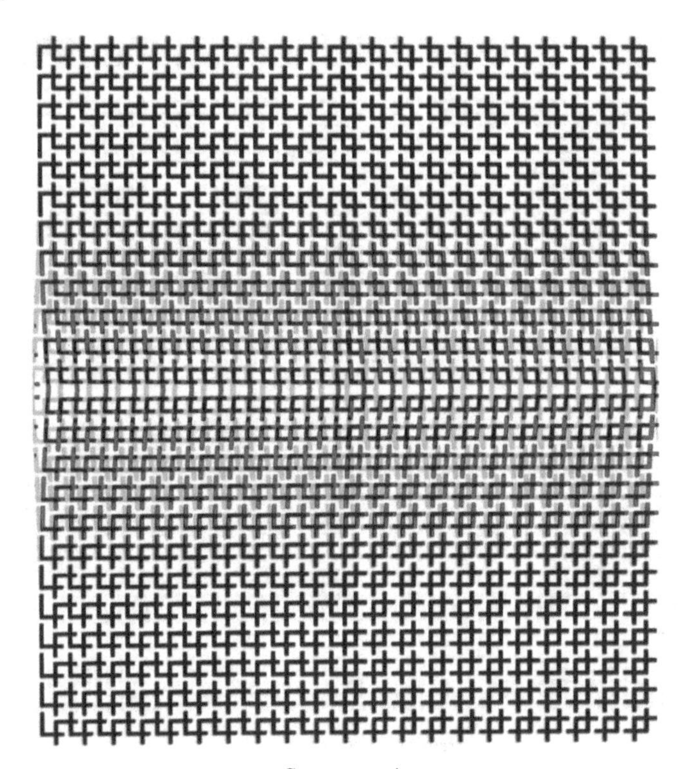

Cusp or crevice

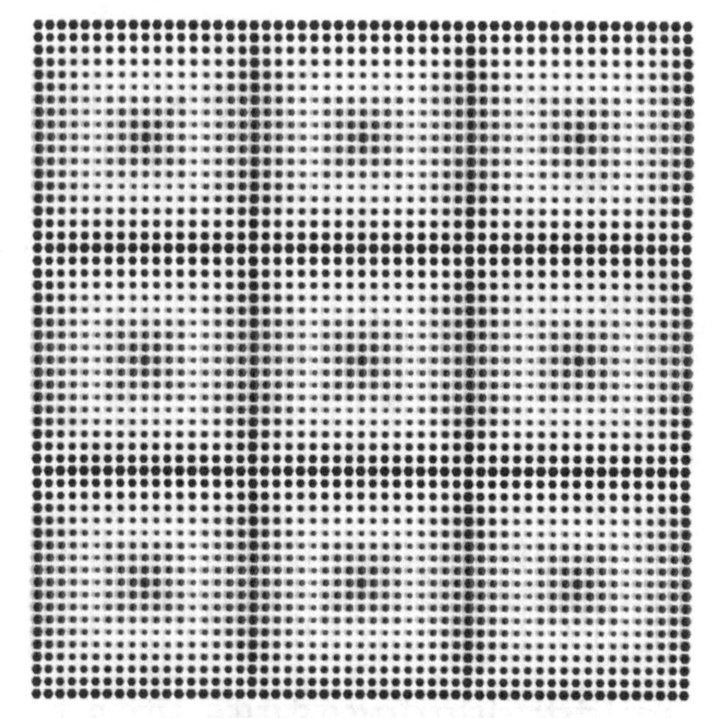

Hemicylindrical hexagons

The stereoscopic depth in designs can be over the whole surface, as in the illustrations above, and depth can also be produced within designs as in the following examples. The first, *Arrows*, contains a simple straight arrow in depth contrasting to those carrying it whereas the depth is more complex in *Facing* and is likely to take longer to see the depth image.

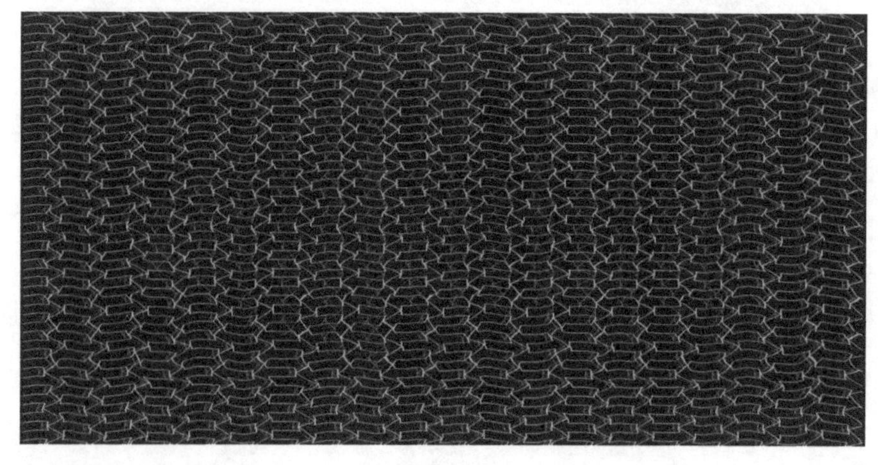

Arrows

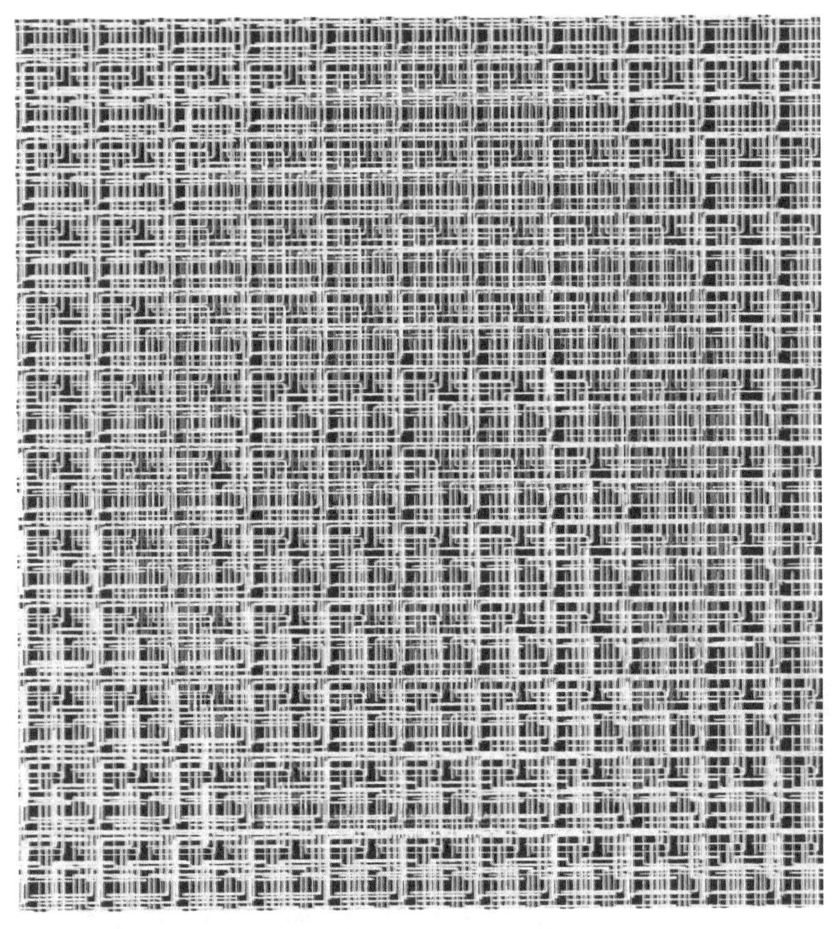

Facing

The upper and lower halves of *Windows on the world* each contain a schematic eye and they can be discerned by their depths with respect to the background. The lower part is the negative of the upper half but the disparities are the same in both. The carrier pattern is derived from a photograph of branches and twigs which produce a network in which the laterally displaced schematic eyes cannot be seen in monocular views. As is often the case with stereoscopic designs, the depth seems more compelling with negative images.

Windows on the world

As is evident in *Windows on the world* the designs employed to create stereoscopic images can be naturalistic; a photograph of tree branches was used to carry two schematic eyes in depth. Branches are a most appropriate carrier because the pattern of retinal blood vessels is referred to as the Purkinje tree! In the case of *Stone well* the symmetrical pattern is derived from a photograph of stones which has been manipulated, multiplied and modified. There are six depth planes in the design but they are not visible until both eyes are used. With the red/LE, cyan/RE combination it looks like different levels of a stone well whereas the reversed arrangement appears tower-like.

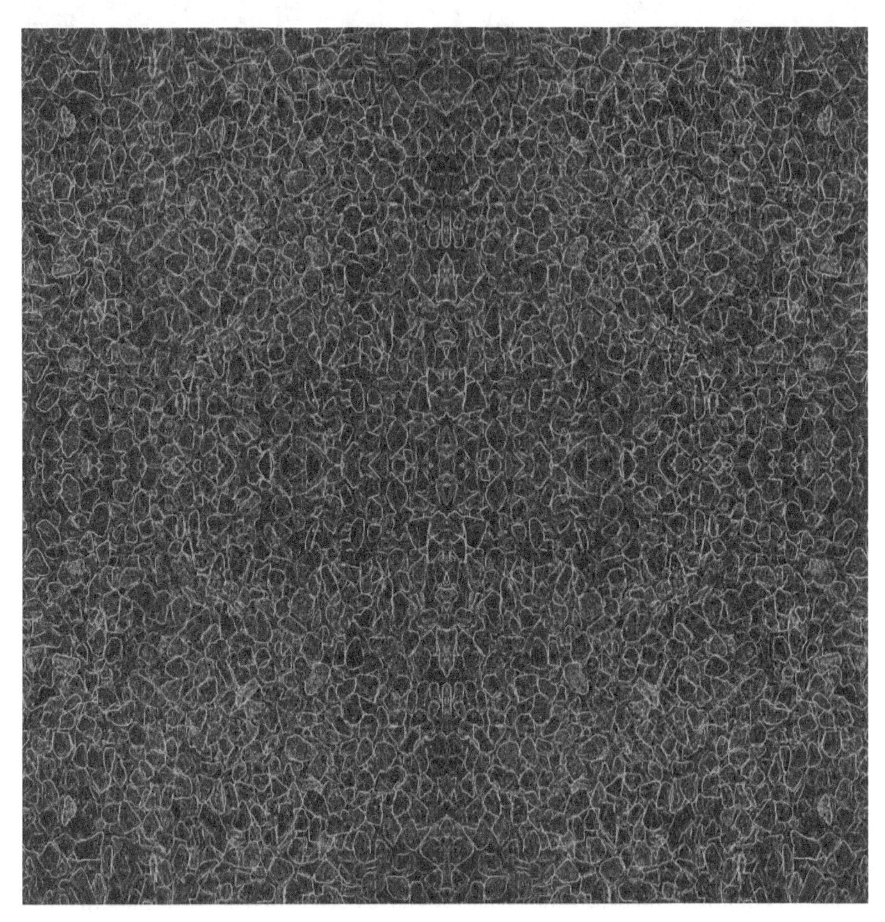

Stone well

A photograph of stones on a beach provided the starting for *Beach petals*. The multiple stereoscopic levels in the pattern create the impression of petals radiating from the centre and circling is at different depths.

Beach petals

The next three designs were derived from natural textures. For *Fronds* the starting point was a photograph of branches and leaves on a pine tree. *Growth and form* started from a photograph of rhododendron flowers. Autumnal leaves provided the basis for *Leaf fall*. In all cases, the disparities have been added after the patterns were designed and the first two involve parts that both approach and recede in depth; there are several depth planes in *Growth and form* and *Leaf fall*.

Fronds

Growth and form

Leaf fall

The next two designs have also been derived from natural forms. In the first example the design is based on a photograph of beech leaves the contours of which have been traced and combined. A single beech leaf can be seen in depth in the centre and the design is framed in a similar leaf shape. The second one was initially a photograph of leaves on a blueberry bush.

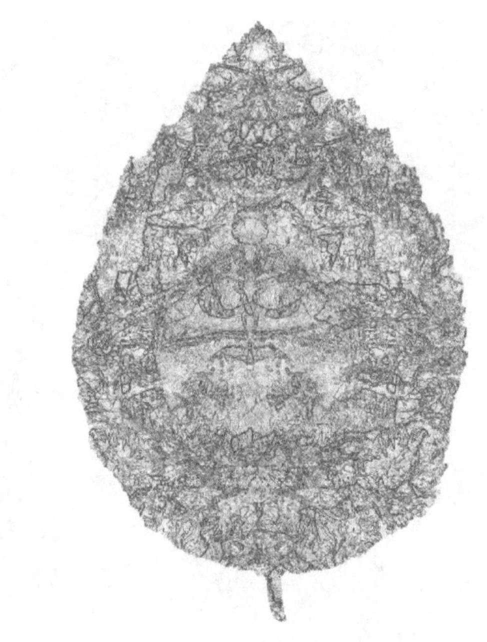

Beech leaves

Blueberry leaves

Continuing the leaf theme, it is possible to play on more subtle aspects of pictorial representation. Magritte drew attention to this in a series of pipe paintings in which beneath the depicted pipe were the painted words (in French) 'This is not a pipe'. He was stating that both written words and images are different representations of objects, one of the category (all pipes) and the other of identity (the particular pipe shown).[5] The theme is echoed in *These are not tobacco leaves*; patterns of tobacco

[5] See Foucault (1982) and Wade (1990, 2011).

leaves initially conceal a picture of a curved briar pipe within them which emerges stereoscopically.

These are not tobacco leaves

The stereoscopic depth in the last set of stereograms has involved disparities between the centrally defined regions of the images, but similar depth can be seen when the centre involves disparity or the surround.

Centre and surround disparities

Many manipulations can be made with stereoscopic designs both in terms of the carrier configuration as well as the form of the apparent depth within it. As is evident from the previous examples in this section, the apparent form of a surface can be modulated as can the relative depths of different parts. Indeed, both of these can be modulated together in the same direction, as is shown in *Hearts*, or the opposite depths as in *Curved ellipses.*

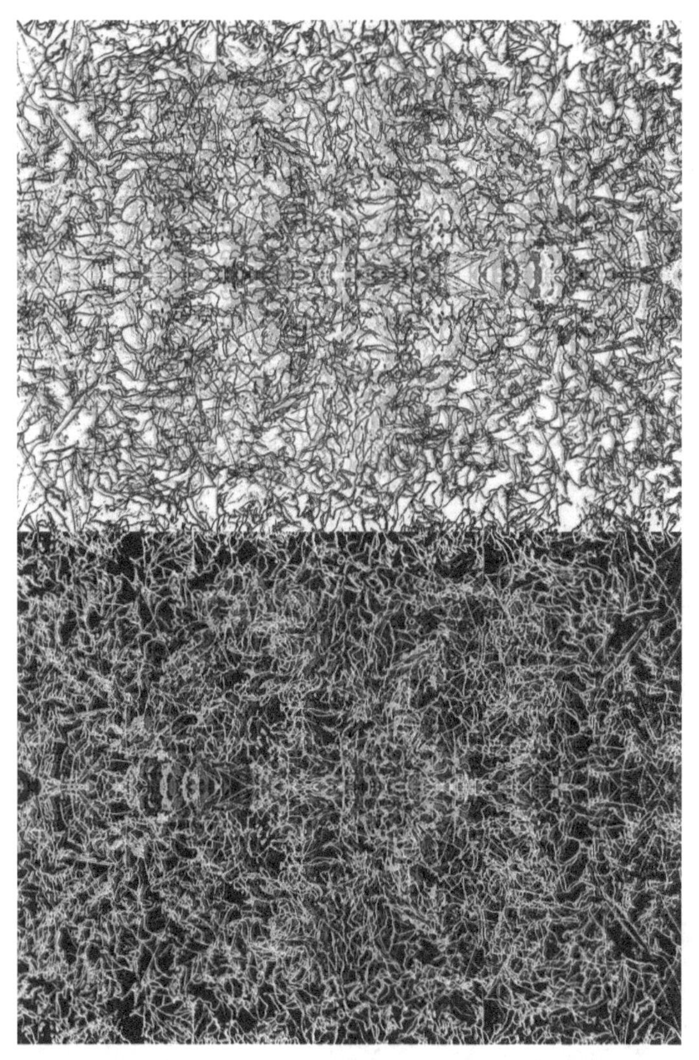

Hearts

Curved ellipses

The final stereoscopic design carries a perceptual puzzle. The carrier design is derived from a photograph of snow-covered branches on a chestnut tree. The shape at the centre of the design is a well-known figure/ground ambiguity. When the components are defined by stereoscopic depth the ambiguity is attenuated, as can be seen by reversing the red/cyan glasses; the apparently nearer part is seen as the figure.

Stereoscopic ambiguity

Rivalry Art

The art of binocular rivalry has not received the attention that has been directed to stereoscopic art and the images that follow attempt to redress this balance a little. Rivalry with anaglyphs presents an additional advantage because the component patterns can interact with one another in monocular vision so that many visual possibilities arise. In all cases the component patterns can be seen by viewing through one coloured filter at a time.

The Romanian visual scientist Liviu Iliescu has produced many examples of what he calls bioptical art; they are paired drawings and paintings which can be viewed by over-convergence thereby generating binocular rivalry.[6] The artist Antonio McAfee[7] enlists anaglyphic images in his examinations of historical photographs of middle-class African Americans. He reprocesses old portraits often combining them with their left–right reversals to produce symmetrical anaglyphs which are viewed with red/cyan glasses. The anaglyphic portraits can be partially masked with yet other images superimposed on them. These techniques have been used to make the double anaglyphic portrait of McAfee; one is centred on the lens of the hand-held camera and the other on a common eye. The left portraits are masked in white and those on the right in black.

Antonio McAfee—Modes of representation

In some ways it is puzzling that so little artistic attention has been applied to an aspect of our binocular vision that is so dynamic. The visual transformations that occur during binocular rivalry are easy to experience but difficult to describe. This

6 http://www.binocular-rivalry.ro/.

7 Https://antoniomcafee.net/.

applies particularly the changes that take place when complex patterns compete with one another. Scientists have tried to overcome this by using very simple patterns like gratings but few artists have reveled in the dynamic variety that is a consequence of processes occurring in the brain rather than on the pictorial surface. Of course, one factor that has inhibited artists from employing rivalry in their work is the need for some binocular viewing device. However, this same argument applies to stereoscopic art which has a vibrant history spanning back to the invention of the stereoscope.

Stereoviewers

In the sections that follow rivalry can be experienced with a bewildering variety of paired patterns; some are graphic designs, others are photographs and yet others are combinations of the two.

Designs

Experimental research on binocular rivalry has tended to use gratings as stimuli. Either achromatic or complementary coloured parallel lines are presented to one eye with the same pattern at right angles in the other, as described in Chap. 6. A more complex pattern involving orthogonal contours is shown below. Each of the four circles contains gratings made up from patterns of painted parallel lines which are in opposite orientations for each eye; the differences are clearly seen when looking through one colour filter and then the other. The preferred orientations employed in experiments are at 45 degrees to vertical because the anisotropy associated with vertical and horizontal is not at play. That is, vertical gratings tend to be visible for longer than horizontals when in rivalry but there is no preference in the case of diagonals.

Quad rangles

Interfering waves

Rotating profiles

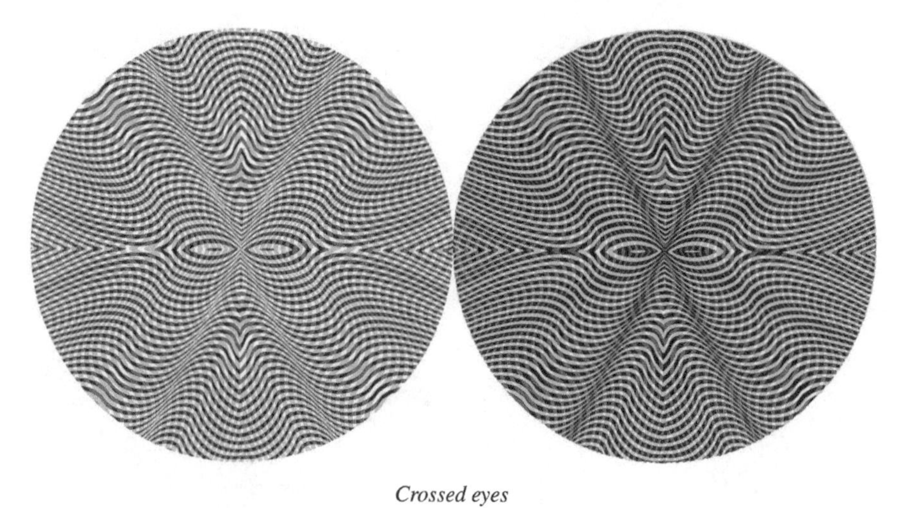

Crossed eyes

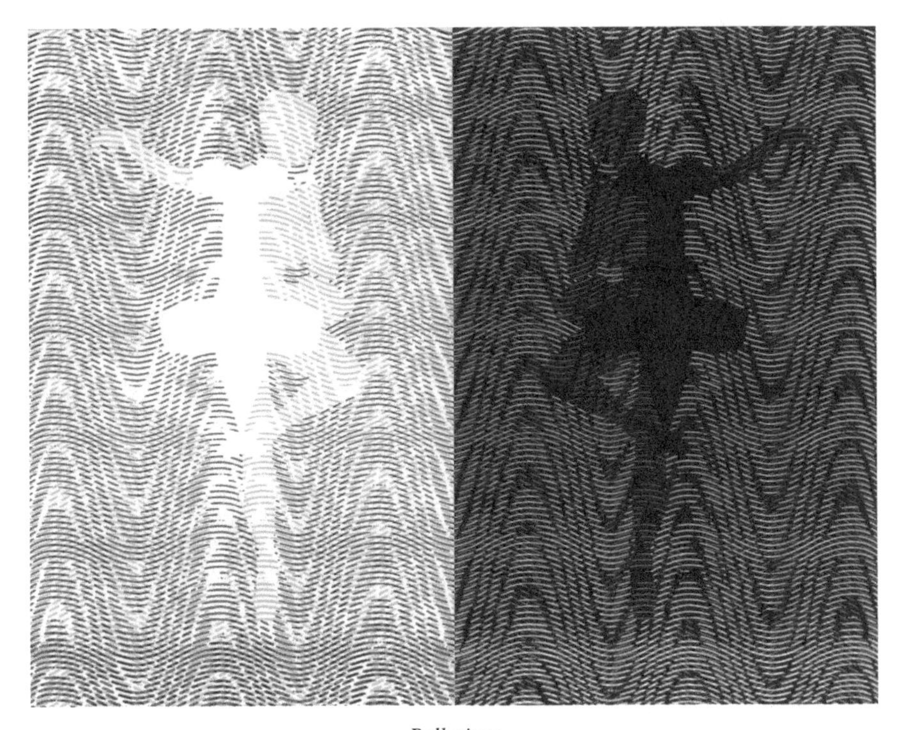

Ballerinas

In the examples above, the patterns in rivalry are graphic designs but it is also possible to use designs derived from naturally occurring patterns following some graphical manipulations as instanced by the following two anaglyphs. In each case a positive and negative combination is shown.

Rivalling leaf structures

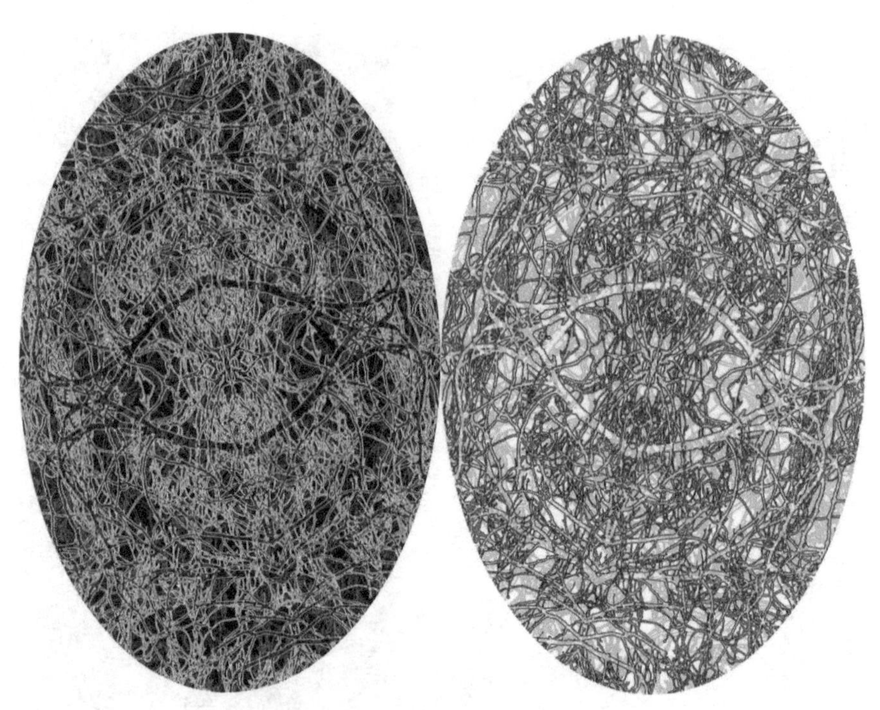

Rivalling branch structures

Rivalry is rarely a subtle phenomenon and the paired patterns that are applied to express it are usually of high contrast. However, it is possible to use lower contrast designs to produce binocular competition but it might take a little longer for the rivalry between the patterns or parts of them to become visible. The components of the following patterns are themselves somewhat intricate. All started as abstract paintings made by flowing paints over a flat surface, then allowing the different colours to intermingle and dry. The paintings or details of them were photographed, digitized and manipulated in the computer. The digital images were usually multiplied and combined in order to produce the symmetries that are evident in most of them. Only at that stage were the anaglyphic images assembled.

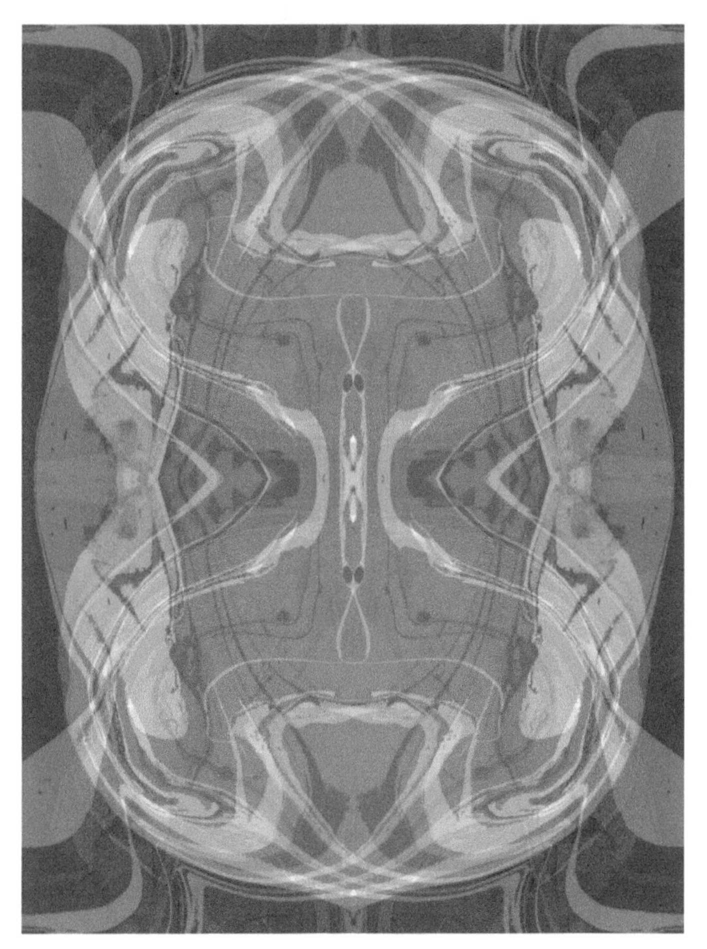

Trace

Enclosures

Inversions

Hidden figure I

Hidden figure II

Photographs

Unlike stereoscopic photographs, where the two components need to retain close correspondence in space and time, rivalling photographs are not so constrained. They do require to be sufficiently different for competition rather than cooperation to take place. Another strategy is to combine two similar photographs with a few features in rivalry; the stability of the overall scene is disturbed by the localized rivalry within it. It is also possible to select the components so that they reflect different aspects of the same subject, as applies to the rivalling photographs below. They are views of the same structure (a road bridge) from the central walkway and from beneath so that the supporting piers are seen. The pair has been combined twice with each component visible to each eye.

Tay Road Bridge from above and below

Feather wall

Tree folly

Distilled wisdom

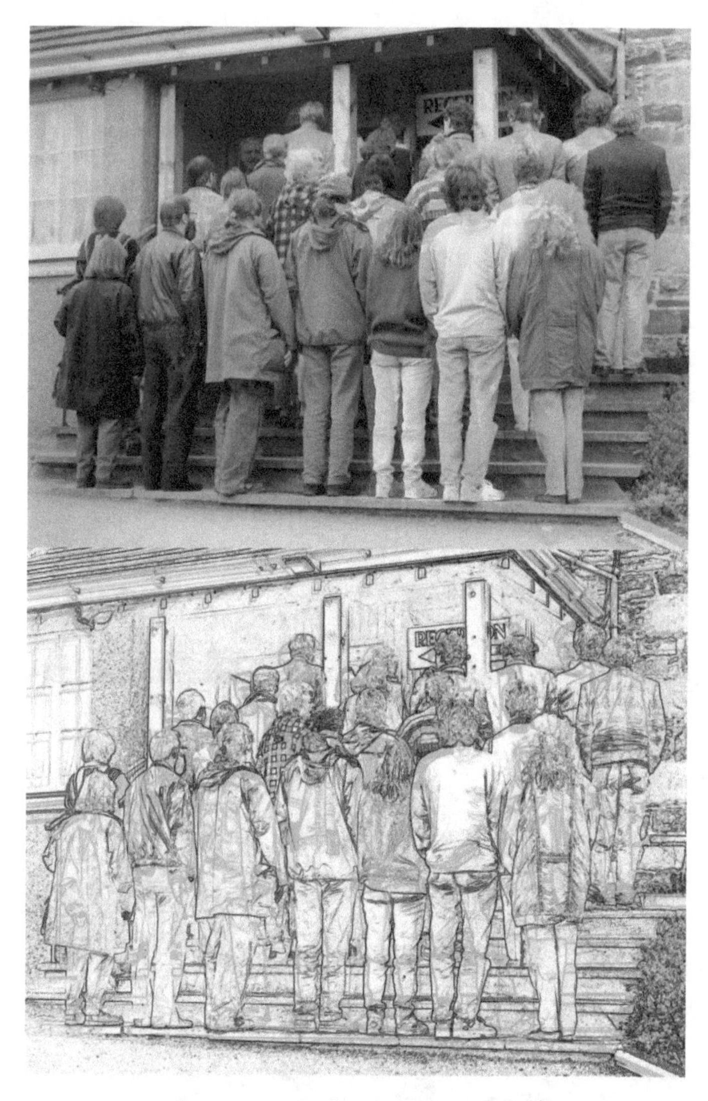

Back to front

Angel of Newcastle

Lille Railway Station

West Light Tayport in winter

Portraits

Portraits that rival with one another can be of the same person either in contrasting postures, at different ages or carried by appropriate graphical or textual motifs. The two components need not be of the same person so that a wide variety of possibilities can be entertained, some of which have been be illustrated earlier and will be amplified in the figures that follow.

Before what we know as photography was invented in 1839 experiments involving fixing images formed by light had been undertaken. In his authoritative history of photography, Beaumont Newhall commenced his chapter on the invention of the process with the statement that: "The first person to attempt to record the camera image by means of the action of light was Thomas Wedgwood".[8] Tom, a son of the potter Josiah Wedgwood, described his experiments thus: "White paper, or white leather, moistened with solution of nitrate of silver, undergoes no change when kept in a dark place; but, on being exposed to day light, it speedily changes colour, and, after passing through different shades of grey and brown, becomes at length nearly black".[9] These are the opening words in the article taken by historians to be one of the foundations of photography. They reflect the experiments conducted by Tom

[8] Newhall (1982), p. 13.

[9] Davy (1802), p. 170. See also Wade (2005).

Wedgwood to copy paintings (on glass) and profiles onto a light sensitive surface. He is shown (twice) in profile in a circular motif that induces impressions of both stereoscopic depth and rivalry.

Tom Wedgwood

It is appropriate to follow Tom Wedgwood with the inventor of the negative/positive photographic process, William Henry Fox Talbot.[10] He is also shown in profile (twice) but the profiles are made up using the words NEGATIVE and POSITIVE. In conventional photography the negative is contact printed and the resulting positive is left right reversed.

[10] See Schaaf (1992).

EGATIVENEGATIVENEGATIVENE
IVENEGATIVENEGATIVENEGATIVENEGATIVE
EGATIVENEGATIVENEGATIVENEGATIVENEGATIVENE
ENEGATIVENEGATIVENEGA IVENEGATIVENEGATIVENEGAT
IVENEGATIVENEGATIVENEGATIVENEGATIVENEGATIVENEGATI
TIVENEGATIVENEGATIV [illegible] ATIVENEGAT VENEGATIV
GATIVENEGATIVENEGAT [illegible] GATIVENEGATIVE ATIV
EGATIVENEGATIVENEGAT POSIT IVEPO [illegible] IVENEGA IVENE V
TIVENEGATIVENEGA POSITIVEPOSI [illegible] VENEGATIV
I POSITIVENEGATIVENE VEPOSITIVEPOSIT [illegible]
TI POS E VEPOSIT GATIVE TIVEPOSITIVEPOSITIV [illegible] IVEPOSIT
ITI P VENEGA IVENEGATIV ITIVEPOSITIVEPOSITIVE [illegible]
ITI IVENEG IVENEGATI SITIVEPOSITIVEPOSITIVE [illegible]
SI GATIVEN GATIVENEGAT POSITIVEPOSITIVEPOSITIVEPO [illegible]
SIT AT VENE GATIVENEGA POSITIVEPOSITIVEPOSITIVEPO [illegible]
SITIVE IT GATIVENEGAT POSITIVEPOS IVEPOSI VEPO [illegible] TIV V
SI VE ATIVENEGAT OSITIVEPOSITIVE TIVE [illegible] ATIV
SI IV VE IVENEGAT OSITIVEPOSITI OSITIV [illegible] ITIV
TIVE OS ATIVENEGATI [illegible] POSITIVEPOS [illegible]
SI IV TIVENEGATIVENE A IVENEGATIVENEGATIV [illegible]
SIT VE OSI VENEGAT IVENEGATIVENEGATIVENEGATIVE [illegible] OSIT
SIT P NEGA IVENEGATIVENEGATIVENEGATIVENE [illegible]
IT IVEPOS GATIVENEGATIVENEGATIVENEGATIV
I EPOSIT TIVENEGATIVENEGATIVENE
VENEGATIVENEGAT
NEGATIVENE
GATI

Talbotypes

One aspect of faces that has long been of interest to scientists and artists concerns the proportions between distinctive features like eyes, nose, and mouth. Certain proportions have historically been considered to be of particular significance mathematically, like the golden ratio. This 'divine proportion' is represented in the portrait of Luca Pacioli (1447–1517), a Pisan mathematician. The three circles are in the golden ratio: the diameter of the outer circle to the larger inner one is the same as the larger to the smaller enclosed circle and its value is approximately 1.62.[11]

[11] See Wade (2007a, b, c).

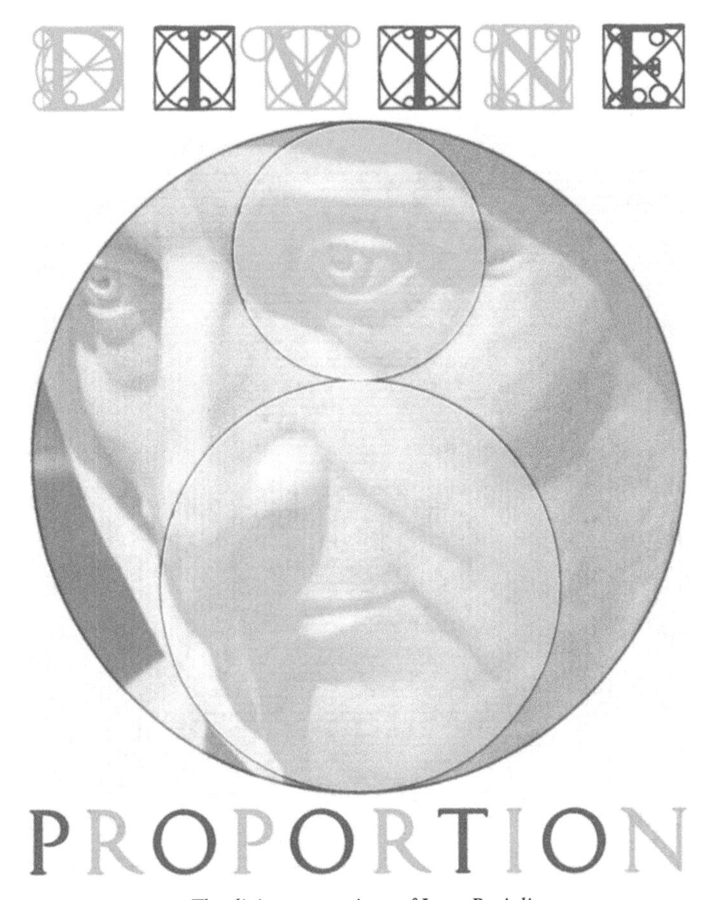

The divine proportions of Luca Pacioli

A constant concern since the origin of geometry has been to marry the circle to the square. That is, to determine the relative dimensions of a square and a circle that have equal areas. The difficulty of this task is enshrined in our language when we refer to squaring the circle as attempting the impossible. The Roman architect Marcus Vitruvius Pollio (ca. 88–26 BC) added an aesthetic dimension to architecture through the harmony of circle and square, and both shapes were in turn related to the symmetry of natural objects like the human body. When the arms are extended horizontally and legs together the extremities of the square are touched, but with raised arms and parted legs they are in contact with the circumference of a circle centred on the umbilicus. Leonardo represented these relationships in a drawing that has become known as 'Vitruvian man' and he is in turn represented within the concentric circles.

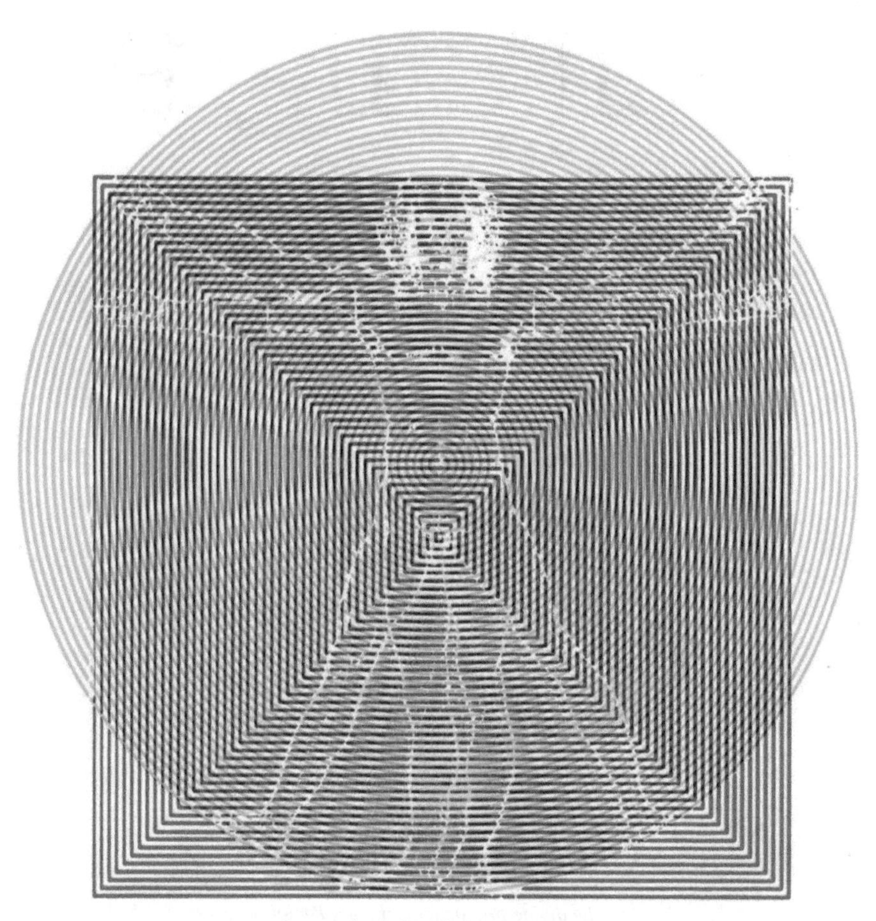

Leonardo da Vinci's Vitruvian man

Until the scientific revolution of the seventeen century the heavenly bodies were thought to be perfect spheres. This geometrical purity was questioned by Galileo Galilei (1564–1642). When he directed his telescope to the moon the impact of his observations reverberated throughout Europe. He described and depicted the mountains and craters on the moon, challenging the received view that heavenly bodies were perfect spheres and that surface imperfections were restricted to earth. The illustration portrays Galileo in the moon he portrayed: the drawing of the moon is derived from a woodcut in *Sidereus Nuncius*, published in 1610.[12] Galileo wrote at length about illusions of the senses, but this concerned the possibility of reaching false conclusions on the basis of observation. He did not apply his optical knowledge to the eye itself but added to the descriptions and analyses of visual phenomena. Galileo was well versed in art and he considered that painting was superior to sculpture because it can allude to depth rather than presenting it directly for the sense of touch.

[12] See Piccolino and Wade (2008, 2014), Wade (2007a, b, c).

Galileo in his moon

Bumps of a different variety were seen as significant to Franz Joseph Gall (1758–1828) and his colleague Johan Gaspar Spurzheim (1776–1832)—those on the skull. They argued that mental and personality characteristics (faculties) could be assessed from the bumps on the head and it later became known as phrenology. Moreover the faculties were localized and Spurzheim listed 35 areas. The stylised phrenological markings on drawings of the head became the stock in trade. Spurzheim's portrait is combined his illustration of a phrenological head taken from one of his books on phrenology.[13] Spurzheim continued spreading the word of phrenology throughout Europe and beyond.

[13] Spurzheim (1832).

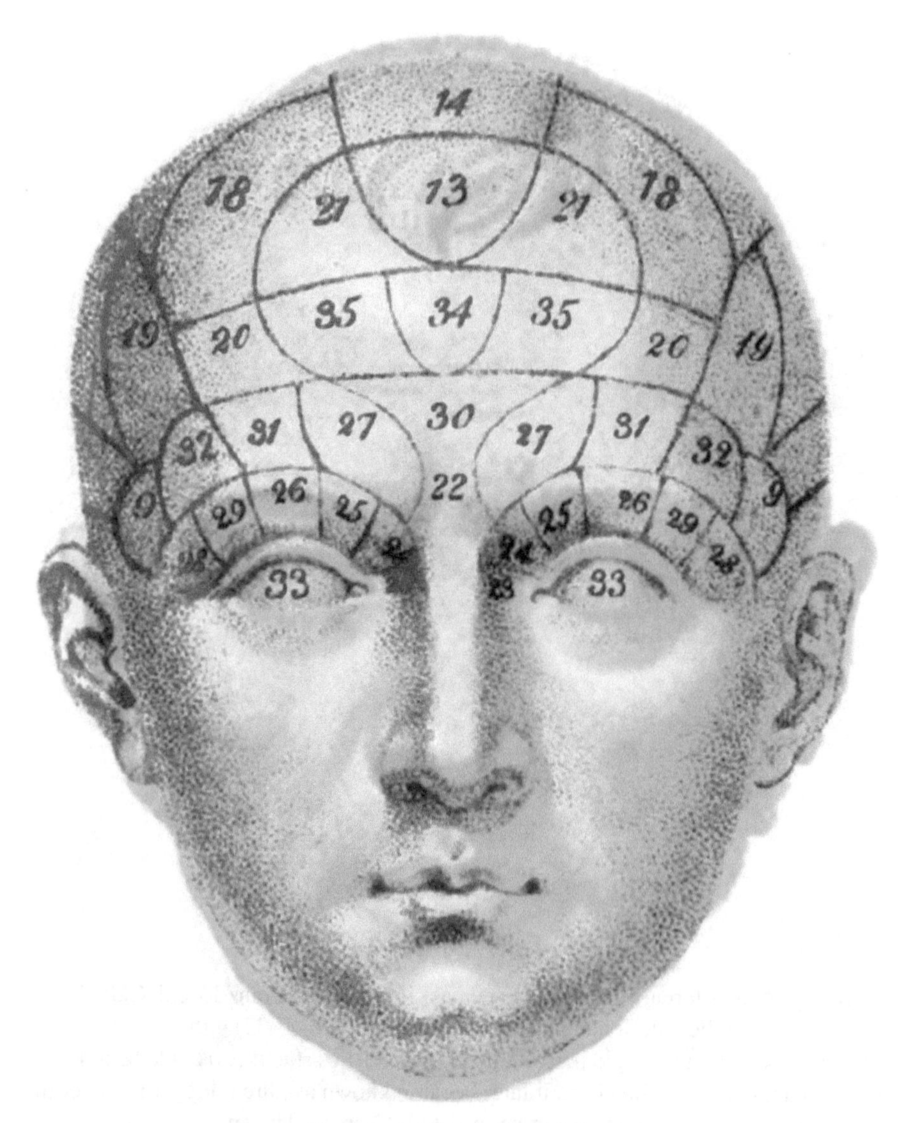

Phrenological head of Spurzheim

Research on vision in the nineteenth century was active before the invention of the stereoscope. Prominent among the researchers was Jan Evangelista Purkinje (1787–1869) after whom several phenomena are named as are microscopic structures in the body: there are Purkinje cells in the cerebellum, Purkinje fibres around the heart, Purkinje images are reflected from the optical surfaces of the eye, a Purkinje tree can be rendered visible, and at twilight we can experience the Purkinje shift. In the 1820s he wrote two books on subjective visual phenomena because he believed that visual illusions revealed visual truths. In his first book[14] he gave detailed descriptions of phenomena such as afterimages, the visibility of the retinal blood vessels (the Purkinje tree), and the distortions that are produced when viewing regular geometrically periodic patterns, like radiating lines and concentric circles. In his second book[15] he described the difference in the visibility of coloured objects when seen in daylight and twilight—blue objects appear lighter and red ones darker in twilight (the Purkinje shift). He is shown twice in two slightly different radiating patterns.

Jan Evangelista Purkinje

Purkinje's first book on subjective vision was partially translated by Wheatstone who was also greatly influences by Thomas Young's studies of vision and hearing. Young was a physicist and a physician by training, but he was also a linguist and a cryptographer. In visual science he explained the process of accommodation, and

[14] Purkinje (1823). This book is translated into English in Wade and Brožek (2001).

[15] Purkinje (1825).

proposed a theory of colour vision based on three receptors, sensitive to different regions of the spectrum. His experiments on interference fringes provided support for a wave theory of light in opposition to Newton's corpuscular theory. Young[16] described the optical aberration of astigmatism, in which lines in different orientations cannot be brought to a focus in the same plane, as will be evident in the pattern carrying Young's portrait.

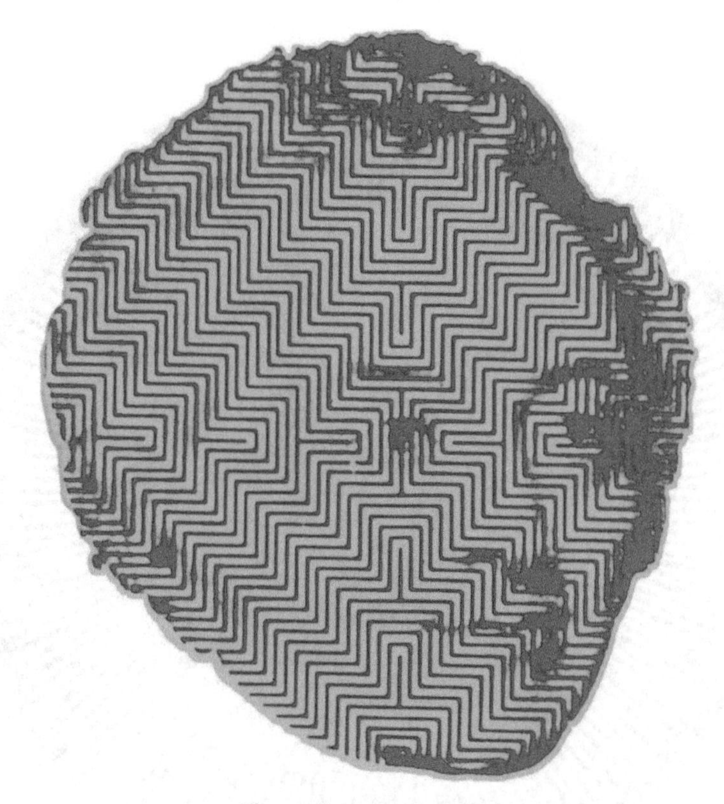

Thomas Young's astigmatism

Like Young, Helmholtz made numerous contributions to vision and audition. The first volume of his *Handbook of physiological optics* concerned physical optics, the second the physiology of vision, and the third was on visual perception. His analysis of perception drew more heavily on British empiricism than on the Kantian philosophy prevalent in Germany. Helmholtz adopted the concept of unconscious inference to account for the way in which we learn to see and how common perceptions can arise from varied patterns of stimulation.[17] His theoretical position was supported

[16] Young (1801).

[17] Helmholtz (1867, 1925), Wade (2021b).

by many experiments, particularly with the stereoscope. Helmholtz is shown twice, as a young man and as an established scientist.

Helmholtz young and old

Artists provide a rich source of rivalry portraiture. Their styles or subject matter can provide the basis for the design competing with their visage. We commence with René Magritte, noted for his paintings that pose perceptual paradoxes. His surreal paintings frequently questioned our conceptions of space and the objects within it. Accordingly, his portrait is dimly discernable within the ambiguous 'cube'. As an additional twist, there are two portraits of Magritte, one to each eye.

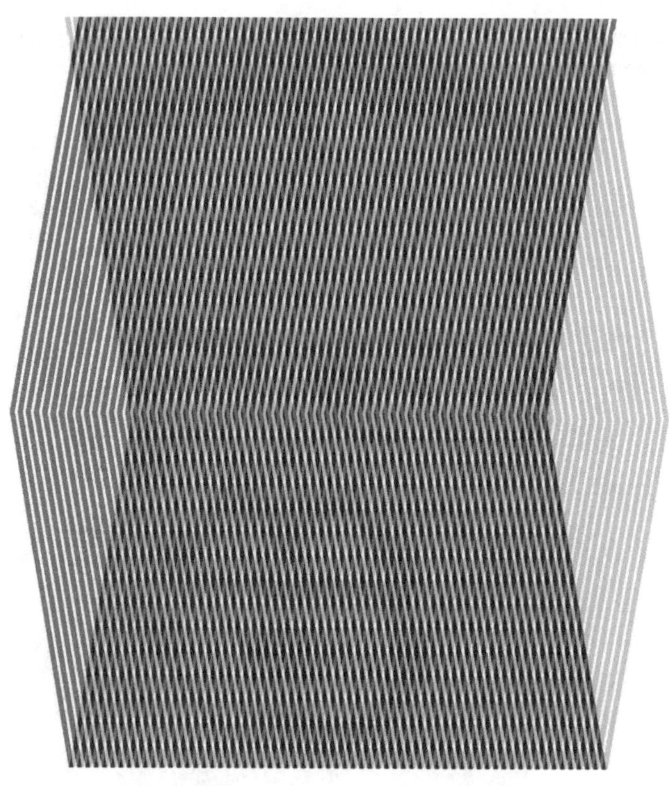

René Magritte—Perceptual puzzler

Op Art has close connections to visual science[18] and several of its practitioners are shown below. Victor Vasarely and Bridget Riley are pioneers of the genre and both manipulate simple geometrical shapes like circles, squares and triangles in order to induce some striking effect on the observer. Repetitions of the geometrical elements often lead to instabilities of the picture surface so that motion is perceived.

[18] See Wade (1978, 1982).

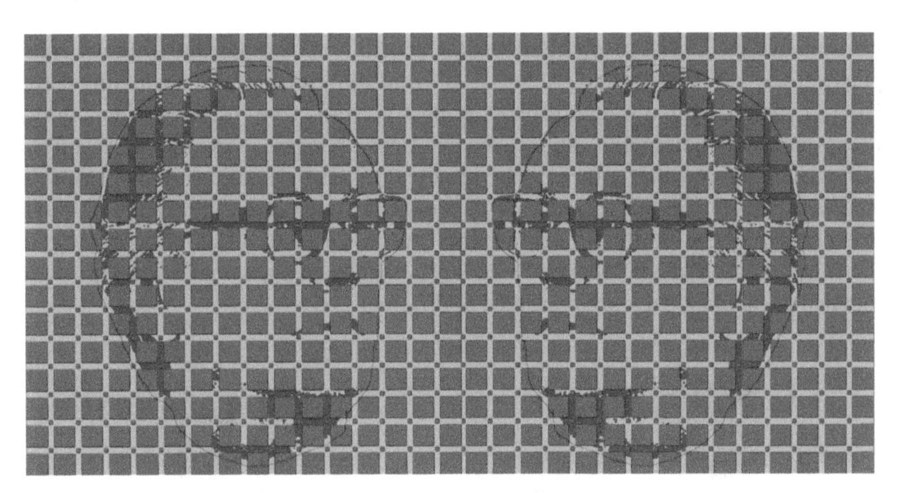

Victor Vasarely—Shimmering squares

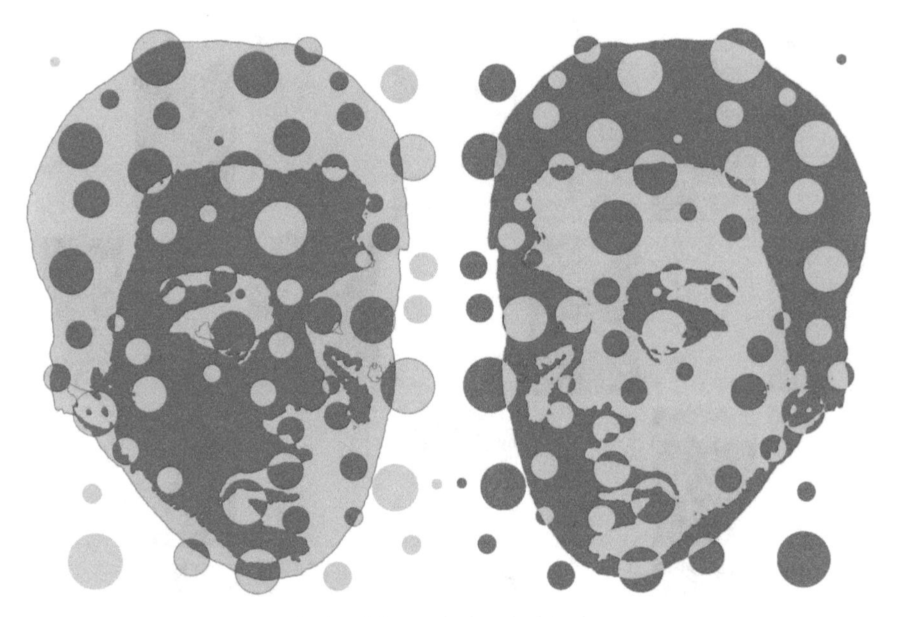

Bridget Riley—Black and white dots

The work of the English artist David Hockney has displayed a constant concern with the representation of space, either historically or culturally. He has embraced new technologies as they emerged and has presented spectators with novel ways of viewing the world. This is especially evident in his *Cameraworks* in which he explores how elements of a bigger picture are influenced by the elements composing it.[19] His

[19] Hockney (1984).

photomontages consist of photographs taken at different times and with different perspectives that are joined to produce a coherent whole despite the conflicting parts. He is portrayed with a range of photographic portraits from different times in his life, some from photographs others from paintings; there are often components from different pictures within a block. The principal unifying features are his spectacle frames and his cigarette. Hockney has not embraced binocular vision in his work but it is deployed here—different components are presented to different eyes. Despite the fragmentation of the blocked components a unity of appearance and identity is retained even when parts are presented to different eyes.

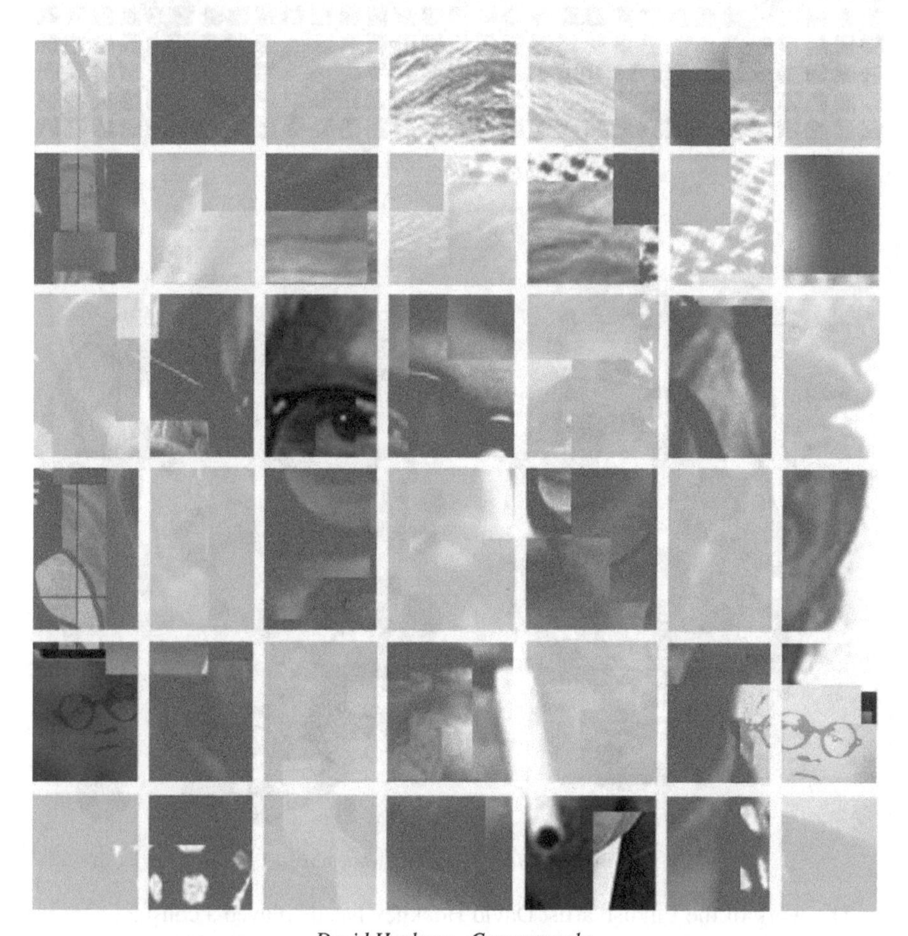

David Hockney—Cameraworks

Derek Besant uses photography and computer graphics to veil the identities of the faces he portrays and then presents them as large scale prints. He has photographed veiled faces of those at rest, photographed bodies and faces under water and superimposed fading text over unfocussed faces.[20] However, these distortions have their limits: whereas it is easy to remove identity from an image, it is very difficult to destroy its detection as a representation of a face. We see faces in almost everything—clouds, smoke, flames, tree stumps—even though we are well aware that they could not possibly be present in those objects. Thus, Besant presents us with another visual intrigue—how do we disentangle the distortions to retrieve a face from all the fragments? It is not vision that Besant plumbs but cognition. He is shown, indistinctly, in rivalry with the pattern of light reflected from the surface of water.

Derek Besant—In other worlds

[20] See Besant (2015).

Binocular Lustre

When a positive image is presented to one eye and its negative to the other the impression of a metallic sheen is called binocular lustre (see Chap. 6) and examples are shown below.

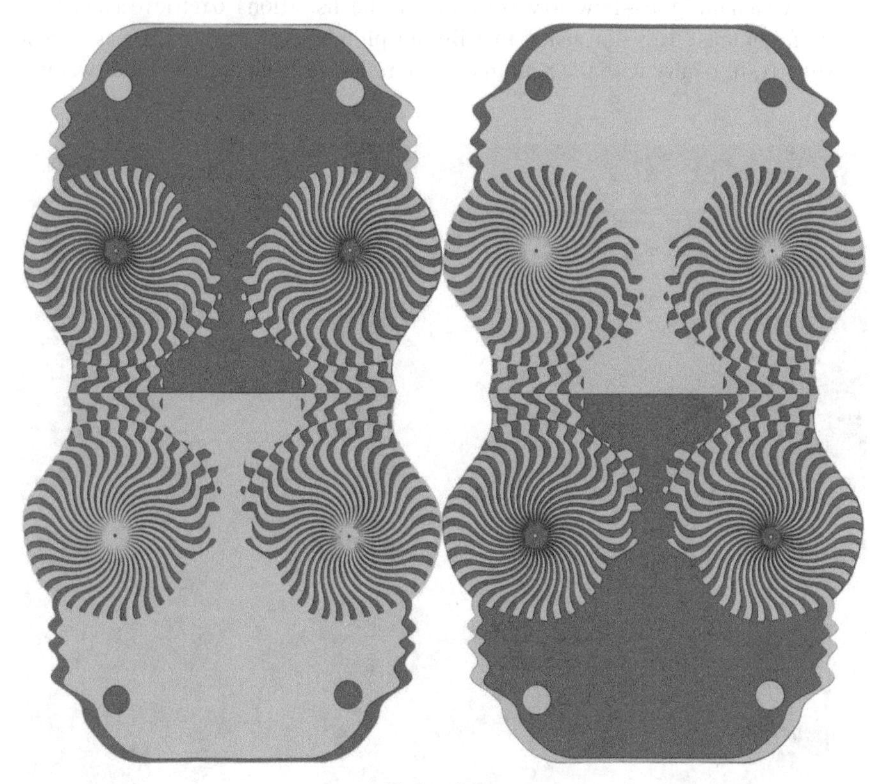

Facing faces

Diametric sides

Golden section arcs

Dundee perspective

Yins and Jungs

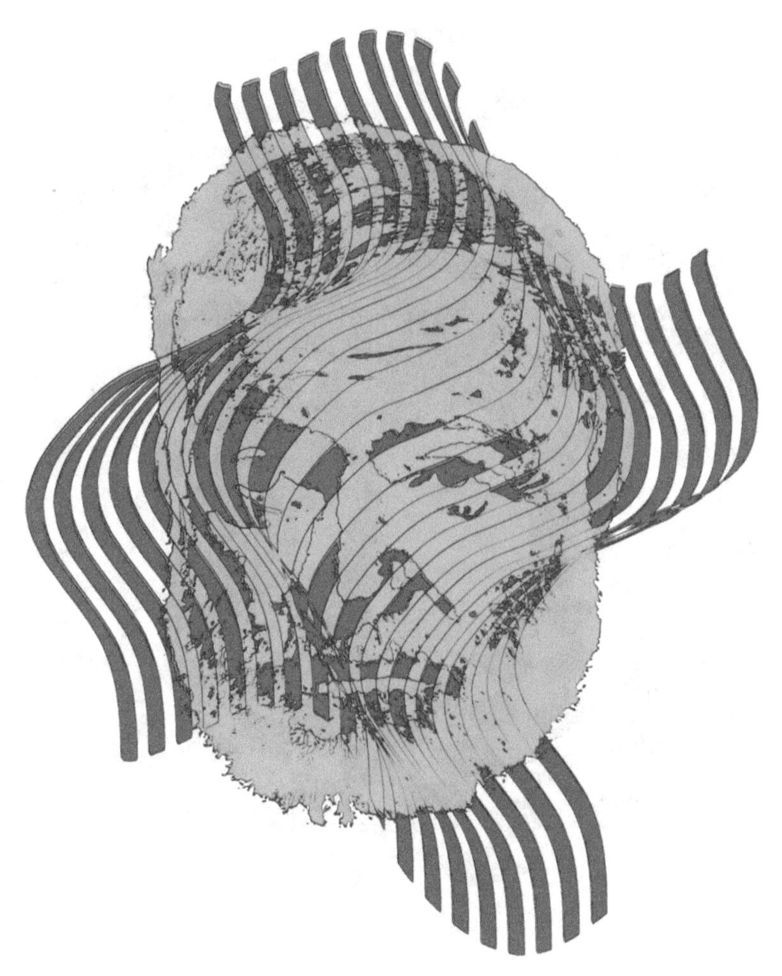

Hemingwaves—the old man and the sea

One of the few artists explicitly espousing rivalry art is Yuki Maruyama, a Japanese artist whose work is based on binocular lustre. She makes large installations painted in red and cyan and provides red/cyan glasses to view them.[21] The juxtaposition of the large areas of the two colours induces a variety of visual effects and these are enhanced with the lustrous reversals seen when wearing the red/cyan glasses.

[21] http://www.adobebooks.com/blog/2020/1/1/yuki-maruyama.

Yuki Maruyama

Partial Rivalry

Many of the portraits in previous chapters have involved partial rivalry. That is, large parts of the portrait or motif were not in competition with one another and they could be seen without the aid of the red/cyan glasses. The examples of partial rivalry that follow involve some common features in both images that do not engage in rivalry.

Wollaston's gaze

Dichoptic Dali

Ames' roominations

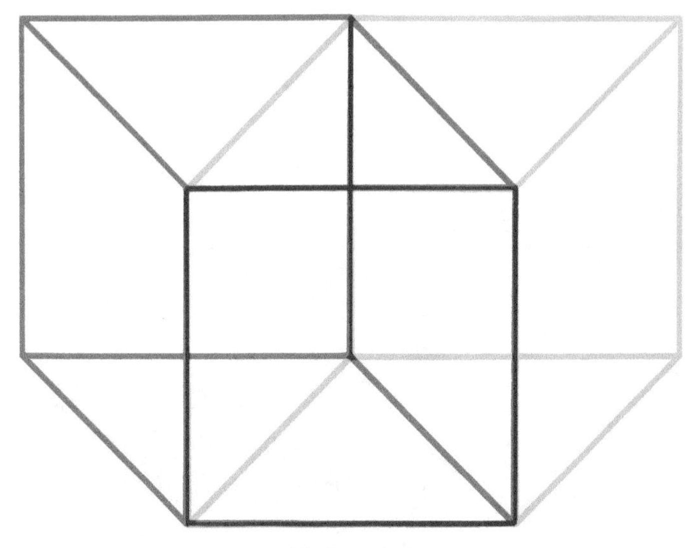

Necker cubed

Certosa Calci

Newport haar

Symmetry

Pisan cloister

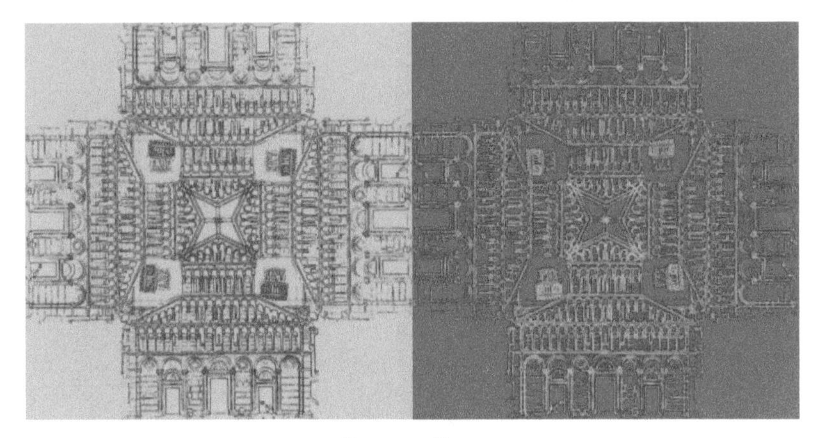

Illustrious Tower

Pisan pillars

Citadelle de Blaye

Le Petarrot plate

Tree lines

West Light, Tayport

Vanishing points

Arches, Lausanne

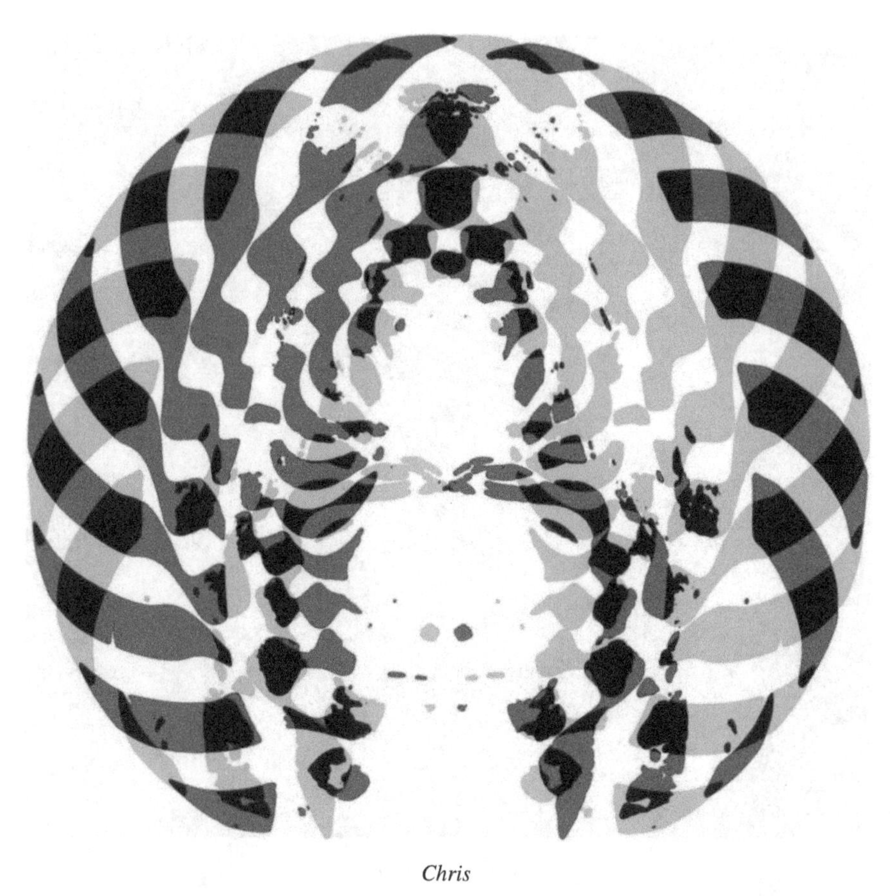

Chris

Literal Binocular Visions

Words can provide the basis for binocular art, as has been evident in many of the images presented earlier in this book. More examples will be shown on the following pages. Rivalling patterns involving words can display the visual phenomena that are described in the text, they can match portraits with literature, words can rival with one another and they can allude to the many subtleties of pictures and prose. In short, they can direct us to the relationships between words and images.

Words and images

Visions and Arts

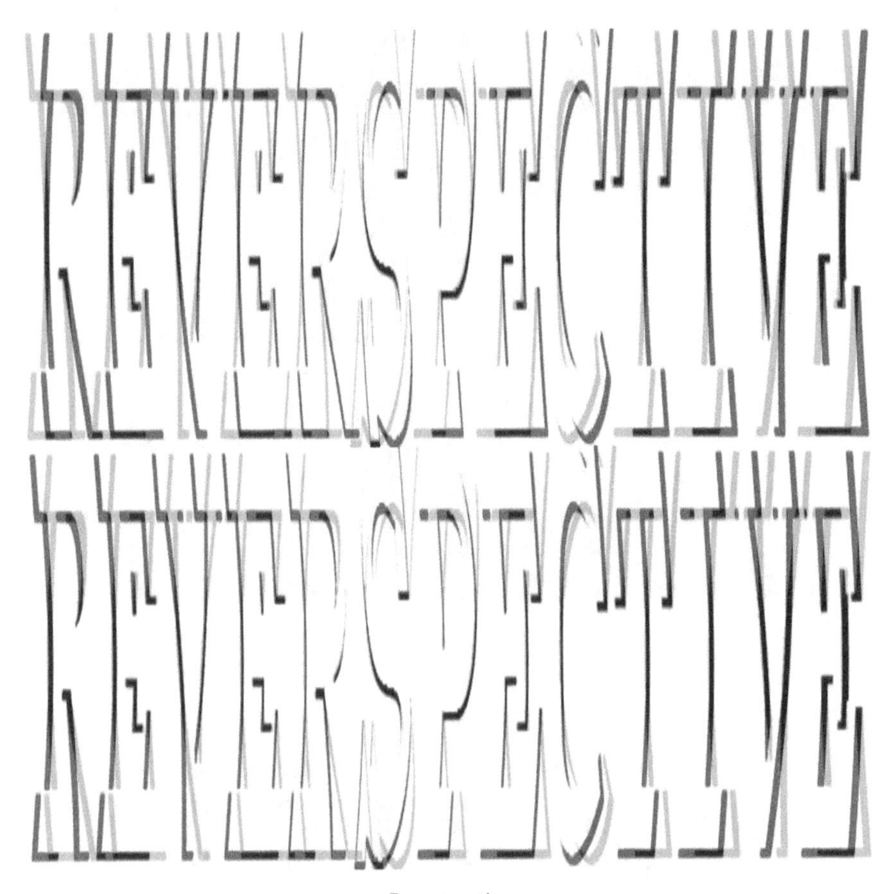

Reverspectives

Reverspector

Illusions of depth

Ocular illusions

The words in the remaining literal pictures should account for the images that carry them or are carried by them. The clues are also in the titles.

Circling the square

Home and away

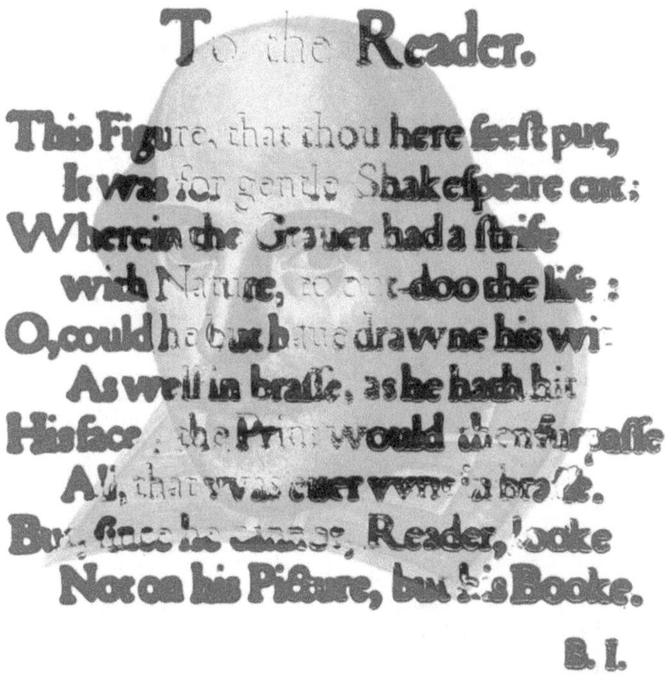

To the Reader.

This Figure, that thou here seest put,
It was for gentle Shakespeare cut;
Wherein the Grauer had a strife
with Nature, to out-doo the life:
O, could he but haue drawne his wit
As well in brasse, as he hath hit
His face; the Print would then surpasse
All, that vvas euer vvrit in brasse.
But, since he cannot, Reader, looke
Not on his Picture, but his Booke.

B. I.

Ben Jonson's Shakespeare

Goethe Dämmerung

Digital images

References

Besant D (2015) In other words. Edinburgh Printmakers Workshop, Edinburgh

Colvin C (2009) Natural magic. Royal Scottish Academy, Edinburgh

Davy H (1802) An account of a method of copying paintings upon glass, and of making profiles, by the agency of light upon nitrate of silver. Invented by T. Wedgwood, Esq. With observations by H. Davy. J Roy Inst Great Britain 1:170–174

Foucault M (1982) This is not a Pipe. University of California Press, Berkeley CA, Trans. Hawkness J

Helmholtz H (1867) Handbuch der physiologischen Optik. In: Karsten G (ed) Allgemeine Encyklopädie der Physik, vol 9. Voss, Leipzig

Helmholtz H (1925) Helmholtz's treatise on physiological optics, vol 3. Optical Society of America, New York, Trans. Southall JPC

Hockney D (1984) Cameraworks. Knopf, New York

Martinez-Conde S, Conley D, Hine H, Kropf J, Tush P, Ayala A, Macknik SL (2015) Marvels of illusion: illusion and perception in the art of Salvador Dali. Front Hum Neurosci. https://doi.org/10.3389/fnhum.2015.00496

Newhall B (1982) The history of photography from 1839 to the present. Secker and Warburg, London

Piccolino M, Wade NJ (2008) Galileo Galilei's vision of the senses. Trends Neurosci 31:585–590

Piccolino M, Wade NJ (2014) Galileo's visions: piercing the spheres of the heavens by eye and mind. Oxford University Press, Oxford

Purkinje J (1823) Beobachtungen und Versuche zur Physiologie der Sinne. Beiträge zur Kenntniss des Sehens in subjectiver Hinsicht. Calve, Prague

Purkinje J (1825) Beobachtungen und Versuche zur Physiologie der Sinne. Neue Beiträge zur Kenntniss des Sehens in subjectiver Hinsicht. Reimer, Berlin

Schaaf LJ (1992) Out of the shadows. Talbot and the Invention of Photography Yale University Press, London Herschel

Spurzheim G (1832) Phrenology, or the doctrine of the mental phenomena. March, Capen and Lyon, Boston

Wade NJ (1978) Op art and visual perception. Perception 7:21–46

Wade N (1982) The art and science of visual illusions. Routledge & Kegan Paul, London

Wade N (1990) Visual allusions: pictures of perception. Lawrence Erlbaum, London

Wade NJ (2005) Accentuating the negative: tom Wedgwood (1771–1805), perception and photography. Perception 34:513–520

Wade NJ (2007) The stereoscopic art of Ludwig Wilding. Perception 36:479–482

Wade N (2007) Circles: science. Dundee University Press, Dundee, Sense and Symbol

Wade NJ (2007) Galileo and the senses: vision and the art of deception. Galilæana 4:259–288

Wade NJ (2009) Natural magicians. Perception 38:633–637

Wade NJ (2011) Eye Contricks. i-Perception 2:486–501. http://i-perception.perceptionweb.com/fulltext/i02/i0442aap.pdf

Wade NJ (2021a) On stereoscopic art. i-Perception 12(3):1–17. https://doi.org/10.1177/20416695211007146

Wade NJ (2021b) Helmholtz at 200. i-Perception 12(4):1–19. https://doi.org/10.1177/20416695211022374

Wade NJ, Brožek J (2001) Purkinje's Vision. The Dawning of Neuroscience. Lawrence Erlbaum Associates, Mahwah, NJ

Young T (1801) On the mechanism of the eye. Philos Trans R Soc 91:23–88

Chapter 9
Conclusion

Vision with two eyes presents us with many marvels but also leaves us with mysteries to solve. The marvels were made evident after the invention of the stereoscope in the 1830s. Depth could be synthesized from two pictures having small horizontal spatial differences. If these differences or disparities were too large then rivalry rather than depth was seen. Thus the stereoscope disclosed the cooperation between the two eyes in seeing depth as well as their competition when rivalry results. How the brain is able to process the signals from each eye to yield these two outcomes is one of the mysteries yet to be resolved. Descriptions of seeing with two eyes stretch back many centuries but the concern was almost always with how the world is seen single despite their different locations in the head. It was only after the invention of the stereoscope that depth from disparity could be demonstrated and manipulated. Scientists were eager to explore the relationship between depth and disparity and photographers were quick to exploit it. Pictorial artists seemed less enthusiastic about espousing stereoscopic vision, largely because of the challenges of producing two pictures that were slightly and systematically different. Binocular competition has a longer descriptive history than stereoscopic cooperation but artists as well as photographers have been reluctant to embrace it in their works. The stereoscope revolutionized our views of vision with two eyes now computer graphics is revolutionizing how we view stereoscopic images. This book attempts to explore the characteristics of both binocular cooperation and competition in photography and graphics. In the concluding chapter some of the issues concerning vision and art with two eyes will be returned to in word and image.

The vision and art of seeing with two eyes was transformed by the invention of stereoscopes, many versions of which were devised from the mid-nineteenth century. The first stereoscopes were made by Charles Wheatstone and a popular model was introduced later by David Brewster. Wheatstone is associated with the reflecting stereoscope and Brewster with one comprised of half-lenses which acted as magnifiers and prisms. The two models are shown below. Stereoscopes facilitated the

N. Wade, *Vision and Art with Two Eyes*, Vision, Illusion and Perception 3,
https://doi.org/10.1007/978-3-030-77995-5_9

observation of different stimuli presented to each eye and they demonstrated the link between perceived depth and retinal disparities.

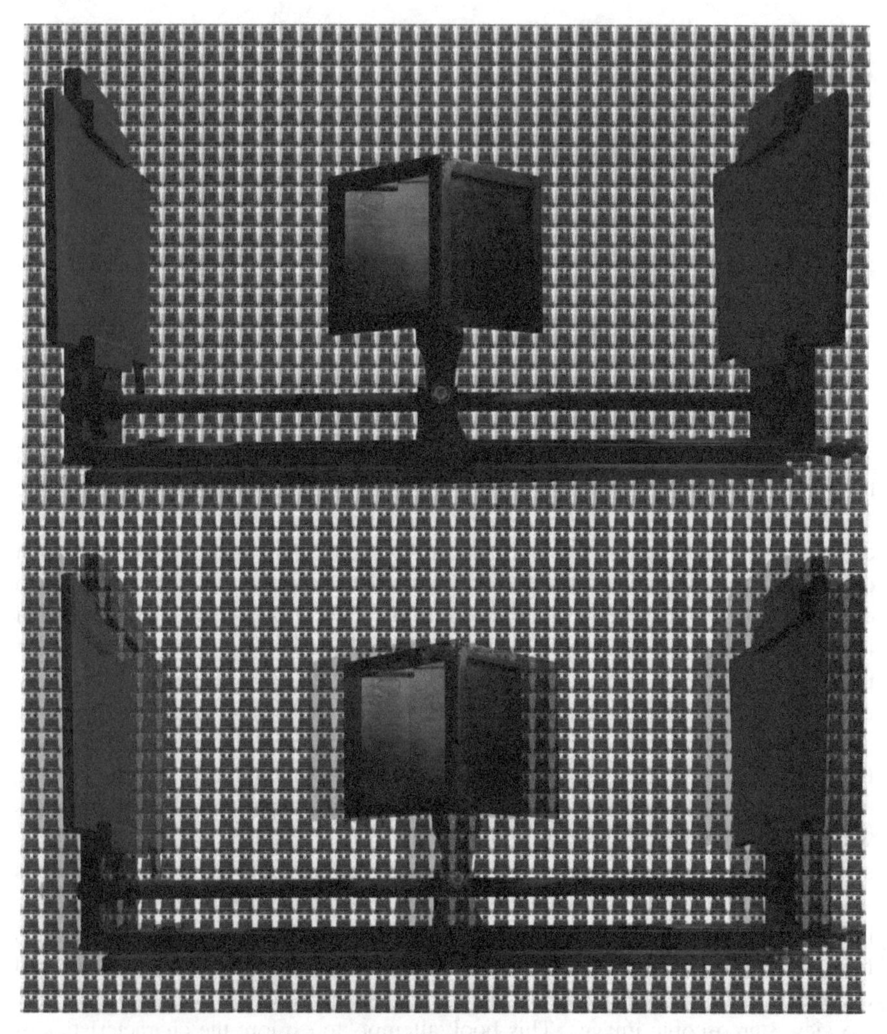

Wheatstone and Brewster stereoscopes

A motif running through this book has been the rivalry between these two pioneers of binocular vision. Their rivalries were expressed at scientific meetings but most particularly in text, as was described in Chap. 7. They are shown below in textual rivalry. Wheatstone can be seen in a page taken from his article in 1838 in which the stereoscope and experiments with it are described. Brewster is represented in his account from his book on the stereoscope of the beautiful instrument presented to Queen Victoria in 1851.

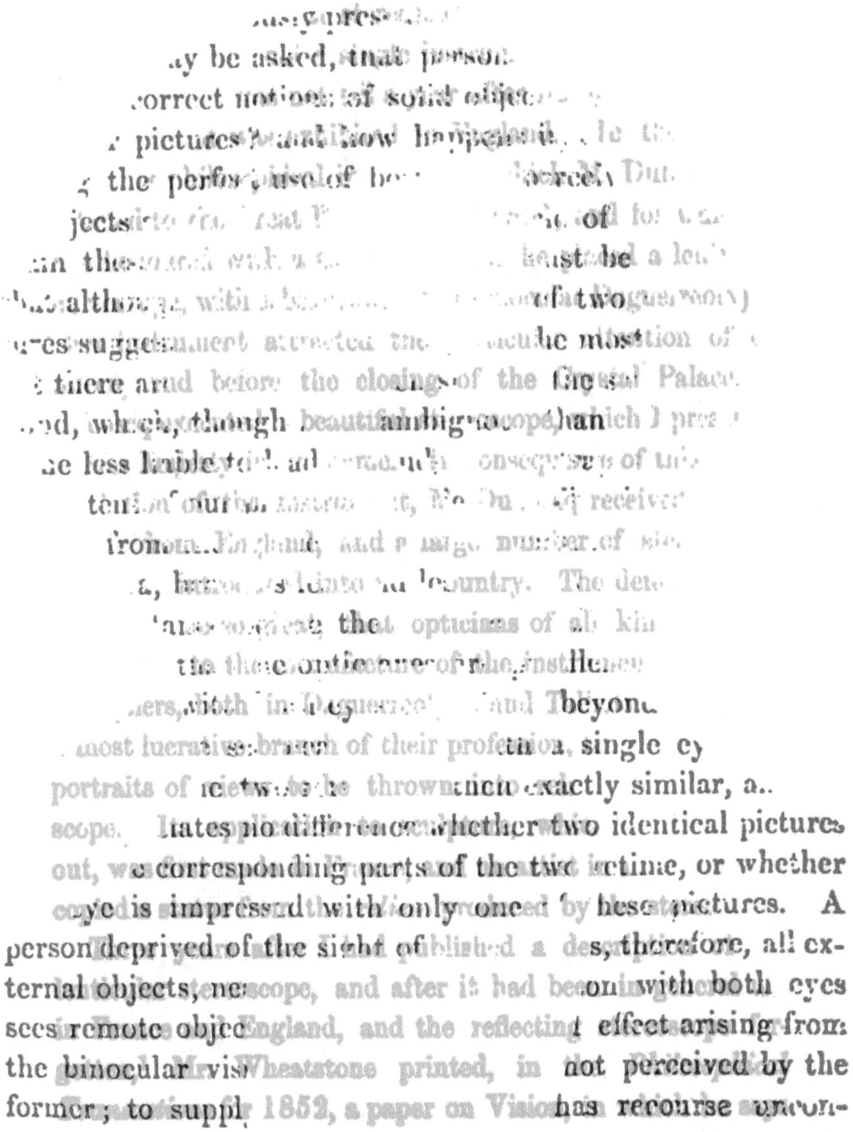

Rivalry between Wheatstone and Brewster

One of the features that facilitated the popularity of the stereoscope was its marriage with photography which was invented at about the same time. Paired photographs for the stereoscope were initially taken by a single camera moved laterally between the two exposures. This technique has been used for the stereoscopic photographs shown in this book and in the example below.

Stereoscopic view of Tayport from Scotscraig

The early paired photographs would have been viewed with optical stereoscopes rather than anaglyphs. Initially anaglyphs were fairly crude patchworks of colour but more subtle combinations of colours and contours are now possible due to the power of computers. This applies to rivalling patterns, too, be they photographs or designs.

Rivalling view of Tayport from Scotscraig

Whereas stereoscopic art has often been concerned with heightening the visual realities of recognizable objects, stereoscopic science has tried to banish them by developing stimuli in which the depth cannot be discerned with a single eye but only in the combined action of two eyes. The carrier patterns for the disparate displacements were initially computer-generated randomly distributed dots. Now more interesting patterns can act as carriers.

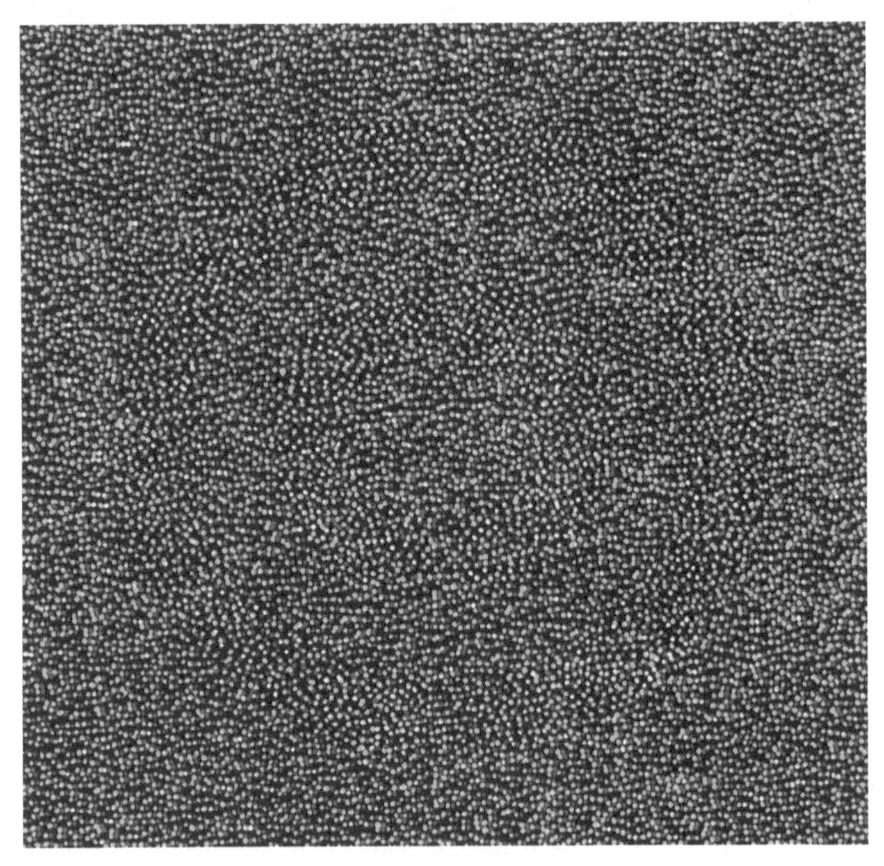

Annuli I

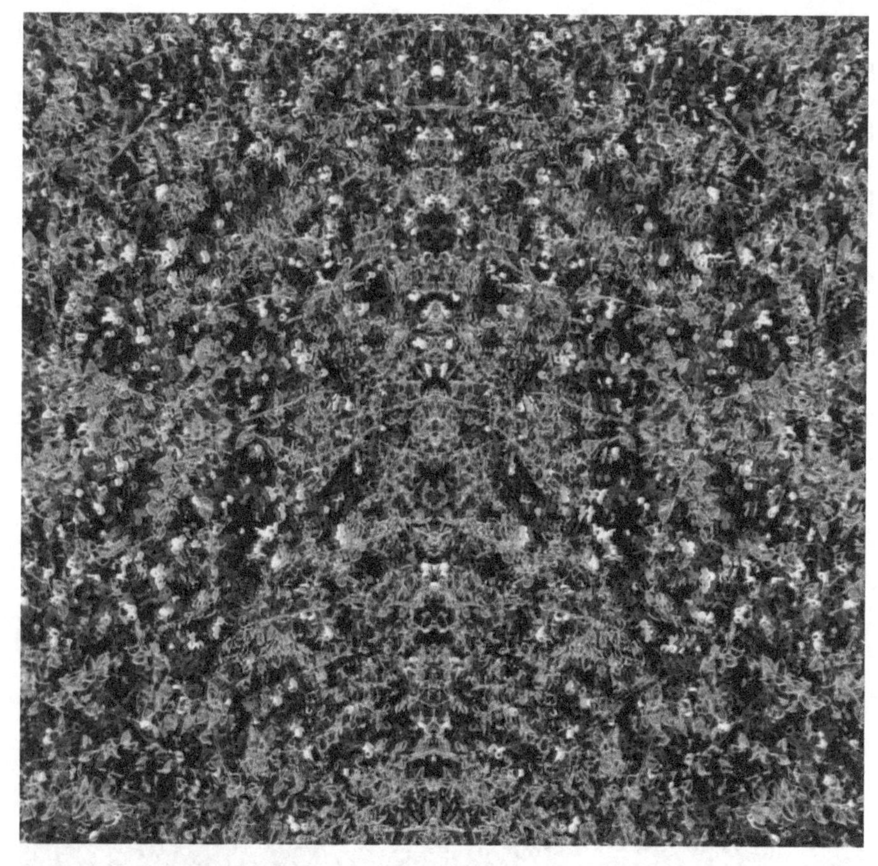

Annuli II

Both designs carry two separated and concentric annuli. Not only is the carrier pattern more interesting in *Annuli II* but the stereoscopic depth is more complex, too; the left and right halves of the annuli are at different depths, one apparently closer and the other further away. By incorporating carrier patterns that have an intrinsic appeal, independently of the depth visible with two eyes when they are combined, this could be called stereoscopic art. In the illustration below this possibility is literally stated. It is in this way that anaglyphs display their advantages over paired images made for optical stereoscopes: the interweavings of the red and cyan contours can create displays that are not otherwise available.

Stereoscopic ART

The carrier pattern in *Stereoscopic ART* was derived from a photograph of leaves on a forsythia bush. Some more examples follow containing different subjects carried by different patterns. The stereoscopic depth can be horizontally or vertically defined depending on the subject matter. For *Sydney Opera House* the majestic sails of the building are defined by depth in the horizontal dimension. While the carrier pattern might give the impression of churning water it is derived from a photograph of leaves. The framing of a design can provide clues to the depth it contains, as in *Divine stereoscopic proportions* which contains vertical semicircles in the golden ratio and they are echoed in depth. The carrier pattern is derived from a photograph of pebbles on a beach. The design *Tom Wedgwood in profiles* returns to aspects of the early explorations of photography when Tom Wedgwood (a son of the potter Josiah Wedgwood) made images from focussed sunlight, but he could not fix them. His profile is repeated ten times, carried by a pattern derived from a painted Wedgwood plate. Thus it is possible to marry a pattern with the depth content that it carries. This is playfully presented in the portrait of Wheatstone that is carried by a pattern derived from combinations of photographs of wheat and stones!

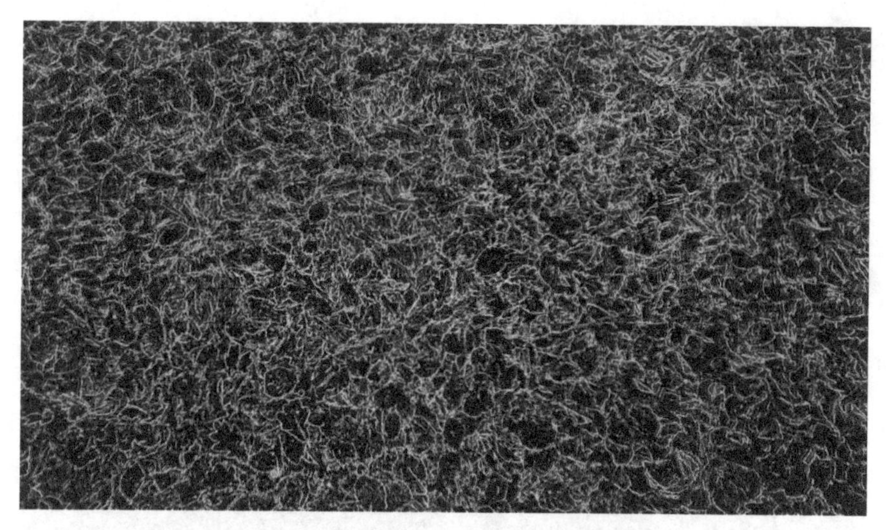

Sydney Opera House

Divine stereoscopic proportions

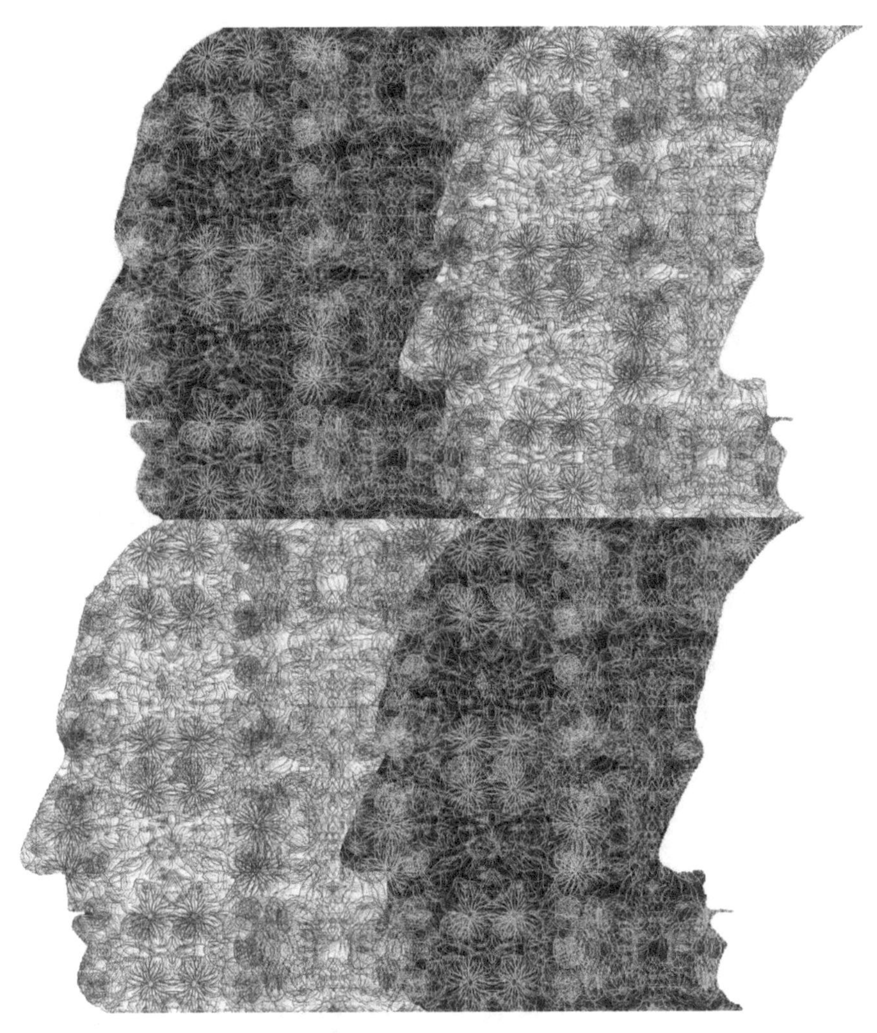

Tom Wedgwood in profiles

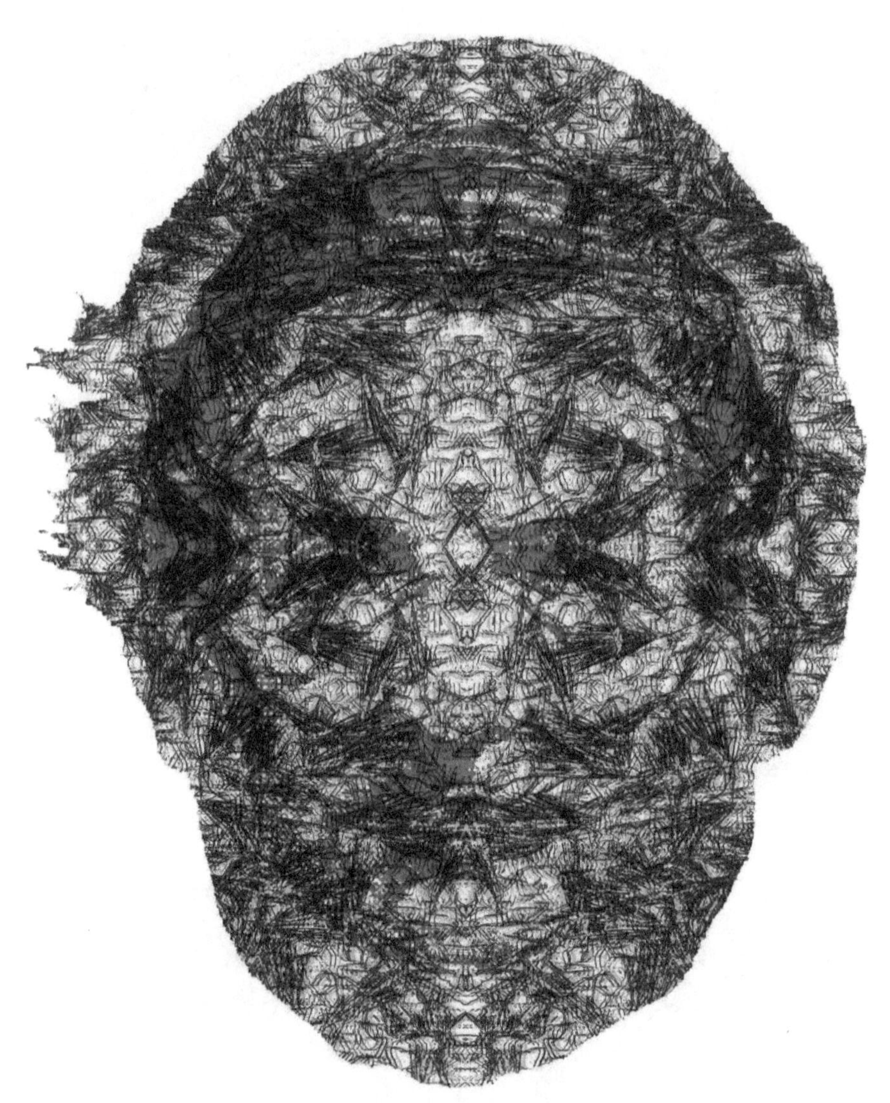

Wheat stones

Many carrier patterns for stereoscopic images can be generated. The ones I have used are generally from photographs of patterns occurring in nature. When they are manipulated and multiplied it is difficult to discern their origins. Simpler combinations of photographed textures can be used to carry disparate patterns and some have been shown in the previous chapters. In the examples below the basic pattern elements can easily be seen but the depth they convey with binocular vision is not available to either eye alone. In each case the pattern is duplicated but the depths are reversed.

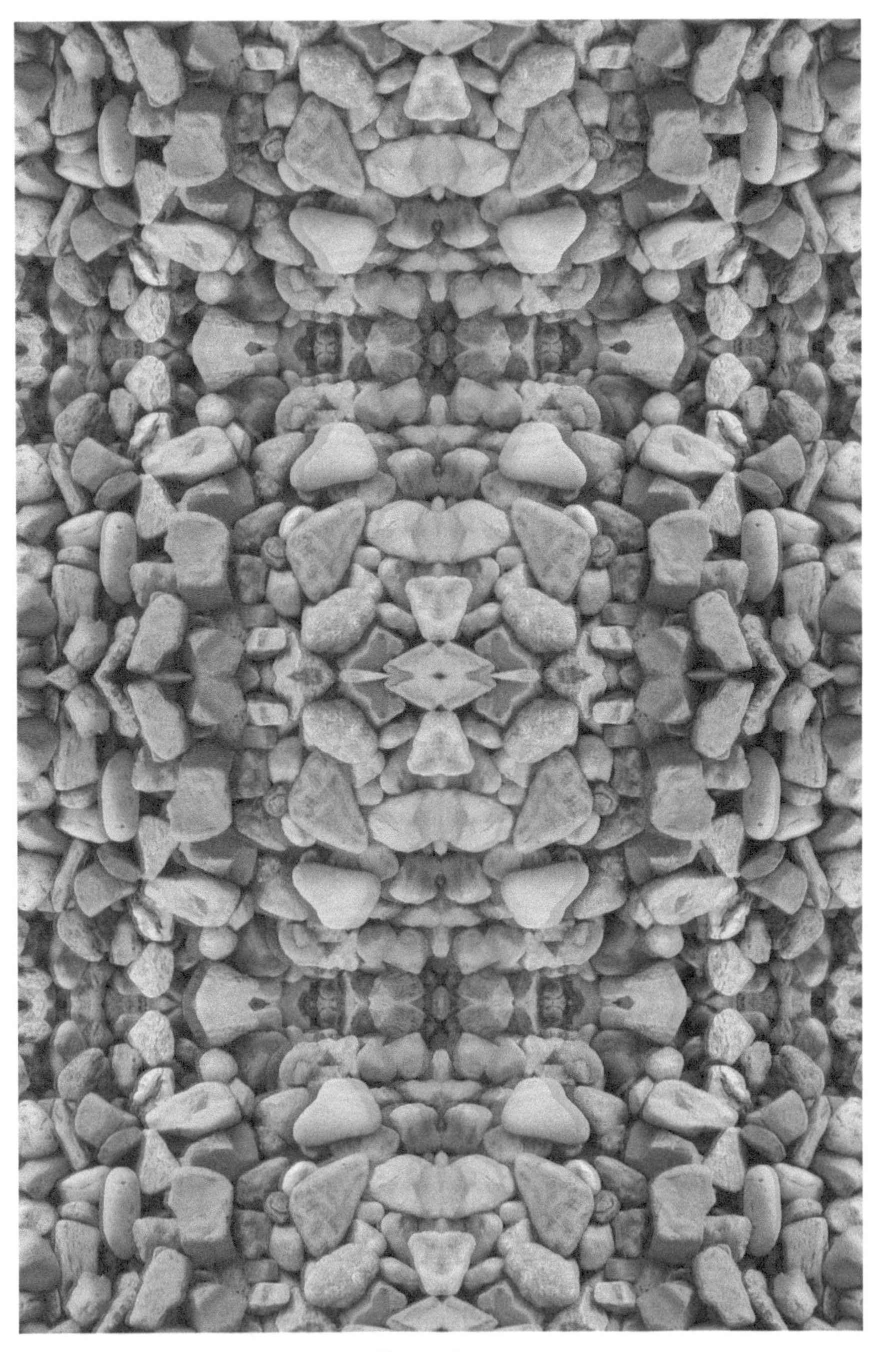

Hearts of stone

Forget-me-nots

An objective in this book is to create patterns that have an intrinsic interest in addition to the stereoscopic depth they carry within them, hence the use of naturally occurring textures rather than computer generated dots. The manipulations of the natural textures can sometimes hide the basic elements from which they are constructed. In *Radiating rhodedendrons* the flowers provide a background as well as acting as the carrier patterns for the stereoscopic shapes within the circles. A similar combination is present in *Petulate* with the stereoscopic petal-shapes surrounded by leaves on a blueberry bush.

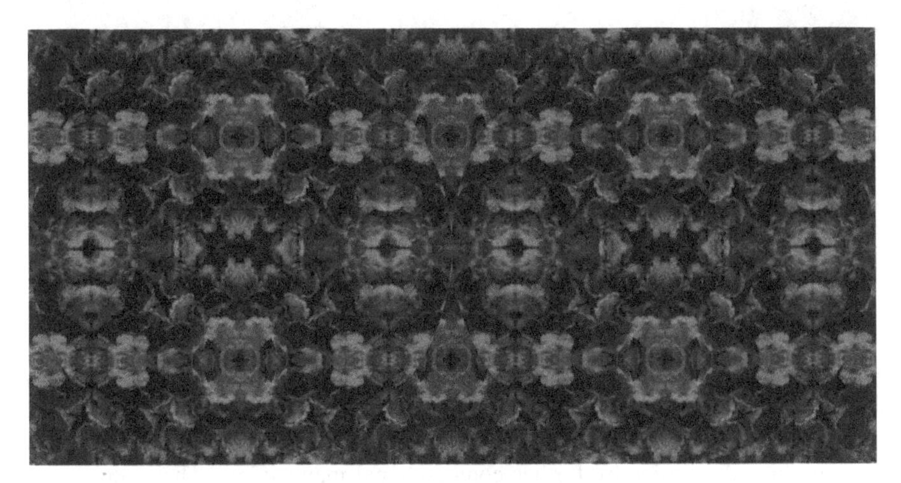

Radiating rhododendrons

Petulate

In this book stereoscopic imagery has been based on photographs and graphics that have been subjected to computer graphical manipulations. An alternative approach that has not been pursued here is to commence with computer generated patterns and the introduce disparities by further computational manoeuvres. These alternative techniques are likely to be adopted increasingly in the future.

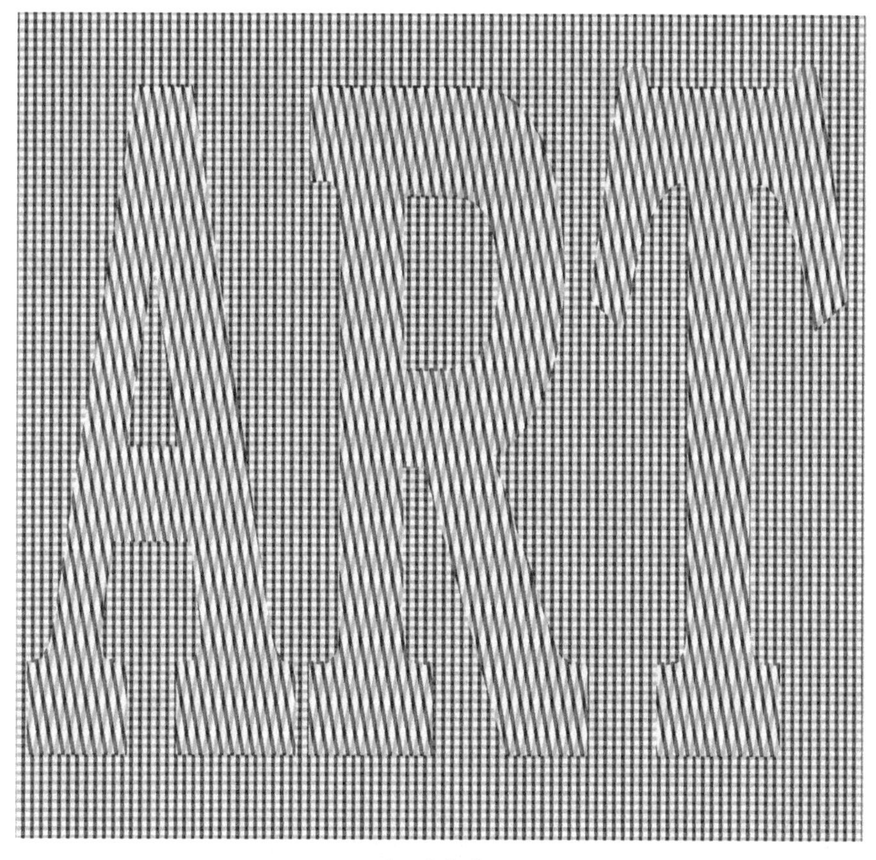

Art of rivalry

Manipulations of rivalling stimuli outside the confines of science are less common than those of stereoscopic stimuli. This is surprising since the scope for manipulating the carrier patterns and what they carry is much broader for rivalry. A good example is provided by portraiture, particularly when antagonists like Brewster and Wheatstone are combined.

Rivalry in action

Binocular rivalry typically involves bold contours that are in different directions for each eye. The ensuing appearances are complex and dynamic with parts of patterns prevailing and also periods in which the stimulus to one eye dominates and that in the other is suppressed. It is less common to present rivalling patterns in low contrast where the sequences of dominance and suppression are slower and more fragmented, as can be seen with *Drawn and quartered*. It is derived from a detail of a painting formed from the flow of pigment over a flat surface; it was photographed and flipped to produce a central symmetry. Each quarter contains the same two patterns at right angles to one another but they vary in terms of either their contrast or orientation.

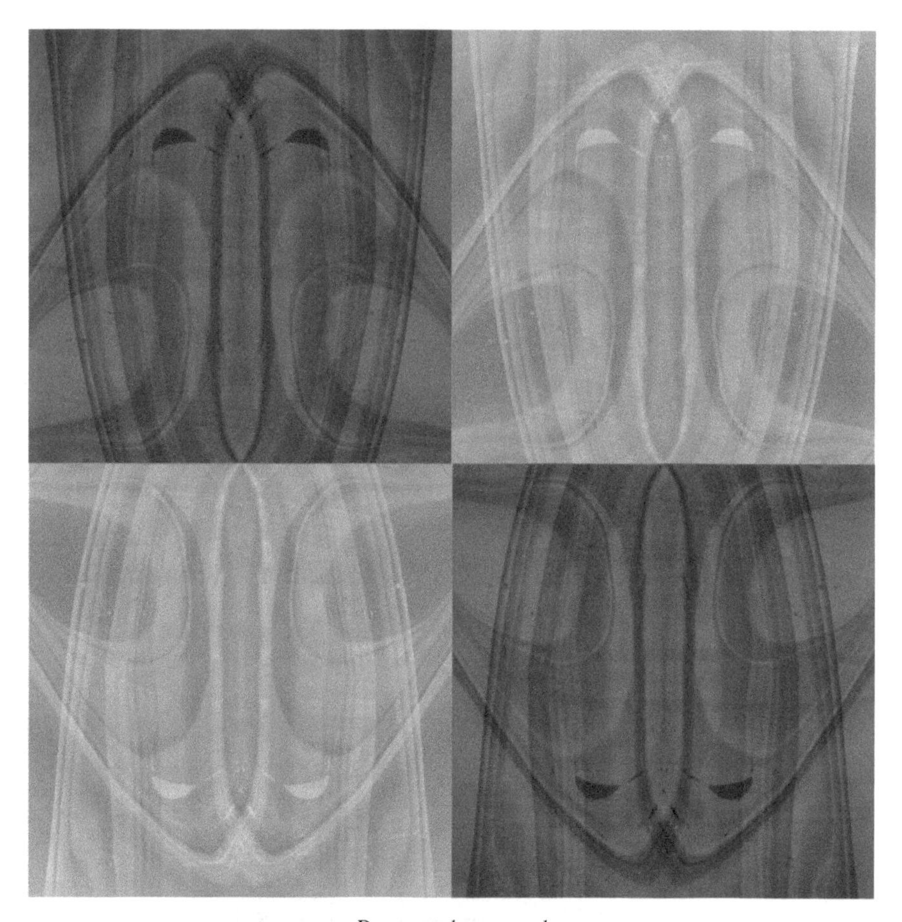

Drawn and quartered

Binocular lustre can be used to combine positive and negative images of individuals in ways which reflect their art or science. Jackson Pollock is represented in *Cigarette smoker.* He was one of the foremost exponents of the Abstract Expressionist movement in America. In the late 1940s he developed his famous action painting (or tachiste) technique, which applied the paint to the horizontal canvas directly rather than via brush strokes. Pollock was smoking a cigarette in most of the photographs and films taken of him, even when he was painting, and it is in this state that he is represented here. The cigarette is drooping from the corner of his mouth, and it corresponds precisely with a white splash of paint in the underlying tachiste picture which is otherwise in lustre. This is followed by positive and negative portraits of the illustrious Albert Einstein together with his equation linking energy, mass and light.

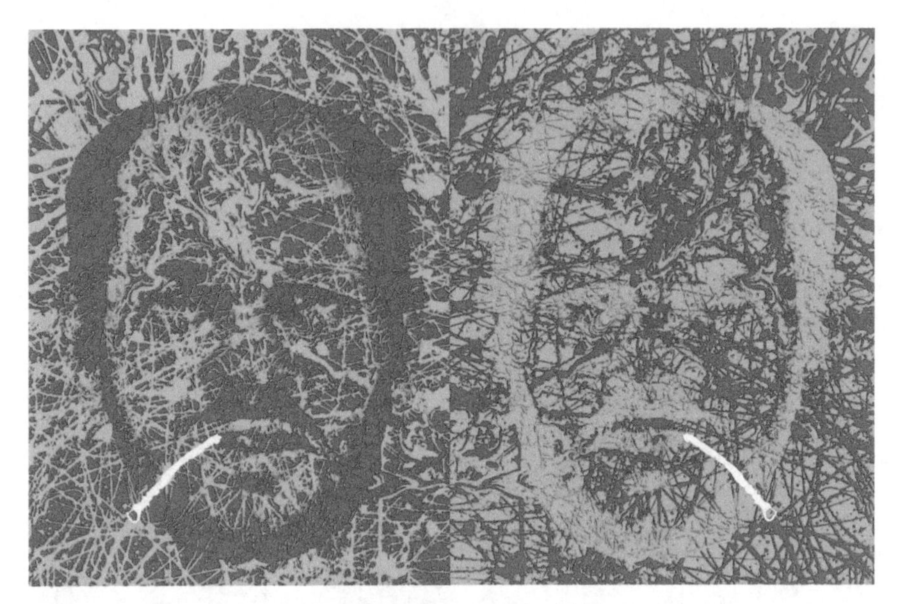

Cigarette smoker–Jackson Pollack

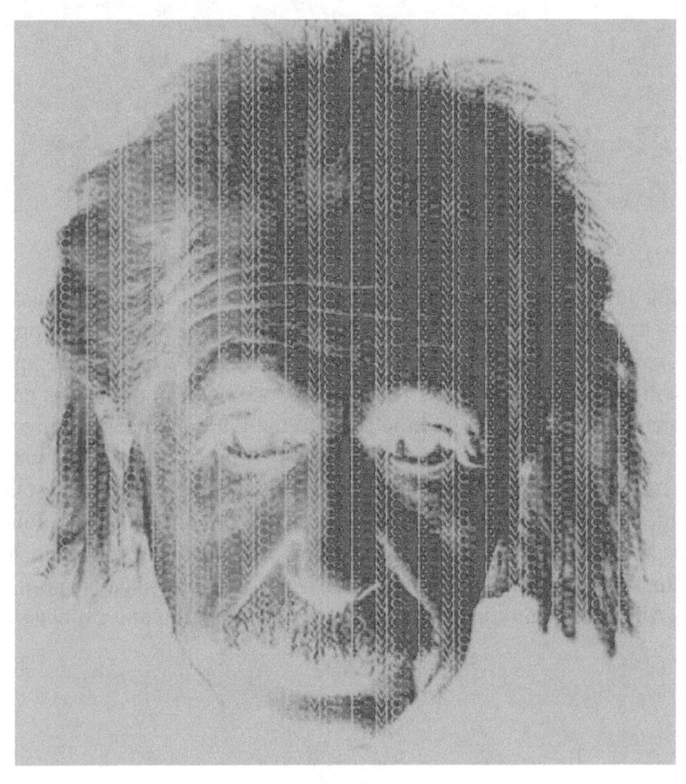

Einstein's equation

Symmetry has been a frequent feature in the images in this book, both in terms of the subjects portrayed and the patterns carrying them. It is also a central aspect of the final two images one of which induces rivalry and the other stereoscopic depth.

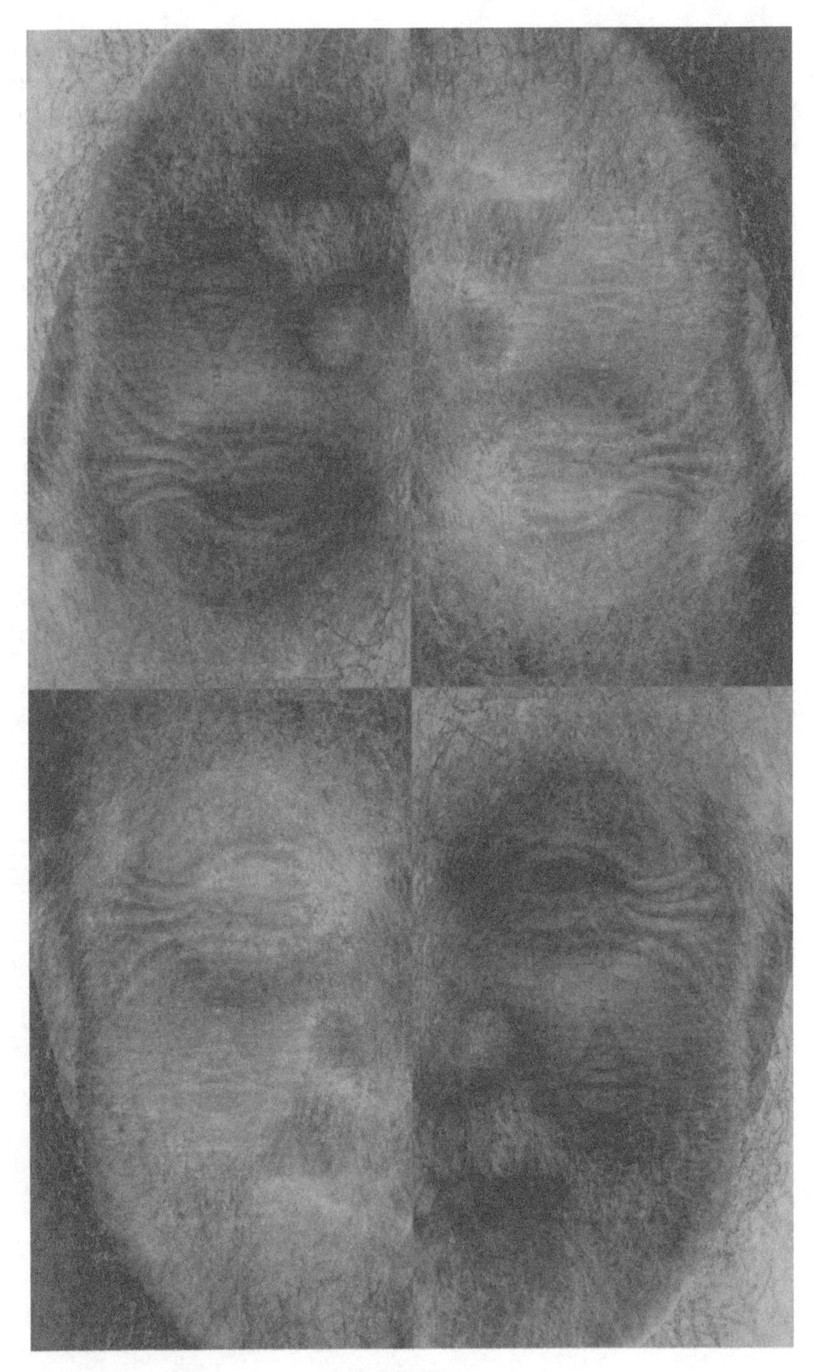

Selbstbildnis

THE END

Author Index

N. Wade, *Vision and Art with Two Eyes*, Vision, Illusion and Perception 3,
https://doi.org/10.1007/978-3-030-77995-5

Subject Index

Page numbers followed by "*f*" indicate figures.

N. Wade, *Vision and Art with Two Eyes*, Vision, Illusion and Perception 3,
https://doi.org/10.1007/978-3-030-77995-5

C

Q

R

SPRINGER NATURE

GPSR Compliance

The European Union's (EU) General Product Safety Regulation (GPSR) is a set of rules that requires consumer products to be safe and our obligations to ensure this.

If you have any concerns about our products, you can contact us on ProductSafety@springernature.com

In case Publisher is established outside the EU, the EU authorized representative is:

Springer Nature Customer Service Center GmbH
Europaplatz 3
69115 Heidelberg, Germany